Network-Centric Naval Forces
A Transition Strategy for Enhancing Operational Capabilities

Committee on Network-Centric Naval Forces
Naval Studies Board
Commission on Physical Sciences, Mathematics, and Applications
National Research Council

NATIONAL ACADEMY PRESS
Washington, D.C.

NOTICE: The project that is the subject of this report was approved by the Governing Board of the National Research Council, whose members are drawn from the councils of the National Academy of Sciences, the National Academy of Engineering, and the Institute of Medicine. The members of the committee responsible for the report were chosen for their special competences and with regard for appropriate balance.

This work was performed under Department of the Navy Contract N00014-96-D-0169/0001 issued by the Office of Naval Research under contract authority NR 201-124. However, the content does not necessarily reflect the position or the policy of the Department of the Navy or the government, and no official endorsement should be inferred.

The United States Government has at least a royalty-free, nonexclusive, and irrevocable license throughout the world for government purposes to publish, translate, reproduce, deliver, perform, and dispose of all or any of this work, and to authorize others so to do.

International Standard Book Number 0-309-06925-4

Copies available from:

Naval Studies Board
National Research Council
2101 Constitution Avenue, N.W.
Washington, D.C. 20418

Copyright 2000 by the National Academy of Sciences. All rights reserved.

Printed in the United States of America

THE NATIONAL ACADEMIES

National Academy of Sciences
National Academy of Engineering
Institute of Medicine
National Research Council

The **National Academy of Sciences** is a private, nonprofit, self-perpetuating society of distinguished scholars engaged in scientific and engineering research, dedicated to the furtherance of science and technology and to their use for the general welfare. Upon the authority of the charter granted to it by the Congress in 1863, the Academy has a mandate that requires it to advise the federal government on scientific and technical matters. Dr. Bruce M. Alberts is president of the National Academy of Sciences.

The **National Academy of Engineering** was established in 1964, under the charter of the National Academy of Sciences, as a parallel organization of outstanding engineers. It is autonomous in its administration and in the selection of its members, sharing with the National Academy of Sciences the responsibility for advising the federal government. The National Academy of Engineering also sponsors engineering programs aimed at meeting national needs, encourages education and research, and recognizes the superior achievements of engineers. Dr. William A. Wulf is president of the National Academy of Engineering.

The **Institute of Medicine** was established in 1970 by the National Academy of Sciences to secure the services of eminent members of appropriate professions in the examination of policy matters pertaining to the health of the public. The Institute acts under the responsibility given to the National Academy of Sciences by its congressional charter to be an adviser to the federal government and, upon its own initiative, to identify issues of medical care, research, and education. Dr. Kenneth I. Shine is president of the Institute of Medicine.

The **National Research Council** was organized by the National Academy of Sciences in 1916 to associate the broad community of science and technology with the Academy's purposes of furthering knowledge and advising the federal government. Functioning in accordance with general policies determined by the Academy, the Council has become the principal operating agency of both the National Academy of Sciences and the National Academy of Engineering in providing services to the government, the public, and the scientific and engineering communities. The Council is administered jointly by both Academies and the Institute of Medicine. Dr. Bruce M. Alberts and Dr. William A. Wulf are chairman and vice chairman, respectively, of the National Research Council.

COMMITTEE ON NETWORK-CENTRIC NAVAL FORCES

VINCENT VITTO, Charles S. Draper Laboratory, Inc., *Chair*
ALAN BERMAN, Applied Research Laboratory, Pennsylvania State University
GREGORY R. BLACKBURN, Science Applications International Corporation
NORVAL L. BROOME, Mitre Corporation
JOHN D. CHRISTIE, Logistics Management Institute
JOHN A. CORDER, Colleyville, Texas
JOHN R. DAVIS, Center for Naval Analyses
PAUL K. DAVIS, RAND and RAND Graduate School of Policy Studies
JOHN F. EGAN, Nashua, New Hampshire
BRIG B. ELLIOTT, GTE Internetworking
EDWARD A. FEIGENBAUM, Stanford University
DAVID E. FROST, Frost and Associates
ROBERT H. GORMLEY, Oceanus Company
FRANK A. HORRIGAN, Raytheon Systems Company
RICHARD J. IVANETICH, Institute for Defense Analyses
WESLEY E. JORDAN, JR., Bolt, Beranek and Newman Co.
DAVID V. KALBAUGH, Applied Physics Laboratory, Johns Hopkins University
ANNETTE J. KRYGIEL, National Defense University
TERESA F. LUNT, Xerox Palo Alto Research Center
DOUGLAS R. MOOK, Sanders, a Lockheed Martin Company
DONALD L. NIELSON, Menlo Park, California
STEWART D. PERSONICK, Drexel University
JOSEPH B. REAGAN, Saratoga, California
CHARLES R. SAFFELL, JR., Titan Technologies and Information Systems Corporation
NILS R. SANDELL, JR., ALPHATECH, Inc.
WILLIAM D. SMITH, Fayetteville, Pennsylvania
MICHAEL G. SOVEREIGN, Monterey, California
H. GREGORY TORNATORE, Applied Physics Laboratory, Johns Hopkins University
PAUL K. VAN RIPER, Williamsburg, Virginia
BRUCE WALD, Center for Naval Analyses
RAYMOND M. WALSH, Basic Commerce and Industries, Inc.
MITZI M. WERTHEIM, Center for Naval Analyses
GEOFFREY A. WHITING, Sanders, a Lockheed Martin Company
DELL P. WILLIAMS III, Teledesic Corporation

Naval Studies Board Liaison

SEYMOUR J. DEITCHMAN, Chevy Chase, Maryland

Staff

RONALD D. TAYLOR, Director, Naval Studies Board
CHARLES F. DRAPER, Study Director
MARY G. GORDON, Information Officer
SUSAN G. CAMPBELL, Administrative Assistant
JAMES E. MACIEJEWSKI, Senior Project Assistant
SIDNEY G. REED, JR., Consultant
JAMES G. WILSON, Consultant

Navy Liaison Representatives

ROBERT LeFANDE, Associate Director, Systems Directorate, Naval Research Laboratory
CDR DAVID L. SPAIN, USN, Office of the Chief of Naval Operations, N513J (through July 1999)
CAPT(S) MARK TEMPESTILLI, USN, Office of the Chief of Naval Operations, N6C3

NAVAL STUDIES BOARD

VINCENT VITTO, Charles S. Draper Laboratory, Inc., *Chair*
JOSEPH B. REAGAN, Saratoga, California, *Vice Chair*
DAVID R. HEEBNER, McLean, Virginia, *Past Chair*
ALBERT J. BACIOCCO, JR., The Baciocco Group, Inc.
ARTHUR B. BAGGEROER, Massachusetts Institute of Technology
ALAN BERMAN, Applied Research Laboratory, Pennsylvania State University
NORMAN E. BETAQUE, Logistics Management Institute
JAMES P. BROOKS, Litton/Ingalls Shipbuilding, Inc.
NORVAL L. BROOME, Mitre Corporation
JOHN D. CHRISTIE, Logistics Management Institute
RUTH A. DAVID, Analytic Services, Inc.
PAUL K. DAVIS, RAND and RAND Graduate School of Policy Studies
SEYMOUR J. DEITCHMAN, Chevy Chase, Maryland, *Special Advisor*
DANIEL E. HASTINGS, Massachusetts Institute of Technology
FRANK A. HORRIGAN, Raytheon Systems Company
RICHARD J. IVANETICH, Institute for Defense Analyses
MIRIAM E. JOHN, Sandia National Laboratories
ANNETTE J. KRYGIEL, National Defense University
ROBERT B. OAKLEY, National Defense University
HARRISON SHULL, Monterey, California
JAMES M. SINNETT, The Boeing Company
WILLIAM D. SMITH, Fayetteville, Pennsylvania
PAUL K. VAN RIPER, Williamsburg, Virginia
VERENA S. VOMASTIC, The Aerospace Corporation
BRUCE WALD, Center for Naval Analyses
MITZI M. WERTHEIM, Center for Naval Analyses

Navy Liaison Representatives

RADM RAYMOND C. SMITH, USN, Office of the Chief of Naval Operations, N81
RADM PAUL G. GAFFNEY II, USN, Office of the Chief of Naval Operations, N91

Marine Corps Liaison Representative

LTGEN JOHN E. RHODES, USMC, Commanding General, Marine Corps Combat Development Command

RONALD D. TAYLOR, Director
CHARLES F. DRAPER, Senior Program Officer
MARY G. GORDON, Information Officer
SUSAN G. CAMPBELL, Administrative Assistant
JAMES E. MACIEJEWSKI, Senior Project Assistant

COMMISSION ON PHYSICAL SCIENCES, MATHEMATICS, AND APPLICATIONS

PETER M. BANKS, Veridian ERIM International, Inc., *Co-Chair*
W. CARL LINEBERGER, University of Colorado, *Co-Chair*
WILLIAM F. BALLHAUS, JR., Lockheed Martin Corporation
SHIRLEY CHIANG, University of California at Davis
MARSHALL H. COHEN, California Institute of Technology
RONALD G. DOUGLAS, Texas A&M University
SAMUEL H. FULLER, Analog Devices, Inc.
JERRY P. GOLLUB, Haverford College
MICHAEL F. GOODCHILD, University of California at Santa Barbara
MARTHA P. HAYNES, Cornell University
WESLEY T. HUNTRESS, JR., Carnegie Institution
CAROL M. JANTZEN, Westinghouse Savannah River Company
PAUL G. KAMINSKI, Technovation, Inc.
KENNETH H. KELLER, University of Minnesota
JOHN R. KREICK, Sanders, a Lockheed Martin Company (retired)
MARSHA I. LESTER, University of Pennsylvania
DUSA M. McDUFF, State University of New York at Stony Brook
JANET L. NORWOOD, Former U.S. Commissioner of Labor Statistics
M. ELISABETH PATÉ-CORNELL, Stanford University
NICHOLAS P. SAMIOS, Brookhaven National Laboratory
ROBERT J. SPINRAD, Xerox PARC (retired)

NORMAN METZGER, Executive Director (through July 1999)
MYRON F. UMAN, Acting Executive Director

Preface

The Chief of Naval Operations (CNO) recently declared that the Navy would be shifting its operational concept from one based on platform-centric warfare concepts to one based on network-centric warfare concepts. This new operational concept can be described as a model of warfare, called network-centric warfare, that derives its power from a geographically dispersed naval force embedded within an information network that links sensors, shooters, and command and control nodes to provide enhanced speed of decision making, rapid synchronization of the force as a whole to meet its desired objectives, and great economy of force.

Realization of a network-centric warfighting capability will depend on a number of factors: development of warfare concepts (and supporting doctrine) that determine how weapons, sensors, and information systems will interact to carry out specific missions; experimentation to test the viability of the new concepts; application of both military and commercial technology, particularly information technology, with essential attention to information and communications security and robustness; timely and effective acquisition of information technology assets; and education, training, and utilization of naval personnel to meet the demands of a network-centric force. This change of operational concept is also part of the Department of Defense (DOD) thrust toward Joint Vision 2010,[1] which encompasses efforts by the four Services to achieve similar objectives DOD-wide.

---

[1] Shalikashvili, GEN John M., USA. 1997. *Joint Vision 2010.* Joint Chiefs of Staff, The Pentagon, Washington, D.C.

Several initial steps have been taken by the Navy and Marine Corps toward achieving network-centric warfare capabilities. These include (1) promulgating the Navy Information Technology 21 (IT-21) initiative, which aims to bring the fleet up to date in information technology and related skills; (2) developing the Navy-Marine Corps intranet, to do the same for the shore establishment; (3) setting up the Navy Warfare Development Command and the Marine Corps Warfighting Laboratory, to develop concepts and doctrine; (4) testing these concepts and doctrines in fleet battle experiments and the Marine Corps "Warrior Series" experiments; and (5) making efforts toward interoperability of battle-group air defense and related command and control systems.

In a larger perspective, network-centric-type concepts have been applied by the Navy in the past, in antisubmarine warfare (ASW) since World War II, in approaches to air defense in the outer air battle in the 1980s, and more recently in the cooperative engagement capability (CEC) now under evaluation.

TERMS OF REFERENCE

At the request of Admiral Jay L. Johnson, USN, CNO (see Appendix A), the National Research Council (NRC), under the auspices of the Naval Studies Board (NSB), conducted a study to advise the Department of the Navy regarding its transition strategy to achieve a network-centric naval force through technology application. The terms of reference for the study call for an evaluation of the following:

• What are the technical underpinnings needed for a transition to network-centric forces and capabilities? Particular emphasis should be placed on assessing the means, the systems, and the feasibility of achieving and delivering data via links with the necessary bandwidth, capacity, and timeliness capabilities. Emphasis also should be placed on establishing and maintaining network security, emissions control when needed, and links with submarines, and on integrating information which may arrive intermittently and with different timescales.

• What near-term program actions need to be taken to begin the transition? What impact will these program actions have on the present platform-centric acquisition strategy? What impact will these program actions have on maintaining a robust industrial base to support the naval forces?

• Recognizing that many areas of technology are evolving faster than the naval forces can develop concepts for their use: What experimental programs need to be put in place to help the forces select needed technologies and systems, develop doctrine, and develop operational concepts that together can support the transition to a network-centric naval force? What organizational adaptations might facilitate rapid progress?

• What are the implications for both the business practices of the Department of the Navy and naval operations of moving away from a platform-centric naval force to network-centric warfare? Implications for the following should be considered especially: resource priorities; force structure; personnel, education, career systems; warfighting doctrine; and coalition building and training with allies.

• Over what period of time can a transition strategy be implemented and in what details will the naval forces be different from today's forces when the strategy is finally implemented?

• What trends, if any, suggest that potential adversaries might move toward a network-centric military capability or exploit its vulnerabilities? What are the implications for U.S. naval forces?

• How will the move toward network-centric forces, if embraced by the Department of the Navy, be accomplished within the joint environment and subject to the likelihood of constrained future budgets?

• What are the implications of network-centric warfare for naval doctrine and for joint operations?

COMMITTEE'S APPROACH

In responding to the CNO's request, the committee organized itself into four ad hoc panels: (1) Panel 1—Concepts, Doctrine, Missions, and Operations; (2) Panel 2—System Architecture, Information Management, Dissemination, Protection, Assurance, and Command and Control; (3) Panel 3—Tactical Networks, Sensor-to-Shooter, Security, Protection, Targeting, Sensor Coordination, and Emission Control; and (4) Panel 4—Resources, Policy, Acquisition, Industrial Base, Career Issues, Education, and Training. In an effort to integrate the work of these four panels, an integration panel was formed with a lead representative from each panel, as well as the committee chair and NSB liaison.

The committee considered network-centric warfare, or better, network-centric operations (NCO), in the context of the Navy's principal missions—strategic deterrence, sea and air control, forward presence, and power projection. Because of its unique characteristics, strategic deterrence was not included in the study. Further, taking a mission-specific approach, the committee decided to focus on NCO in the power projection mission, since power projection must also encompass sea and air control (as well as a degree of forward presence), and, in anticipated littoral operations, the land-attack aspect of power projection was considered to be less developed with respect to NCO than sea and air control, with which the Navy has considerable experience.

The following report attempts to treat in as much detail as was feasible the issues raised in the terms of reference listed above. As often happens, once the study's directions of inquiry developed and results began to emerge, the committee found that its discussions of the issues raised in the terms of reference tended to group in a contextual and logical order different from the order initially anticipated. The next few paragraphs therefore sketch briefly where in the report discussions of the issues may be found.

The technical underpinnings needed for the transition to network-centric forces, capabilities, and operations are treated in detail throughout the report. Implications for naval force doctrine and joint operations are reviewed, directly and indirectly, in Chapters 1 and 2, while implications for joint operations in designing and creating NCO systems, in designing and creating a common information infrastructure (i.e., the Naval Command and Information Infrastructure, the NCII), and in undertaking network-centric combat operations are treated in detail at many points in Chapters 3, 4, 5, and 6 in connection with the overall topics of those chapters.

Presented in the Executive Summary is a short list of recommended near-term program, management process, and organizational actions that must be undertaken to begin the transition from platform-centric to network-centric naval forces. The list was developed from the more detailed sets of recommendations given in Chapter 1, which were, in turn, taken from the fully developed findings and recommendations in the body of the report.

The implications for Department of the Navy business practices and organizational responsibilities needed to better transition to network-centric operations are considered in detail in Chapter 7. Management and technical aspects of some business practices and acquisition strategy are covered further in parts of Chapters 2, 4, 5, and 6 in discussions of the need for a new approach to thinking about the naval forces under the NCO concept and in descriptions of the many aspects of NCII design, operation, and information assurance. Needed experimental programs are described as part of these discussions, in Chapter 2 and also in Chapter 3, in connection with the technical details of subsystems and components needed to complete the NCO orientation of the naval force systems.

The committee believes that NCO will rely on a dual industrial base. The purely military aspects of such systems will draw on the base that currently furnishes the platforms and the specialized sensors and weapons that will enter NCO subsystems and components. Much commercial off-the-shelf technology will also support these subsystems and components. The NCII will draw largely from the huge commercial technology base that is developing to support civilian communication and computer-based information networks (e.g., the Internet) and the exponentially increasing commercial activity that their presence is fostering. This commercial base is as much a driver of the U.S. military's movement to network-centric forces and warfare as it is an enabler for that movement.

The committee did not fully examine the capability of allies and potential coalition partners in the information and networking technology and systems areas relevant to network-centric operations. Similarly, it was not possible to investigate in depth, from the intelligence viewpoint, the possibility that potential adversaries could engage in network-centric conflict as defined in this report. The United States is so rapidly outpacing every other significant power in the

PREFACE   xiii

world in the area of linking military forces in large, computer-based information networks that it is difficult for intelligence to estimate where the rest of the world stands relative to the United States in this area.

This does not mean that U.S. network-centric operations capability is now or will in the future be safe from attack or interference. As detailed in Chapter 5, U.S. information and combat networks and the NCII have, because of their inherent design and by virtue of their reliance on the commercial technology base, many vulnerabilities. Anyone with modern computing and communications capability can wage information war or cyber war against the United States, often in ways that have no easy counter. Approaches to mitigating this risk are discussed in detail in Chapter 5.

Overall, the committee believes that it has assembled a relatively complete picture of the significance of the movement toward NCO for the naval forces in the joint environment. The menu of needed actions to achieve the capability is large and will require a dedicated and extended effort throughout the Department of the Navy, building on and greatly extending actions currently under way.

COMMITTEE MEETINGS

The committee first convened early in 1999 and met for approximately 8 months. During that time, it held the following committee and panel meetings:

- January 26-28, 1999, in Washington, D.C (Plenary). Organizational meeting. Navy, Marine Corps, Joint Chiefs of Staff, and Defense Advanced Research Projects Agency (DARPA) briefings on network-centric warfare.
- February 16-17, 1999, in Washington, D.C. (Representatives, Panels 1 and 3). Office of the Chief of Naval Operations concepts of operations and tactical data links briefings.
- February 18, 1999, in Washington, D.C. (Integration Panel).
- March 4-5, 1999, in Washington, D.C. (Panel 2). Defense Information Systems Agency (DISA), DARPA, Joint Chiefs of Staff, and Office of the Secretary of Defense information infrastructure and interoperability briefings.
- March 9 and 11, 1999, in Washington, D.C. (Panel 4). Joint Requirements Oversight Council, Navy, and Marine Corps assessment and requirements briefings.
- March 23, 1999, in Washington, D.C. (Plenary). Air Force Battlespace Infosphere, Army Digital Battlefield, Defense Science Board Integrated Information Infrastructure, and DARPA Discover II briefings.
- March 24, 1999, in Washington, D.C. (Representatives, Panels 1 through 4). DARPA, DISA, Military Satellite Communications Joint Program Office, and National Imagery and Mapping Agency information dissemination and management briefings. Naval Air Systems Command weapons, Navy Warfare Devel-

opment Command concepts of operations, and Office of the Secretary of Defense acquisition and technology briefings.
- March 25, 1999, in Washington, D.C. (Integration Panel).
- April 15-16, 1999, in Washington, D.C. (Panel 2). CitiGroup, DARPA, Naval Research Laboratory, and Office of Naval Research information assurance and security briefings.
- April 19, 1999, in Alexandria, Virginia (Representatives, Panels 2 and 3). National Reconnaissance Office briefings.
- April 20-21, 1999, in Washington, D.C. (Representatives, Panels 1, 3, and 4). Office of the Secretary of Defense and Marine Corps C4ISR requirements briefings. Air Force Rivet Joint and U2 briefings.
- April 27-29, 1999, in San Diego, California (Panel 2). Site visit to Space and Naval Warfare Systems Command. Briefings on information assurance and infrastructure programs, as well as related network-centric topics.
- May 19-20, 1999, in Washington, D.C. (Representatives, Panels 1 through 4). Air Force Expeditionary Force Experiment, DARPA information assurance, Joint Theater Air and Missile Defense Organization single integrated air picture, naval intelligence threat, and Naval Sea Systems Command battle force interoperability requirements briefings.
- May 21, 1999, in Washington, D.C. (Integration Panel).
- June 8-9, 1999, in Crystal City, Virginia (Panel 4). Navy and Air Force briefings on DD-21 and Joint Strike Fighter, respectively.
- June 16-17, 1999, in Washington, D.C. (Panel 2).
- June 21, 1999, in Washington, D.C. (Panel 4).
- June 23, 1999, in Washington, D.C. (Panel 1).
- June 22-23, 1999, in Washington, D.C. (Panel 3).
- June 24, 1999, in Washington, D.C. (Plenary). Status from panels.
- June 25, 1999, in Washington, D.C. (Integration Panel).
- July 13-14, 1999, in Washington, D.C. (Panel 4).
- July 19-23, 1999, in Woods Hole, Massachusetts (Plenary).
- August 31 to September 1, 1999, in Washington, D.C. (Integration Panel).
- September 29 to October 1, 1999, in Washington, D.C. (Integration Panel).
- November 8-10, 1999, in Washington, D.C. (Integration Panel).
- January 11-12, 2000, in Washington, D.C. (Integration Panel).

Acknowledgments

The Committee on Network-Centric Naval Forces extends its gratitude to the many individuals who provided valuable information and support during the course of this study. Special acknowledgment goes to VADM Arthur K. Cebrowski, USN, president, Naval War College, who formulated the concept of network-centric warfare. His knowledge and insights made an invaluable contribution to the success of the study.

The committee extends a special thanks to the Navy liaisons to the committee, CAPT(S) Mark Tempestilli, USN, CDR David Spain, USN, and Dr. Robert LeFande, who responded to the committee's numerous requests for information throughout the stages of the study.

The committee also thanks Mr. Kin Searcy, who helped arrange a visit to the Space and Naval Warfare Systems Command. He and his staff were gracious in hosting members of the committee on its 4-day site visit to learn more about ongoing Navy information technology investments.

In addition, the committee wishes to thank Mr. Paul Blatch, who serves as the Navy's action officer for Naval Studies Board activities and assisted with this study from its inception to completion.

The committee is grateful to the staff of the Naval Studies Board for its assistance, support, and guidance throughout the course of the study and to the CPSMA editorial office for help in editing the manuscript.

Finally, the committee thanks the many men and women throughout the Armed Services, as well as government, academic, and industry leaders who provided the committee with insightful discussions throughout the course of this study. Without their combined efforts, the committee's report would not have been possible.

Acknowledgment of Reviewers

This report has been reviewed by individuals chosen for their diverse perspectives and technical expertise, in accordance with procedures approved by the National Research Council's (NRC's) Report Review Committee. The purpose of this independent review is to provide candid and critical comments that will assist the authors and the NRC in making the published report as sound as possible and to ensure that the report meets institutional standards for objectivity, evidence, and responsiveness to the study charge. The contents of the review comments and draft manuscript remain confidential to protect the integrity of the deliberative process. The committee wishes to thank the following individuals for their participation in the review of this report:

William F. Ballhaus, Jr., Lockheed Martin Corporation,
Ruth M. Davis, Pymatuning Group, Incorporated,
John S. Foster, Jr., TRW, Incorporated,
Robert A. Frosch, Harvard University,
Charles M. Herzfeld, Silver Spring, Maryland,
Anita K. Jones, University of Virginia,
David A. Richwine, Fairfax, Virginia,
John P. Stenbit, TRW, Incorporated,
Jerry O. Tuttle, ManTech Systems Engineering Corporation,
Andrew J. Viterbi, QUALCOMM, Incorporated, and
Larry Welch, Institute for Defense Analyses.

Although the individuals listed above provided many constructive comments and suggestions, responsibility for the final content of this report rests solely with the authoring committee and the NRC.

Contents

EXECUTIVE SUMMARY   1

1   OVERVIEW OF STUDY RESULTS   11
    1.1   Mission Effectiveness: What Is Required, 11
    1.2   Leading the Transformation to Network-Centric Operations, 17
    1.3   Integrating Force Elements: A Mission-Specific Study of Power Projection, 23
    1.4   Designing a Common Command and Information Infrastructure, 31
    1.5   Adjusting the Department of the Navy Organization and Management, 41

2   NETWORK-CENTRIC OPERATIONS—PROMISE AND CHALLENGES   52
    2.1   Introduction, 52
    2.2   Basic Capabilities Required in a Common Command and Information Infrastructure, 63
    2.3   The Need for System Engineering, 65
    2.4   The Critical Role of Leadership in Network-Centric Operations, 66
    2.5   A Proposed Process for Developing CONOPS for Network-Centric Operations, 71
    2.6   Summary of Findings and Recommendations, 85
    2.7   Bibliography, 86

3   INTEGRATING NAVAL FORCE ELEMENTS FOR
    NETWORK-CENTRIC OPERATIONS—A
    MISSION-SPECIFIC STUDY   88
    3.1   Introduction, 88
    3.2   Weapons, 94
    3.3   Sensors, 96
    3.4   Navigation, 107
    3.5   Tactical Information Processing, 116
    3.6   System Engineering, 127
    3.7   Summary and Recommendations, 133

4   DESIGNING A COMMON COMMAND AND INFORMATION
    INFRASTRUCTURE   140
    4.1   The Naval Command and Information Infrastructure
          Concept, 140
    4.2   Tactical Networks, 151
    4.3   Architectural Guidance and Development Processes, 156
    4.4   Recommendations, 170

5   INFORMATION ASSURANCE—SECURING THE NAVAL
    COMMAND AND INFORMATION INFRASTRUCTURE   175
    5.1   Introduction, 175
    5.2   Threats to the Naval Command and Information
          Infrastructure, 176
    5.3   Vulnerabilities of the Naval Command and
          Information Infrastructure, 178
    5.4   Defense in Depth, 181
    5.5   Assessment of Current Information Assurance Activities, 190
    5.6   Research Products Suitable for Near-term Application, 201
    5.7   Information Assurance Research, 206
    5.8   Recommendations, 215

6   REALIZING NAVAL COMMAND AND INFORMATION
    INFRASTRUCTURE CAPABILITIES   219
    6.1   Baseline Naval Systems, 219
    6.2   Functional Capabilities Assessment, 236
    6.3   Recommendations, 280

7   ADJUSTING DEPARTMENT OF THE NAVY ORGANIZATION
    AND MANAGEMENT TO ACHIEVE NETWORK-CENTRIC
    CAPABILITIES   289
    7.1 Key Decision Support Processes and Their
         Interrelationships, 289
    7.2 Requirements Generation: Clearly Stating Operators'
         Mission Needs, 291
    7.3 Mission Analyses and Resource Allocation: Aligning
         Program and Budget Resources to Meet Mission
         Needs, 301
    7.4 System Engineering, Acquisition Management, and Program
         Execution, 308
    7.5 Personnel Management: Acquiring Personnel and
         Managing Careers to Meet Network-Centric Needs, 317
    7.6 Organizational Responsibilities for Effective Network-
         Centric Operations Integration, 324
    7.7 Recommendations, 331

APPENDIXES

A Admiral Johnson's Letter of Request   351
B Current Sensor Capabilities and Future Potential   352
C System Requirements to Hit Moving Targets   384
D Weapons   404
E Tactical Information Networks   429
F The Organizational View of the Recommended Operations
   Information and Space Command   462
G Committee Biographies   464
H Acronyms and Abbreviations   474

Executive Summary

ES.1 WHAT ARE NETWORK-CENTRIC NAVAL FORCES?

ES.1.1 Network-Centric Operations Defined

This report responds to a request from the Chief of Naval Operations to help the Navy "[realize] . . . the full potential of network-centric warfare. . . ."[1] The committee received many briefings on the subject, none of which defined "network-centric warfare" in the same way. Thus, the committee deemed it important to establish a common basis of understanding regarding what is meant by the "network centric" concept and its characteristics within the Department of the Navy and from there into the joint arena. Further, it concluded that once adopted as an organizing principle the concept must apply to *all* military force operations, in peace as well as in war. The committee therefore defined network-centric operations (NCO) as *military operations that exploit state-of-the-art information and networking technology to integrate widely dispersed human decision makers, situational and targeting sensors, and forces and weapons into a highly adaptive, comprehensive system to achieve unprecedented mission effectiveness.*

ES.1.2 The Promise and Significance of Network-Centric Operations

In network-centric operations naval force assets are linked together to carry out a mission in ways that were not previously possible, through the application of modern means of acquiring, processing, disseminating, and using information

---

[1]See Appendix A.

and information networks. The gathering, exploitation, and transmission of information about the enemy and the environment have always been of critical importance in guiding military operations. The means for doing so have become so powerful in recent times that they have overtaken the capabilities of individual platforms and weapons as primary drivers of global naval force capability.

Network-centric operations thus represent a new force design and operational paradigm for the naval forces. In network-centric operations, naval force and other Service elements, organized as a single, joint, networked system, will be able to achieve mission objectives far more rapidly, decisively, and with greater economy of force than was possible earlier. However, the entire, joint system will be more intricate than any the naval forces and joint forces have ever dealt with in the past. For the Navy and the Marine Corps, the transition to NCO will require that many of the traditional approaches to development and operations be transformed into new methods and concepts of operation.

ES.1.3 Attributes of Naval Forces in Network-Centric Operations

The key attribute of NCO is the unprecedented ability to support well-informed and rapid decision making by naval force commanders at all levels, within a system of flexible and adaptable command relationships. The information network and infrastructure in which the naval force elements will be embedded will enable dynamic adjustment and adaptation to battlespace situations and needs as they emerge. Multiple platforms separated by great distances will be able to work as closed-loop systems with the same speed and assurance that have characterized single platform-weapon combinations. Within the physical limits of time required for movement and weapon range and speed, the force commanders operating in the network-centric mode will be able to concentrate widely dispersed forces' fire and maneuvers at decisive locations and times. The forces will be able to achieve the precision needed to identify and engage opposing forces and specific targets with minimal casualties and the least civilian damage. And they will be able to do so at a pace that overwhelms the opposition's ability to prevent the actions or to respond in time to avoid defeat.

To develop these attributes of NCO, information and networking technology will have to be applied to achieve the following, to the greatest extent possible:

• Knowledge of where all U.S., allied, neutral, and opposition installations, forces, and platforms are, in terms of common space and time coordinates, in time to use the knowledge to desired military effect;
• Sharing of processed information throughout the force as and when needed by the decision makers at various command levels;
• Coordination of all (possibly widely dispersed) assets—sensors, weapons, platforms, Marine units—to operate as a common whole; and

- Assurance that the information that is gathered and distributed is timely, accurate, and not subject to disruption, corruption, or exploitation by the opposition.

ES.1.4 The Inevitability of Network-Centric Operations

The committee believes that development of the naval forces in the direction of network-centric operations is inevitable, because of both the push of developing threats worldwide and the pull of opportunities that the information and networking technology offers.

All of the following are becoming available to potential opponents of U.S. naval forces: stealth in antiship missiles; quieter submarines; long-range air defenses with counterstealth characteristics; battlefield ballistic missiles that may have chemical, biological, and eventually nuclear warheads; hiding of organized criminal, terrorist, and irregular forces in civilian populations and difficult terrain; cell phone and satellite communication and navigation; and cyber-warfare capability. A concatenation of such threats can be met only by sharing, among all friendly force elements, information gathered by widely dispersed assets and fused to make a coherent operational and tactical picture for the force's decision makers, so as to enable an effective response or preemptive action, all in less time than it takes the threat to strike. Information and networking technology makes such sharing possible.

In addition, current and, it is expected, future U.S. superiority in exploiting the technology presents the opportunity to build naval forces that will be able to undertake the decisive operations basic to success in missions as far into the future as can be foreseen.

ES.2 TRANSITION TO NETWORK-CENTRIC NAVAL FORCES

To achieve naval forces able to perform as described will require leadership from the top levels of the Navy Department; new concepts of operation; a common information infrastructure with assured reliability and integrity of the information that passes through it; and an integrated approach to shaping the Navy and the naval forces.

ES.2.1 Leadership

The Department of the Navy's top leadership must convey understanding, acceptance, and their continuing support of the concept of network-centric operations throughout the naval forces, including their anticipation of and support for the NCO-induced changes in command relationships that will inevitably come about as the command and information structure of the naval forces evolves.

Recommendation 1: The Secretary of the Navy, the Chief of Naval Operations (CNO), and the Commandant of the Marine Corps (CMC) should agree on the basic concepts essential to transforming today's naval forces to network-centric forces, including:

  a. Integrating all the naval force elements involved in a mission into an adaptive, comprehensive, information-driven NCO system;

  b. Adopting the spiral development process that is described in this report[2] as the primary development and procurement mechanism for creating such NCO systems;

  c. Constructing a common command and information infrastructure (the Naval Command and Information Infrastructure; NCII[3]) as the framework that enables the creation and effective utilization of effective NCO systems; and

  d. Making the attending adjustments and enhancements in organization and management.[4]

They should promulgate those concepts throughout the naval forces as top-level policy.

ES.2.2 Concepts of Operation

Operations in which all force elements are closely coupled and function as a single system within a common command and information network will differ in speed and character of execution from those familiar in the past. New kinds of operations will be possible, as illustrated by the recent development of the cooperative engagement capability for fleet air defense. The flow of information from many sources to multiple command levels will tend to flatten the combat command hierarchy within agreed mission plans and rules of engagement. All future military operations, in peace and in war, will be joint, and will occur most often in coalitions. Even when the Navy and Marine Corps are the only military forces at a point of action, the information network and the sensors that the forces rely on will be interconnected with information assets from other Services and National[5] agencies. Command and information links with coalition partners will also have to be assured.

The CNO and the CMC have assigned to the Navy Warfare Development Command (NWDC) and the Marine Corps Combat Development Command (MCCDC), respectively, the responsibility for developing new concepts of operation in the joint and combined environment. Each of these organizations is

---

[2]See Chapters 1 and 2.
[3]See Chapters 1, 4, and 6.
[4]See Chapters 1 and 7.
[5]The term "National" refers to those systems, resources, and assets controlled by the U.S. government, but not limited to the Department of Defense.

EXECUTIVE SUMMARY 5

devising concepts for its parent Service. However, the naval forces as a whole cannot function in the NCO mode unless they share common concepts of operation involving both Services.

Recommendation 2: The CNO and the CMC should assign NWDC and MCCDC the responsibility to work *together* to devise joint concepts and doctrine for network-centric operations of the naval forces as a whole. Joint and coalition aspects of such operations should be incorporated in the concepts developed.

ES.2.3 Common Command and Information Infrastructure

Network-centric operations require an infrastructure that supports not only the manipulation and transport of information but also the actual functions of command, to hold the elements of the network together and guide their operation in concert as an integrated system according to the NCO concept. That infrastructure, the NCII, will include the communications trunk lines, the terminals, the central processing facilities, the common support applications, connectivity to tactical networks, and the Department of Defense (DOD)-wide and commercial standards, rules, and procedures that will enable the flow of raw and processed information and commands at all levels of command among units that are involved in an action. The NCII will be connected to, and will essentially have to become a part of, the joint National and coalition information infrastructures to the extent that all will function as a single infrastructure to ensure consistency and interoperability among all the parts.

Recommendation 3: The Secretary of the Navy, the CNO, and the CMC should arrange for assembly, augmentation, and interweaving of all related ongoing efforts[6] to begin creating the NCII as a common command and information infrastructure to provide the global framework for networked naval force operations.

Recommendation 4: The Secretary of the Navy, the CNO, and the CMC should develop a comprehensive and balanced transition plan to aid realization of the functional capabilities necessary for the NCII (as described in the detailed recommendations in the body of this report[7]).

---

[6]As discussed at length in Chapters 4 and 6, these efforts include the Navy's IT-21 strategy, the Global Command and Control System–Maritime, common-user long-haul communications, tactical networks, common support application software, and sensor and intelligence feeds, including as necessary other joint and National assets.

[7]See Chapter 6.

ES.2.4 Information Assurance

Many threats[8] will arise from the very structure of the NCII, and also from the need to rely heavily on civilian systems for the transport of data and processed information, the need to share information and techniques with coalition partners, and the potential for damaging actions by malicious insiders who may also be enemy agents. There is currently no single individual within the Department of the Navy who has the responsibility and authority to ensure the integrity of the NCII and the information that flows through it, and the timeliness and continuity of the information flow.

Recommendation 5: The Secretary of the Navy, the CNO, and the CMC should assign responsibility for information assurance at a high enough level within the Navy and the Marine Corps, and with sufficient emphasis, to ensure that adequate and integrated attention is paid to all aspects of information assurance in the design and operation of the NCII.

Recommendation 6: The CNO and the CMC should take steps to ensure that fleet and Marine training encompasses situations with impaired information and NCII functionality, and that fallback positions and capabilities are prepared to meet such eventualities.

ES.2.5 Integrated Approach to Shaping the Navy and the Naval Forces

Network-centric operations will span all Navy and Marine Corps activities. Since the force components, the people in the force, and the information network in which they are embedded will be treated as a complete system, the new approach to shaping the Navy and the naval forces will entail performance and economic trade-offs among *all* the parts of the system—weapons, platforms, people, command, control, and information assets—not simply *within* the parts as has been customary heretofore. And there will have to be corresponding organizational and business practice adjustments in the Navy and the naval forces to suit the new conditions. The committee examined alternative approaches to achieving these changes but concluded that the best Department of the Navy strategy to meet these needs would be to build on existing organizations with some changes in emphasis.

The following needs were identified, and recommended approaches to meeting these needs are given. It is, of course, recognized that internal and external considerations that were not known to the committee may lead the Navy Department to reach other solutions to the problems posed.

---

[8]Described in detail in Chapter 5.

- In the current fleet/Office of the Chief of Naval Operations (OPNAV)/ Systems Command (SYSCOM) organizational relationships, there is no mechanism for integrating cross-platform/cross-mission needs of the battle force in operations information—including terrestrial and space assets; command, control, communications, computing, intelligence, surveillance, and reconnaissance (C4ISR); and the NCII. The lack of a functional type commander[9] resource for C4I who can interact with the platform type commanders exacerbates this cross-platform integration problem.

Recommendation 7: The Secretary of the Navy and the CNO should create a new functional type commander, the Commander for Operations Information and Space Command, to be the single point of information support to all the fleets. Responsibilities for the new functional type commander and related other changes in Navy organizational responsibilities are described in the detailed recommendations in the body of this report.[10]

- A mechanism is needed to integrate various competing and complementary requirements presented by the fleets to ensure rapid improvement of at-sea operational capabilities in the NCO mode through the spiral development process.

Recommendation 8: The CNO should establish a requirements board[11] under the chairmanship of the Vice Chief of Naval Operations to deal with operations information and to integrate requirements presented by the fleets as the NCII is assembled and other NCO plans and acquisitions take shape.

- An authority is needed to make funding, scheduling, and program adjustments, trade-offs, and decisions in relevant areas, based on review, oversight, and prioritization of the acquisition, installation, and program execution aspects of NCO systems treated in an integrated fashion.

Recommendation 9: The Secretary of the Navy, the CNO, and the CMC should establish a board of directors[12] under the chairmanship of the Undersecretary of the Navy to provide coordinated guidance and ensure the integration and interoperability of all the Navy and Marine Corps NCO acquisition and program execution activities.

- Decision support and program execution mechanisms are needed to improve and enhance implementation of the decisions made by the above authority.

---

[9]The flag officer who has responsibility for all ships of a certain type in the fleet.
[10]See Chapters 1 and 7.
[11]See Chapters 1 and 7.
[12]See Chapters 1 and 7.

Recommendation 10: The CNO should strengthen mission analysis and component trade-off evaluations by (1) providing staff and resources for the integrated warfare architecture (IWAR) process to enable continuous assessments from requirements generation through programming, budgeting, and execution; (2) developing output-oriented measures of effectiveness and measures of performance for network-centric operations; and (3) developing a comprehensive set of design reference missions across all mission areas. Resource planning should support the spiral development process.

    a. The Secretary of the Navy and the CNO should appoint a designated SYSCOM Commander to be a deputy to the Assistant Secretary of the Navy (Research, Development, and Acquisition) (ASN (RDA)) for Navy NCO integration.

    b. The Secretary of the Navy should adjust the responsibilities of the Chief Information Officer, the Chief Engineer, and the N6, with due account for authorities and responsibilities established in law, to enable the implementation and operation of the NCII, including interaction and collaboration with the other Services, the joint community, and defense agencies.[13]

- There is a need to ensure that all missions are given balanced emphasis in the naval force planning and acquisition processes. In particular, the committee found that the power projection mission is not as well represented in the planning process as other naval force missions. Special attention is needed to the planning and design of end-to-end (surveillance and targeting through effectiveness assessment) fleet-based land-attack (strike and fire support) subsystems for network-centric operations.[14]

Recommendation 11:

    a. The Office of the ASN (RDA), in conjunction with other interested Navy and Marine Corps elements, should review the Navy's overall planning and acquisition processes and if necessary and as appropriate adjust the program executive office structure to orient it toward the integrated design and acquisition of systems suited to network-centric operations.

    b. The CNO should review and if necessary and as appropriate adjust the N8 structure and assignments within his staff to ensure balanced attention to all missions, including the mission of power projection from the sea.

- Without effective, appropriately educated and trained people the NCO concept cannot be made to work. To be fully effective in implementation over

---

[13] See Chapters 1, 4, and 7.

[14] See Chapters 1, 3, and 7. There were some differing views within the committee regarding the following recommendations, as indicated in related discussion in these chapters.

the long term, NCO concepts must pervade the Navy and Marine Corps training and education system. This approach includes identifying the qualifications for billets critical to network-centric operations (including both domain and infrastructure experts); identifying training and education needs for those billets; developing career paths for both military personnel and civil service employees to retain and reward those with information technology expertise; and orienting the education of naval officers toward NCO concepts from the beginning of their schooling.[15]

Recommendation 12: The CNO and the CMC should review NCO education and training at all levels across the Navy and the Marine Corps, and institute changes as necessary and appropriate to achieve the objectives outlined above.

- Research and development is needed to meet the challenges of creating an advanced NCII, including providing for information assurance, and to meet the new challenges of network-centric operations, including especially support of the power projection mission in NCO.

Recommendation 13: The Office of the ASN (RDA), in conjunction with other interested Navy and Marine Corps elements, should join with the other components of DOD to sponsor a vigorous, continuing research and development program aimed at the objectives noted above.

The above recommendations, and related ones, are expanded and discussed more fully in the overview that follows this summary. Many additional recommendations for actions to reorient the naval forces toward NCO, involving many areas of naval force endeavor, emerged from this study. All the recommendations, including those above and many others, are developed in detail and presented in the main body of the report.

---

[15]See Chapters 1 and 7.

1

Overview of Study Results

1.1 MISSION EFFECTIVENESS: WHAT IS REQUIRED

1.1.1 Joint Vision 2010

In one way or another all military operations will be joint. That is, systems and forces from all the Services and from National agencies will contribute to the U.S. Armed Forces' operations in ways that vary with the circumstances. Developed in the past few years by the Joint Chiefs of Staff, Joint Vision 2010[1] envisions how the Armed Forces will channel the vitality and innovation of the nation's people and use the leverage offered by advancing technology to achieve unprecedented levels of power, timeliness, and decisiveness in joint operations and warfighting. The Navy and Marine Corps have also developed conceptual descriptions of their own future warfighting strategies—"Forward...From the Sea"[2] and "Operational Maneuver From the Sea"[3]—that have themes in common with Joint Vision 2010. Most importantly, all of these concepts have recognized

---

[1] Shalikashvili, GEN John M., USA. 1997. *Joint Vision 2010*. Joint Chiefs of Staff, The Pentagon, Washington, D.C.

[2] Department of the Navy. 1997. "Forward...From the Sea." U.S. Government Printing Office, Washington, D.C.

[3] Headquarters, U.S. Marine Corps. 1996. "Operational Maneuver From the Sea." U.S. Government Printing Office, Washington, D.C., January 4.

the fundamental role that information superiority will play in the forces' ability to prevail over adversaries.[4]

Focusing on achieving dominance across the range of military operations through the application of new operational concepts, Joint Vision 2010 provides a joint framework of doctrine and programs within which the Services can develop their unique capabilities as they prepare to meet an uncertain and challenging future. The scope and complexity of the challenges and the capabilities required to meet them were projected in a recent Naval Studies Board report (the TFNF—Technology for Future Naval Forces—study;[5] see Box 1.1), an effort from which this current study follows naturally.

1.1.2 Network-Centric Operations

The implications of Joint Vision 2010, future naval operational concepts, and the spread of advanced technology and commercial information systems worldwide make it inevitable that joint forces, and particularly forward-deployed naval forces, must move toward network-centric operations. The committee defines such operations as follows: *Network-centric operations (NCO) are military operations that exploit state-of-the-art information and networking technology to integrate widely dispersed human decision makers, situational and targeting sensors, and forces and weapons into a highly adaptive, comprehensive system to achieve unprecedented mission effectiveness.*

Forward deployment of naval forces that may be widely dispersed geographically, the use of fire and forces massed rapidly from great distances at decisive locations and times, and the dispersed, highly mobile operations of Marine Corps units are examples of future tasks that will place significant demands on networked forces and information superiority. Future naval forces must be supported by a shared, consolidated picture of the situation, distributed collaborative planning, and battle-space control capabilities. In addition, the forces must be capable of coordinating and massing for land attacks and of employing multisensor networking and targeting for undersea warfare and missile defense.

In network-centric operations, the supporting information infrastructure, ideally, will deliver the right information to the right place at the right time to achieve the force objectives. Also, although rules of engagement (ROEs) are

---

[4]*Joint Vision 2010* (p. 16) defines information superiority as "the capability to collect, process, and disseminate an uninterrupted flow of information while exploiting or denying an adversary's ability to do the same." Information superiority will therefore require "both offensive and defensive information warfare" capabilities.

[5]Naval Studies Board, National Research Council. 1997. *Technology for the United States Navy and Marine Corps, 2000-2035: Becoming a 21st-Century Force,* 9 volumes. National Academy Press, Washington, D.C.

usually determined politically and morally, accurate information delivered rapidly to a commander may affect how ROEs are applied, for example, by providing input to decisions for preemptive attack in primarily defensive situations. Network-centric operations must also ensure that when forces move and weapons are delivered according to the information furnished, they arrive at the right places and times to achieve the force objectives. Thus, the command relationships, the information systems and networks, implementations of ROEs, and the combat forces themselves must all evolve toward network-centric operations together.

The trend toward network-centric operations is inevitable. There are many reasons why this is so. One reason is the pull of opportunity: The anticipated effectiveness of joint, networked forces is compelling. A second is the push of necessity: Threats are becoming more diverse, subtle, and capable. If they are to be discerned, fathomed, and effectively countered in timely fashion, increasingly complex information gathering and exploitation will be required. Also, the diversity and geographic spread of potential threats and operations, many of which will occur simultaneously or nearly so, demand that forces of any size be used to their maximum effectiveness and efficiency. Another reason derives from the relentless advance of U.S. and foreign technology in both the civilian and military spheres: There will be no other way for U.S. forces to develop. Only a force that is attuned to and capable of harnessing the power of the information technology that drives modern society will be able to operate effectively to protect that society.

The naval forces are already moving toward network-centric operations. Joint task force commands afloat are being established to direct ongoing operations and are the subjects of fleet battle experiments. Elements of network-centric forces and operations are both in place and in the making, in the Aegis system and its extensions to theater missile defense, and in the cooperative engagement capability (CEC) for fleet defense against cruise missiles and its shoreward extensions.[6] The Navy's information technology thrust is becoming

---

[6]It is remarkable that in World War II the U.S. Navy's Tenth Fleet exercised network-centric antisubmarine warfare (ASW) operations in the Battle of the Atlantic against German submarines, characterized by Morison as ". . . a contest between systems of information . . ." (as quoted by Cohen, Eliot A., and John Gooch. 1990. *Military Misfortunes: The Anatomy of Failure in War,* The Free Press, A Division of Macmillan, Inc., New York and Collier Macmillan Publishers, London, p. 75). The Tenth Fleet integrated information from distant direction-finding fixes with data from local high-frequency direction finder and radar contact from forces in the action area with decrypted messages and other intelligence from vessels attacked, and with the help of a strong operational analysis group directed the coordinated efforts of warships, aircraft, and convoy commanders, with time delays from initial detection to action orders of minutes to hours. The Tenth Fleet also shared its operational picture and coordinated actions with the British in charge of the Eastern Atlantic ASW operations and conducted information warfare in the form of psychological warfare messages directed specifically to the enemy submarines at sea.

Box 1.1 Future Naval Operations

Technology for the United States Navy and Marines Corps, 2000-2035 (the TFNF study)[1] projected that future naval forces would continue to be required to perform tasks such as the following (Vol. 1, *Overview*, p. 3):

- Sustaining a forward presence;
- Establishing and maintaining blockades;
- Deterring and defeating attacks on the United States, our allies, and friendly nations, and, in particular, sustaining a sea-based nuclear deterrent force;
- Projecting national military power through modern expeditionary warfare, including attacking land targets from the sea, landing forces ashore and providing fire and logistic support for them, and engaging in sustained combat when necessary;
- Ensuring global freedom of the seas, airspace, and space; and
- Operating in joint and combined settings in all these missions.

These tasks are not new for the naval forces and have changed little over the decades. However, advanced technology is now spreading around the world, and burgeoning military capabilities elsewhere will, in hostile hands, pose threats to U.S. naval force operations. The most serious are as follows (pp. 4-5):

- Access to and exploitation of space-based observation to track the surface fleet, making surprise more difficult to achieve and heightening the fleet's vulnerability;
- Increased ability to disrupt and exploit technically based intelligence and information systems;
- Effective antiaircraft weapons and systems;
- All manner of mines, including "smart" minefields with networked sensors that can target individual ships for damage or destruction by mobile mines;
- Antiship cruise missiles with challenging physical and flight characteristics;
- Accurately guided ballistic missiles able to attack the fleet;
- Quiet, modern, air-independent submarines with modern torpedoes; and
- Nuclear, chemical, and biological weapons.

Future naval forces must be designed to meet these threats while maintaining the forward presence and operational flexibility that have characterized U.S. naval forces throughout history. This capability must be achieved in a world of ever advancing technology (particularly information technology) available globally through the commercial sector and sales to foreign military users.

The TFNF study described the characteristics of future naval force operations as follows (p. 6):

- Operations from forward deployment, with a few major, secure bases of pre-positioned equipment and supplies;
- Great economy of force based on early, reliable intelligence; on the timely acquisition, processing, and dissemination of local, conflict-, and environment-related information; and on all aspects of information warfare;
- Combined arms operations from dispersed positions, using stealth, surprise, speed, and precision in identifying targets and attacking opponents, with fire and forces massed rapidly from great distances at decisive locations and times;

- Defensive combat operations and systems, from ship self-defense through air defense, antisubmarine warfare, and antitactical ballistic missile defense, always networked in cooperative engagement modes that extend from the fleet to cover troops and installations ashore;
- Marine Corps operations in dispersed, highly mobile units from farther out at sea to deeper inland over a broader front, with more rapid conquest or neutralization of hostile populated areas, in the mode currently evolving into the doctrine for Operational Maneuver From the Sea;
- Extensive use of commercial firms for maintenance and support functions; and
- Extensive task sharing and mission integration in the joint and combined environment, with many key systems, especially in the information area, jointly operated.

The TFNF study concluded that these future threats and operational requirements would demand the development of new naval force capabilities, which would in turn necessitate a complete transformation of future naval forces. These breakthrough capabilities included the following (p. 5):

- Sustained information superiority over adversaries;
- Major ships operated effectively by fewer people, through the use of networked instrumentation and automated subsystems [with high maintainability and reliability];
- A family of rocket-propelled, guided missiles, significantly lower in cost than today's weapons, that will greatly increase the responsiveness, rate of fire, volume of fire, and accuracy of strike, interdiction, and supporting fire from surface combatants and submarines;
- STOL [short takeoff and landing] or STOVL [short takeoff and vertical landing], stealth, and standoff in combat aircraft;
- Cooperative air-to-air engagement at long range using networked multistatic sensor, aircraft, and missile systems;
- Use of unmanned aerial vehicles (UAVs) for both routine and excessively dangerous tasks;
- Greatly expanded submarine capability to support naval force operations ashore;
- Recapture of the antisubmarine warfare advantage that has been eroded by quieting of Russian nuclear submarines and by advanced air-independent non-nuclear submarines that are being sold by other nations on world markets;
- The ability to negate minefields at sea, in the surf, and on the beaches much more rapidly than has been possible heretofore;
- Novel weapons, systems, and techniques for fighting in populated areas, against organized military forces, irregulars, and terrorist and criminal groups; and
- Logistic support extensively based at sea that will provide needed materiel on time with far less excess supply in the system than has been the case in the past.

---

[1]Naval Studies Board, National Research Council. 1997. *Technology for the United States Navy and Marine Corps, 2000-2035: Becoming a 21st-Century Force,* 9 volumes. National Academy Press, Washington, D.C.

evident in the fleet and its support operations. During the Cold War, networked antisubmarine warfare (ASW) systems were devised to overcome the Soviet submarine threat. As the TFNF report points out, networked operations will become necessary to achieve an effective defense against quiet submarines in the littoral environment and against mine warfare; effective fleet fire and logistic support of Marines ashore in Operational Maneuver From the Sea (OMFTS); and effective protection against growing air defense capabilities of potential adversaries that will demand engagements at very long ranges.

Today, however, all of these network-centric operations and capabilities, existing and under development, are evolving in an essentially fragmented and stand-alone manner. The focus is still on the subsystems or components of the total naval force combat system, and they are not yet fully coordinated with one another. It has become clear that unless networked naval forces are treated as a total system, a great deal of money will be wasted and opportunities to enhance warfighting capabilities will be lost. Beyond optimizing individual sensors, weapons, and command, control, communications, and intelligence (C3I) systems, it is essential to achieve overall optimization of the total system of networked combat assets, including the information that ties them all together and makes them fully effective.

Network-centric operations with fully networked forces will provide the significant advances demanded for success in future warfighting and in countering the capabilities of future adversaries. They will enable better and faster battlespace decisions, providing time and direction for rapid, integrated execution of tasks with flexible use of both dispersed and concentrated (and other joint and combined) assets. At the same time, however, network-centric operations will present significant new vulnerabilities that must be actively managed through the application of technology and doctrine. Both aspects of network-centric operations are treated in this report.

1.1.3 Approach and Emphasis in This Report

This report describes the operational concepts, command and control relationships, and information systems architecture necessary to support the networked naval forces. Many requirements for sensor and weapon systems assets in the future systems are also discussed, as is information assurance, which is critical to achieving true information superiority.

In keeping with the definition of network-centric operations given above, the committee considered more than just the design of information and communication systems, a critically important topic in itself. Since the point of network-centric operations is to empower the entire naval force to maximize the effectiveness of its operations, this examination of network-centric operations has been extended to include the entire naval force system encompassed by the committee's definition of network-centric operations, and network-centric operations are treated in terms of mission accomplishment by that system.

When the committee examined the naval forces' mission spectrum from this point of view, it realized that the force capability has not developed rapidly enough in all mission areas since the end of the Cold War to keep up with the ensuing profound change of emphasis in overall mission orientation (see discussion in Box 1.2). As a consequence, attention is devoted in several parts of this report to the power projection mission, and network-centric operations are discussed in terms of the subsystems and components that will enable the naval force network to succeed in that mission.

Finally, as requested in the terms of reference, attention is also given to the demands that the move to network-centric operations will make on the business practices and organization of the Department of the Navy, including the problems associated with the training, retention, and promotion of naval personnel in the developing network-centric operations environment, as well as the unprecedented opportunities offered by the new information and networking technologies.

In the following overview of study results, the recommendations associated with each major topic are presented following the discussion of that topic. Additional recommendations are offered in Chapters 2 through 7.

1.2 LEADING THE TRANSFORMATION TO NETWORK-CENTRIC OPERATIONS

1.2.1 Integrated Systems for Operations

Network-centric operations represent a new approach to warfighting. When that approach and its elements are discussed, familiar terms come to be used in new ways to deal with new concepts.

In network-centric operations, a set of assets, balanced in their design and acquisition so as to be integrated with one another, must operate together effectively as one complete system to accomplish a mission. The assets assembled in such a *network-centric operations (NCO) system* encompass naval force combat, support, and command, control, communications, computing, intelligence, surveillance, and reconnaissance (C4ISR) elements and subsystems, integrated into an *operational and combat network*. Such subsystems will be designed and acquired to meet specific requirements of their tasks in the overall mission. For example, a fleet and amphibious force assembled for an expeditionary operation along the littoral will comprise subsystems designed for power projection but will also include antiair, antimissile, and antisubmarine subsystems to protect the naval force while it is projecting power ashore, as well as logistics subsystems to support the forces at sea and ashore.

The subsystems' *components* will be ships, aircraft, missiles, communications, and other parts of the C4ISR network. These components will continue to

Box 1.2 Network-Centric Operations for Power Projection

The naval forces have always had the missions of deterrence, forward presence, sea and air control, and power projection. During the Cold War the emphasis was on strategic deterrence, protection of the sea transit of reinforcements to the European theater, and the ability, under the maritime strategy of the 1980s, to bring naval aviation within striking distance of the Soviet Union. Because the Soviet threats to the fleet were severe enough to keep it from carrying out those missions, defensive operations were of critical importance and led to networked operations in antisubmarine warfare (ASW) and Fleet Air Defense. The ASW network included fixed arrays such as the Sonar Ocean Surveillance Underwater System, as well as sensor and attack capabilities by maritime patrol aircraft, carrier-based aircraft, and ship- and submarine-based ASW systems, all operated in a cooperative manner to find and neutralize Soviet submarines. The Fleet Air Defense system included the Outer Air Battle systems, Aegis, and ultimately the cooperative engagement capability to counter low, stealthy, or supersonic antiship cruise missiles.

Since the end of the Cold War the naval forces have turned their attention to expeditionary warfare and military operations other than war in the world's littoral zones, especially those of the Eurasian and African land masses. As threats against the fleet and movement over the seas have diminished, emphasis has shifted to the forward presence and power projection missions. In the words of the Chief of Naval Operations, Admiral Jay Johnson, USN, "The purpose of Naval Forces is to influence directly and decisively, events ashore from the sea—anytime, anywhere."[1] Although much work remains to be done in the other mission areas, it became apparent to the committee during its study that elements of the power projection mission have lagged significantly and now require renewed emphasis. These mission elements may be grouped according to the following phases of a campaign:

- Preparing the battlespace: This involves integrated battlespace sensing and sea- and air-launched strikes against inland targets using fleet firepower and information warfare;
- Landing the force: This includes countermine warfare, landing the Marines ashore in their developing Operational Maneuver From the Sea mode of operation, and providing them with close air support during the landing;
- Engaging the enemy; and
- Supporting the force ashore: This entails supplying fire support and logistic support from the sea.

---

[1]Sestak, RADM Joseph A., Jr., USN, Director of Strategy and Policy Division, "A Maritime Concept for the Information Age," briefing to the Naval Studies Board on November 18, 1999, Office of the Chief of Naval Operations (N51), Washington, D.C.

involve research, development, and acquisition efforts involving extensive resources.

Although this characterization of the NCO system might imply a classical system-subsystem-component hierarchy, it must be recognized that NCO systems may differ in composition, but not in concept, depending on the mission or the circumstances. Thus there can be different NCO systems for various purposes—e.g., for forward presence and deterrence, or for fighting a major theater war—sometimes operating simultaneously within a global network.

To support such adaptations of the overall system concept, different stages of system design and acquisition will require different types of system-oriented analyses. Development and experimentation in the field to perfect various NCO concepts require *operational analyses*. System planning, programming, and budgeting, as well as making trade-offs among mission-oriented subsystems of what will become NCO systems, require *systems analyses*. Building the components and subsystems to work together satisfactorily requires *system engineering*.

Network-centric operations represent a new paradigm for the naval forces, which no longer will be considered in terms of assemblages of ships, aircraft, Marine units, and weapons drawn together to fight battles. Rather, the platforms, Marine units, and weapons will be part of a network integrated into a system to carry out a mission, supported by a common command and information infrastructure. All the naval forces, at all command levels, will be involved in and affected by this change.

Network-centric operations are characterized by the rapid and effective acquisition, processing, and exchange of mission-essential information among decision makers at all command levels, enabling them to operate from the same verified knowledge base, kept current according to the temporal needs of the commanders at the different levels. This approach will enable the naval forces to perform collaborative planning and to achieve rapid, decentralized execution of joint actions, based on the most accurate and timely situational and targeting knowledge available. It will enable them to focus the maneuvers and fire of widely dispersed forces to carry out assigned missions rapidly and with great economy of force.

Network-centric operations systems include, in addition to the people who use the information in the network to direct operations, the naval forces' platforms, weapons, Marine units, and all the parts of the command and information structure within which they fit and that binds them together and guides their operations. Joint Service elements or forces and coalition forces operating with the naval forces must also be included. In any mission assignment, from peacetime engagement to combat in a major theater war, NCO systems encompass, as appropriate, all operations from a single weapon engaging a single target to a regional force including one or more fleets and Marine expeditionary forces that might be operating anywhere in the world.

The command and information parts of NCO systems include all the sensors and their platforms, from shore-based installations through ships, manned and unmanned aircraft, and spacecraft; processing and display subsystems; communication links; common supporting software; the standards, rules, and procedures that lend structure to the network and enable seamless, integrated functioning of all its parts; and the people at all levels, in joint and combined forces, who use the information in carrying out their tasks and missions and who maintain and operate the system's infrastructure. The Naval Command and Information Infrastructure (NCII), meshed with and functioning as part of a joint and national infrastructure, must provide a functional framework for establishing and maintaining the relationships and for transferring information among all the system parts, and for coordinating functions across all the platforms and force units in the joint and combined environment.

Figure 1.1 summarizes the comprehensive nature of network-centric operations systems; that view guided the committee's deliberations.

1.2.2 Creating Network-Centric Operations Systems

Transforming the naval forces from platform-centric to network-centric design and operations will require a disciplined approach to developing very-large-scale integrated systems. New concepts of operation embodying new technical capabilities will have to be developed and then tested in the field, with the test results used to refine the concepts continually and adapt them to changing conditions of threat, environment, and technological advance. This means using up-front, empirically founded operational and system analyses to set system performance, cost, and schedule requirements based on emerging concepts of opera-

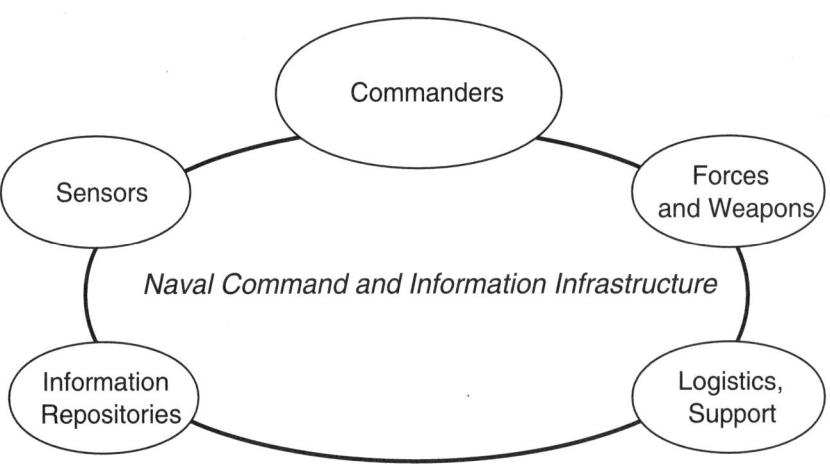

FIGURE 1.1 An NCO system structure.

tion; performing studies of the trade-offs in alternative approaches to system design; selecting and documenting a baseline approach; managing the design and implementation of the system according to the planned schedule and cost targets, while being adaptable to unforeseen contingencies; verifying that the design meets requirements; and maintaining meticulous documentation of the entire process.

To implement the system, responsible organizations must first devise joint concepts of how network-centric operations would work. These concepts will form the starting point for the spiral development process described below. Within the naval forces, the Chief of Naval Operations (CNO) and the Commandant of the Marine Corps (CMC) have assigned such development responsibility to the Navy Warfare Development Command (NWDC) and the Marine Corps Combat Development Command (MCCDC), respectively. To date, these organizations have been functioning more or less independently, each devising concepts for its parent Service. However, since network-centric operations will be joint and will most likely involve coalition partners, the NWDC and MCCDC must work together and must incorporate the inputs from other Services and agencies, as well as from potential coalition partners, into their work.

The implementation of network-centric operations does not start from a zero base. The naval forces are faced with transforming today's systems—including "legacy" subsystems, new ones entering service or under development for future service, and also elements of subsystems of other Services, National agencies, and possibly coalition partners—into new, all-inclusive systems. All of these subsystems and their components must be able to operate together, even if they were not originally designed to do so. All must be accounted for in devising network-centric concepts of operation and in designing the systems that will support them.

One of the greatest problems in shifting from today's platform-centric operational concepts to tomorrow's network-centric operational concepts is to ensure interoperability among the subsystems and components of the fleet and the Marine forces as well as of joint and coalition forces. The forces can operate to their full potential only if all subsystems and information network components can operate smoothly and seamlessly together. In the current context "interoperability" does not necessarily mean that the characteristics of all subsystems and components must match at the level of waveforms and data formats. Interoperability means that the subsystems must be able to transfer raw or processed data among themselves by any means that can be made available, from actually having the common waveforms and data formats to using standard interfaces or intermediate black boxes enabling translation from one to another.

Ensuring interoperability will be a very complex, technically intensive task involving network protocols, data standards, consistency algorithms, and many other aspects of network design, as well as numerous procedural matters. The subsystem mix will evolve and will be different from the one that exists today.

Eventually, today's legacy subsystems, most of which were not designed for interoperability, will give way to subsystems that are so designed, but only if the networks are configured appropriately now. Even so, different subsystem and component upgrades or replacements will have different time frames for development and installation, so that interface standards will have to ensure their proper meshing into overall systems as they are created. As network-centric operations systems are constituted, all will have to be based on the same command and information framework (the NCII) and all will have to be interoperable.

Network-centric operations must be based on the transformation of both raw and processed data into "knowledge." That is, the masses of information from often dispersed sources must be integrated, interpreted, and presented to combat leaders in a common operational picture that will enable them to discern meaningful patterns of enemy activity in conditions that are disordered and confused, and to act effectively on that information. This knowledge, coupled with their own experience, judgment, and intuition, will allow well-trained leaders to adapt to the situation at hand, identifying and exploiting enemy vulnerabilities while guarding against exploitation of their own. All the design concepts, equipment, and supporting elements of NCO systems must support this capability.

Essential as they are, analytical methods alone are insufficient for the design of systems of this complexity. Actual experimentation by the fleet and Marine force elements is required, to learn how legacy subsystems and their components will operate together with existing or testbed versions of new subsystems and components and to devise concepts of operation using the new and the legacy subsystems and components in the actual operational environment. When such a development process, part of what has been called spiral development, is used, new equipment and concepts can be incorporated into the fleet and the Marine forces based on validated concepts of operation.

In spiral development, equipment and operational concepts are designed, tested, and then refined or redesigned based on the results of real-world experiments. Concepts and components whose effectiveness is demonstrated in the experiments are incorporated into the operational forces, while those requiring improvement enter the next phase of the development spiral. This process will ensure that NCO systems remain vital and current, evolving continuously to incorporate new technology in a constantly changing environment. The process of spiral development can be expected to converge on successive versions of NCO systems that incorporate major force elements far more rapidly than do traditional processes that call for the full development of subsystems and components before outfitting the forces. Also, it will help to identify and resolve interoperability problems in time to avoid large and expensive retrofit programs.

The shift from platform-centric to network-centric thinking and operation of naval forces will require a shift in the mind-sets, culture, and ways of doing business of all the naval forces (and, indeed, in their connections to other Services and National agencies). To shorten the interval between learning about situations and opposition activity from a variety of information sources both within and outside the naval forces, and taking necessary action, command relationships will have to adapt to the exigencies of operations. Achieving the required speed of action will require flattening of the command hierarchy at certain times and preserving the familiar hierarchy at others. Such profound transformation can only be effected through continuous commitment, attention, and guidance from the top levels of naval force leadership.

1.2.3 Recommendations Regarding the Transformation to Network-Centric Operations

1. Network-centric operations planning, design, and management should emphasize mission success in the network-centric operations mode, not the physical aspects of the C4ISR network per se.

2. The Department of the Navy and its component Services should take a mission-driven, integrated approach with a total-system view to achieve success in transforming the naval forces from platform-centric to network-centric operations. Specific steps to achieve this are included in Section 1.5.

3. The CNO and the CMC should give the Navy Warfare Development Command and the Marine Corps Combat Development Command the responsibility of working together to devise joint concepts and doctrine for network-centric operations for the naval forces as a whole, and to incorporate joint and coalition aspects of such operations in their concepts.

4. The spiral development approach involving the design-test-design of new software and equipment and model-test-model to devise new joint concepts and their testing in fleet and Marine units should be adopted as a standard mechanism for achieving network-centric operations systems.

1.3 INTEGRATING FORCE ELEMENTS: A MISSION-SPECIFIC STUDY OF POWER PROJECTION

1.3.1 Mission Orientation

Network-centric operations systems comprise a number of subsystems, each designed and engineered to accomplish a military purpose. These subsystems are networks of components such as sensors, weapons, command elements, and mission-specific communications, tied together by the NCII. First it is necessary to understand the characteristics of the components and the interdependencies of component performance and subsystem performance.

The four missions of the U.S. Navy are illustrated in Figure 1.2, which summarizes the major components in the Navy's integrated warfare architecture (IWAR) process. The subsystems for strike and fire support missions against land targets are used here as example subsystems to accord with the selected

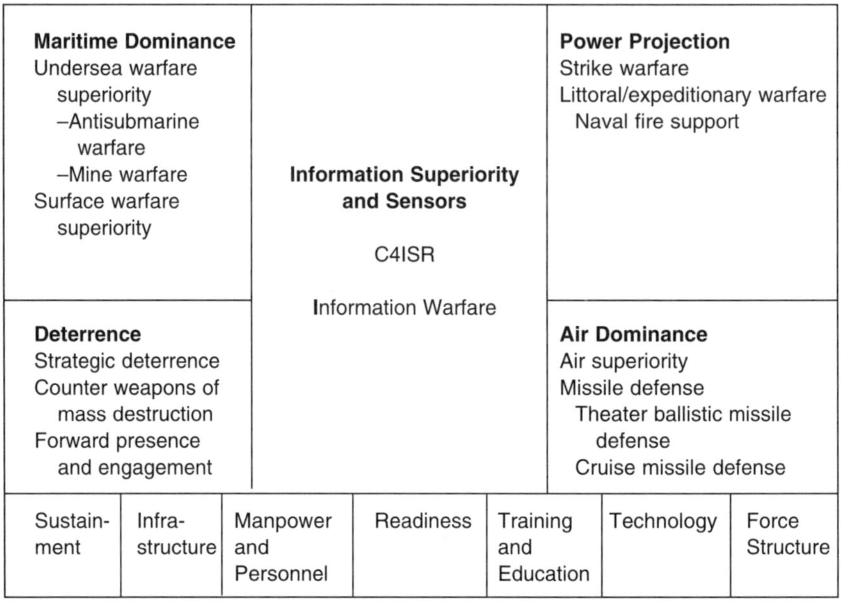

Maritime Dominance
Undersea warfare
  superiority
  –Antisubmarine
    warfare
  –Mine warfare
Surface warfare
  superiority
	Information Superiority and Sensors
	
	C4ISR
	
	Information Warfare
		Power Projection
		Strike warfare
		Littoral/expeditionary warfare
		Naval fire support

Deterrence
Strategic deterrence
Counter weapons of
  mass destruction
Forward presence
  and engagement
		Air Dominance
		Air superiority
		Missile defense
		  Theater ballistic missile
		    defense
		  Cruise missile defense

Sustainment	Infrastructure	Manpower and Personnel	Readiness	Training and Education	Technology	Force Structure

FIGURE 1.2 Naval forces integrated warfare architecture (IWAR) structure. The four fundamental naval force missions are listed in the side columns; all the remainder are essential for carrying them out. (Information superiority and sensors are enablers of all four missions.)

emphasis on the power projection mission described in the introduction to this overview and integrated into the detailed discussion in Chapter 3.

1.3.2 Critical System Needs

Developing the capability for effective power projection by the Navy and Marine Corps requires that the mission-specific networked operations that have already been developed must be integrated into a comprehensive NCO system structure (see Figure 1.1). Under OMFTS, landing (and supporting) forces expands the battlespace deeper into opposition territory and more widely along the littoral. Elements of the total force may be widely dispersed, requiring that they be firmly and effectively linked through the command and information infrastructure. Barriers to landing, such as minefields and proliferated shoulder-fired surface-to-air missiles, must be overcome rapidly. Greater involvement with civilian populations and the need for rapid closure and success and for minimization of U.S. and collateral casualties increase the criticality of accurate, timely fire support from the sea.

Such performance cannot be achieved unless intelligence, targeting, launch platforms, weapons, and postattack assessment are integrated into a fully connected, robust operational and combat network covering every phase of an expeditionary campaign: preparing the battlespace (strike warfare); landing the force (mine clearing, suppression of enemy air defenses (SEAD), amphibious and air-landed operations); engaging the enemy (fire support to forces on the ground); supporting the force ashore (logistics from sea and land); consolidating the position (civic and psychological operations, defending against counterattack); and handing off to follow-on forces and debarking the Marines who were the landing force. Moreover, since naval force operations involve the simultaneous execution of many activities in many mission areas, networked capabilities in other areas, such as ASW and CEC, must be integrated with those for power projection to achieve network-centric operations for an entire force in a total operational context. Creation of such a force-wide NCO capability requires multiple lines of research and development (R&D), procurement, and organizational effort, including the spiral development process described above.

A critical aspect of power projection is the delivery of accurate and timely firepower from the sea on targets ashore, either for strike or for fire support of Marines (and other forces) there. In the past, weapons were typically developed largely independently of the targeting means and of the means for penetrating defenses to deliver the weapons or to assess their effects once delivered. Network-centric operations will require effective integration of sensors and target acquisition, navigation, and weapons to account for all the factors shown in Figure 1.3 and for multiple feedback loops (which have been eliminated from the diagram for simplicity of illustration).

Some specific component needs are discussed below.

1.3.2.1 Sensors and Target Acquisition

To provide all the information needed for force movement and weapon delivery, sensors will have to be linked, as, for example, distributed radars are used in CEC, electronic intelligence sensors are used to guide SEAD attacks, and the Joint Surveillance and Target Attack Radar System (JSTARS) is used to cue specific weapon-targeting sensors against ground forces. Figure 1.4 illustrates how triangulation can reduce target location uncertainty, provided that the sensor positions are precisely known and the observations are synchronized. Coherent processing of detailed sensor observations can produce identified tracks in situations where no single sensor could perform an unambiguous detection, identification, or track. (Although two radar sensors are shown in Figure 1.4, sensors in different frequency domains that meet the above conditions can yield similar results, or better if they contribute to more positive target identification.)

The importance of real-time fusing of multiple sensor outputs as a driver for the target engagement architecture cannot be overemphasized; it is fundamental

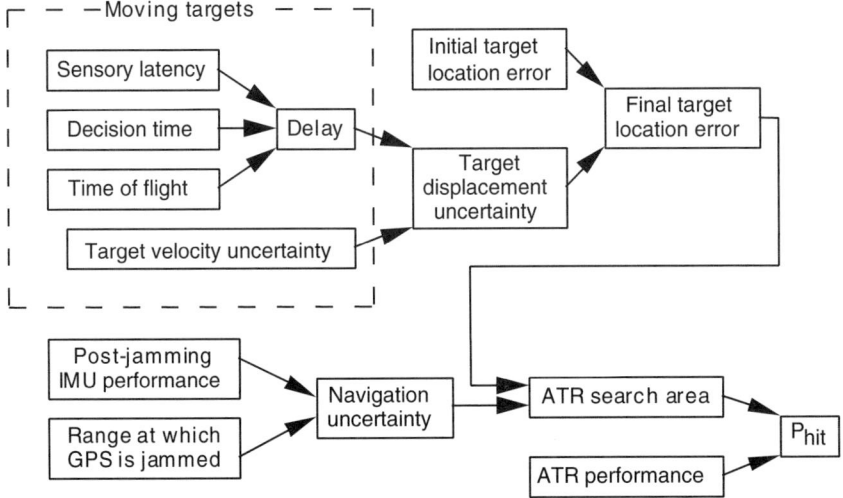

FIGURE 1.3 System factors in delivering firepower ashore without in-flight links to targeting sensors. ATR, automatic target recognition; IMU, inertial measurement unit; GPS, Global Positioning System; P_{hit}, probability of hit.

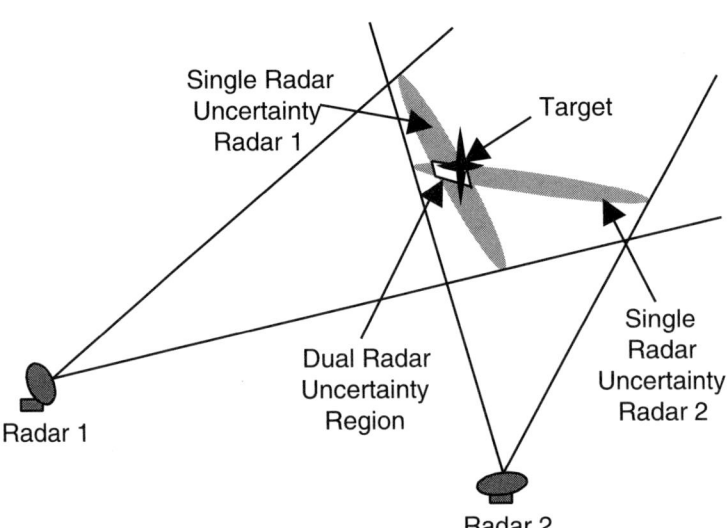

FIGURE 1.4 Reducing target location error with linked cooperating sensors in common coordinates.

to bringing network-centric operations to the point where U.S. forces meet the enemy. The change in architecture brought about by linked sensors is illustrated schematically in Figure 1.5.

The implications for change in the nature of combat engagement as illustrated in Figure 1.5 are profound. On a single platform it is relatively easy to close the observe, orient, decide, and act (OODA) loop. The challenge in network-centric operations is to enable OODA loops that span space and time as effectively and as rapidly for dispersed force elements as for a single platform, particularly when some sensors may be involved in multiple loops. Any sensor and processor with useful data or information will provide it for anyone who can use it, and the provider may not know who the user is nor the user who the provider is. In the large, however, the operation of the network will remain a closed loop in that information will lead to action, and the mission decision maker—the one who decides what the target is—will have to know that the target was engaged and the outcome of the engagement, as a condition for deciding on further action.

In addition to having to be linked, sensors require continual improvement. Phenomenology in all spectral domains must be explored to exploit multiple sensing paths to the greatest extent possible, both physically and economically, and the quest must continue for automatic recognition of targets that are detected. Automatic target recognition (ATR) will, when it is achieved, aid not only in finding targets in noisy backgrounds but also in defeating the effects of countermeasures to accurate navigation and guidance of weapons. It will also reduce the number of personnel needed for the information-processing parts of the NCII and other information operations.

The use of unmanned aerial vehicles (UAVs) for surveillance; target detection, recognition, and location; and postattack reconnaissance for effectiveness

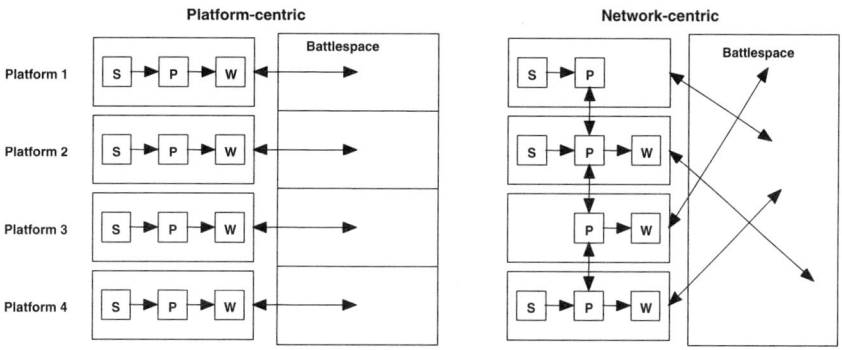

FIGURE 1.5 Platform-centric versus network-centric architecture.

assessment has been accelerated during military operations in the former Yugoslavia, where UAVs were effectively utilized as joint and coalition-based assets. In addition, the Marines will have a continuing need for short-range, organic UAVs for close-in targeting and to elevate communications relays.

1.3.2.2 Navigation

The problem of Global Positioning System (GPS) jamming will become more acute as weapon ranges and times of flight increase and will have to be overcome. No single technique will make GPS-aided weapon guidance invulnerable to GPS jamming. Practical solutions are likely to involve a combination of cheaper, precise inertial measurement units (IMUs), better target acquisition (including ATR), improved satellite signals and receiver signal processing, directive arrays of antenna elements, and the correlation of multiple signals and sources. Shorter times of flight, achieved by increasing weapon speed, together with improved, low-cost IMUs, can reduce reliance on GPS in the endgame against targets whose locations have been determined a priori. For moving targets being attacked by weapons without update links, time of flight and ATR go hand in hand; shorter delays make the ATR task easier. However, update links to enable the "forward pass" mode, in which weapons are given continually updated target location information after they are launched, are preferred for attacking moving targets, and when such links (and the sensors behind them) are available, ATR becomes much less important. Also, targeting and weapon delivery must be locked in the same reference grid to minimize the error due to target location inaccuracies.

1.3.2.3 Weapons

Naval force weapons are being made more accurate to reduce the need to reattack targets and to reduce collateral damage. The combination of greater accuracy and improved warhead lethality will allow lighter warheads, thereby increasing the range of weapon delivery systems. Weapons will need shorter times of flight to engage fleeting, moving, or highly threatening targets, despite the longer standoff needed to enhance the safety of launching platforms. This will be achievable by launching from advanced aircraft, often at supersonic speed, and by rocket propulsion of air- and sea-launched weapons. For the most effective results in some parts of the strike and fire support missions (e.g., attacks against concentrated targets embedded in population centers or very close to U.S. ground forces), accuracy at the target will have to be improved from the currently specified 13-meter circular error of probability (CEP) to 1 or 2 meters, including target location error. Additionally, the much greater use of precision weaponry will require that, notwithstanding all the weapon improvements called for, weapon costs be reduced significantly to achieve sustainability in a campaign.

In considering the design conditions for an overall subsystem, the performance goals for the components of the subsystem must be traded off against one another on the basis of mission performance. For example, GPS jam resistance can be traded off against ATR performance, guidance accuracy can be traded off against warhead radius of lethality, sensor latency can be traded off against weapon time of flight, and the reduced sensor latency afforded by data links to weapons in flight can be traded off against target location and guidance accuracy.

Network-centric operations require an intimate connection among all the sensing, processing, navigational, and weapon components of the NCO power projection system. Thus, all must conform to the compatibility and interface standards of the NCII. Currently there is no mechanism to coordinate the development of Navy and Marine Corps doctrine and apparatus for joint littoral operations or to coordinate such functions as tracking and network control.

Success in the power projection mission will require that all the areas touched on above and elaborated in Chapter 3, and many of the related areas discussed there, be supported with resources and worked on simultaneously in a fully integrated fashion.

The fielding of improved subsystems will have to be integrated in any NCO system by continually improving subsystems to support the force. Also, the United States may have more than one NCO system or force operating simultaneously in different parts of the world, or even in the same theater of operations. There must be an overall infrastructure—the NCII—with joint and coalition connections, to ensure consistency and interoperability among such far-flung assets, from local tactical networks to major commands, in a global naval force network.

1.3.3 Recommendations Regarding the Integration of Force Elements for the Power Projection Mission

1. In all Department of the Navy planning and acquisition activities, the integration of components for the power projection mission, as well as the integration of the power projection subsystems with the subsystems for other naval force missions such as air and maritime dominance, should be considered as the combination of related parts of a total NCO system, including all the component functions and equipment described above. This includes the naval forces' continuing efforts in the areas of countermine and amphibious warfare, and other efforts.

2. The Department of the Navy should engineer the strike and naval fire support subsystems of NCO systems in an end-to-end fashion. This includes the capability to sense, track, and hit high-priority relocatable or mobile targets with ad hoc or on-call fire and then to assess the results of strike and naval fire support operations in near-real time. Engineering studies and tests should be conducted to define effective, affordable, and balanced major subsystems in all mission areas.

3. System engineering should be performed to determine what combinations of improvements would be required to overcome the effects of foreseeable GPS jamming. Technology base funding and demonstration funding should be made available to determine whether these improvements are attainable.

4. A number of technology directions should be pursued in furtherance of the power projection mission:

a. Diversity of sensor phenomenology and locations should be sought; new sensors should provide for cooperative behavior and participation in ad hoc networks;

b. Organic airborne moving-target indicator (MTI) sensors should be considered for guiding precision weapons fired from over the horizon toward moving targets in the forward-pass mode; it should be ensured that closed-loop control in the forward-pass mode is not foreclosed in the design of sensors and weapons or by the concepts for their targeting;

c. Technology for better long-range identification of targets (including ATR) should be sought; in this regard, the Department of the Navy should interact more strongly with Defense Advanced Research Projects Agency (DARPA) programs; and

d. Technology to achieve affordable antennas with adequate gain, bandwidth, and flexibility, and that maintain low observability of the platform, should be sought. A particular challenge is to provide multiple-beam, directional, shared large-aperture antennas on major Navy platforms to serve the needs of the NCII as well as weapon systems.

5. The Department of the Navy should move more urgently toward providing the naval forces the capability to acquire data from theater and National sensors.

6. As part of the assignment to NWDC and MCCDC to jointly devise NCO concepts for the naval forces as a whole, the relationship between the two organizations should be formalized and institutionalized to encompass NCO innovation; tactics, techniques, and procedures; and doctrine for operations in the littorals. In particular, they should reach agreement on the need for a family of short-time-of-flight, over-the-horizon weapons and concepts for their targeting.

Many additional recommendations are included in the main body of this report, at a more detailed level than is appropriate for this overview. Those recommendations aim at improving specific sensor and weapon technologies, thereby greatly enhancing the naval forces' ability to carry out effective sea-launched strike missions and to provide highly responsive, long-range, affordable, sustainable, accurate, high-volume ship- and aircraft-launched supporting fire. These detailed recommendations are as essential to successful achievement of the aims of NCO systems as are the higher-level recommendations included in this overview.

1.4 DESIGNING A COMMON COMMAND AND INFORMATION INFRASTRUCTURE

1.4.1 The Naval Command and Information Infrastructure Concept

The Naval Command and Information Infrastructure will become the enabling framework for network-centric operations. The NCII includes the communications trunk lines, terminals, tactical networks, central processing facilities, common support applications, and Department of Defense (DOD)-wide and commercial standards, rules, and procedures that will enable the flow of raw and processed information and commands among units at all levels of command. Its attributes are listed in Figure 1.6, an expansion of Figure 1.1.

All the Services are striving to achieve the capability to share information, based in large measure on the Internet paradigm. The Internet's robust, networked communications base enables rapid, ready, and flexible access to information and supports the applications that provide information and services to a widely dispersed user population. Some top-down principles and standards are necessary for the communications base so that the applications can easily use it and so that users can interoperate with applications. In the Internet applications are developed from the bottom up by a diverse developer population. Thus there is a broad base for innovation, an important factor contributing to the utility of the Internet. The point for the NCII is that it should use standards that will permit its applications to come from diverse sources to serve a diverse set of users. In this respect, the Internet is the best model available to describe the design approach for the NCII.

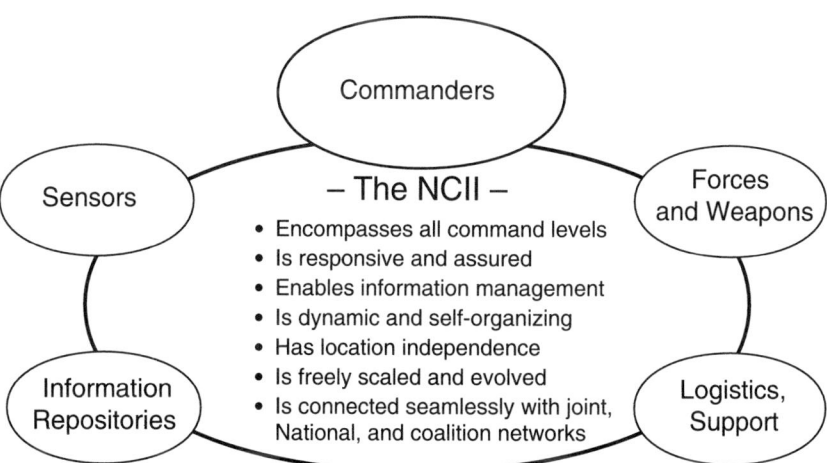

FIGURE 1.6 Attributes of the Naval Command and Information Infrastructure.

As with the Internet, users of the NCII will not be satisfied with, nor will their needs be met by, some fixed, predefined set of information. The uncertainty as to the type and location of future military operations ensures that. Relatedly, different operators may vary in their approach to a situation, and hence in their information needs. Furthermore, the manner in which information is used in the NCII will change continually as operational concepts are refined and new technologies introduced. For all these reasons, a central notion of the NCII is that it be flexible, adaptable, and evolvable in meeting the needs of its users.

While the NCII includes tactical networks and allows for widespread dissemination of information, it must also accommodate the need of commanders for some degree of control over such dissemination, for, among other things, security purposes and bandwidth management. This management of information dissemination facilitates and allows for decentralization of command, but at the same time it allows for the centralized collection of information and hence for greater centralization of authority. There is no one generally appropriate point to aim for on this centralization-decentralization spectrum; it will depend on the nature of the military operation. The NCII must be able to support varying modes of command.

The NCII is conceived not only as carrying long-haul traffic but also as enabling short-haul and tactical information acquisition, processing, and transfer. The acquisition of raw information and its processing into an accurate understanding of the current details of environments, forces, targets, and maneuvers must be treated separately from the transport (communication) of the information and the commands based on it. The NCII provides for the integration of the acquisition and processing mechanisms and provides the transport for information and command at all levels, from major force operations to single target-shooter engagements.

The mechanisms for transporting information for many services and functions will rely heavily on civilian, commercial systems. Purely military functions will appear more in the information processing and command parts of the NCII, where security and the special characteristics of military operations are driving factors, although purely military functional capabilities will be built in good measure from commercial sources and technology.

The NCII should be recognized as the naval force portion of an information infrastructure that is interwoven with, shares common components with, and adheres to the same set of standards as other Service, National, and, when appropriate coalition networks, such that all function as a global whole. Thus, the NCII will have to be built to standards established by others, although the Department of the Navy should play a part in developing some of the standards. Since the network will have commercial components, the standards will also have to be compatible with and often the same as commercial standards. These standards, and the rules and coordinated operational procedures that go with them, will be the only means by which full interoperability can be achieved. Full interoperability will be essential to bring all the benefits and advantages of network-centric operations to fruition.

Tactical networks are of special concern since they pose the greatest challenge to the goal of using standard, Internet-based networking technology throughout the naval infrastructure. The Navy and the Assistant Secretary of Defense for C3I (ASD (C3I)) have argued that this class of radio networks must necessarily be based on nonstandard, military-developed technology to meet the tight time constraints and extreme reliability that tactical communications require. Accordingly, the current Navy networking architecture defines two special-purpose tactical radio networks in addition to the standards-based Joint Planning Network: the Joint Data Network (actually, the Joint Tactical Information Distribution System (JTIDS)) and the Joint Composite Tracking Network (actually, CEC).[7] Although the Navy and ASD (C3I) argument has merit, the committee concluded that there are greater advantages in extending a uniform, open, standards-based network architecture across the entire naval infrastructure, including the tactical networks. The committee envisions a network in which tactical data communications are provided via the NCII standards, including a standardized naming and addressing scheme and data transport using the Internet Protocol (IP). The committee believes that advancing commercial technology will make it possible to remove technical impediments to allowing any type of data to be conveyed across any type of radio link.[8] If an Internet-based architecture is adopted, new types of tactical services can be rapidly deployed across in-place radios.

It is important to note that the committee does not believe that all types of traffic should be allowed to cross any tactical radio network freely. Quite the contrary: Strict controls will be necessary at the connection points between the tactical and nontactical portions of the NCII. These controls will ensure that only authorized types of traffic are allowed onto the tactical networks, and hence they will provide continued guarantees that the tactical networks can provide highly reliable, low-latency data services. These controls will also aid in providing security boundaries (i.e., firewalls) within the NCII as part of the network defense in depth discussed in Section 1.4.2.

In the end, it is likely that a few tactical networks will remain outside the NCII for some combination of technical and economic reasons. Such outlying tactical networks can be connected into the Internet-based NCII via IP-capable

---

[7]Furthermore, as far as the committee can tell, this focus on the Joint Data and Joint Composite Tracking networks omits consideration of all other tactical communications networks currently employed by the Navy that are part of the overall information transfer capability. These include various sensor links—e.g., for MTI and synthetic aperture radar data—and links to weapons control systems—e.g., ultrahigh frequency satellite communications target location updates for Tomahawk.

[8]Approaches to this are described in some detail in Appendix E.

gateways so that they can still enjoy the advantages of being part of the overall, seamless naval network infrastructure.

It is important to understand that the NCII itself does not represent a major new investment. Rather, it requires an investment of resources sufficient to integrate the many subsystems and components, some of which exist, some of which are being developed, and some of which are or may be planned, in a way that provides guidance and structure for an overarching concept for information support to network-centric operations.

The general composition of the NCII is illustrated by its functional architecture, shown in Figure 1.7 and discussed in detail in Chapter 4. The starting premise of this functional architecture is the need to support the warfighting decision process extending across all levels of command, to include those engaged in actual weapons delivery. Shown in the top half of the figure are the functional capabilities (collection management, etc.) that gather and generate information to support the decision process, and then see that the decisions are conveyed to their appropriate recipients. Across the bottom of the figure are the supporting resources (communications, etc.) used by the "upper level" functional capabilities.

Collection management determines the tasking of sensors to collect data. The information exploitation and integration function takes the initial data and refines the information by correlating, fusing, and aggregating it. Information request and dissemination management provides information based on user-specified requests for a given type of information. Its operation is transparent in that users do not have to know the details of where the information is located. This function will also provide information to users based on the directions of any other authorized party. Information presentation and decision support includes the graphical means for displaying information to users and the set of automated

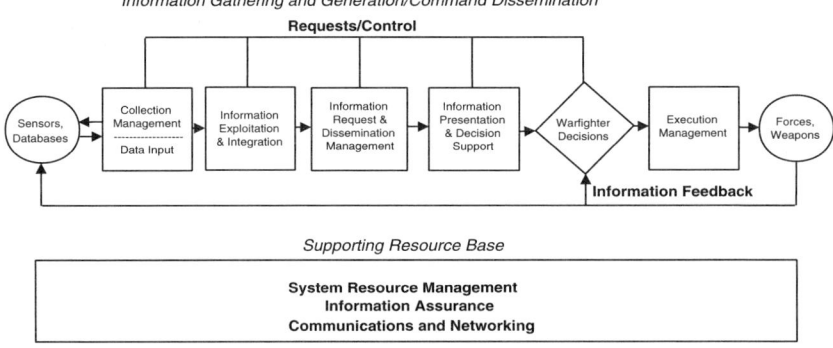

FIGURE 1.7 Functional architecture of the Naval Command and Information Infrastructure.

tools that allows users to manipulate information for the purposes of making a decision. Execution management supports delivery of decisions to the intended recipients and allows for dynamic adaptation of those decisions in the light of rapidly changing events. The other functions illustrated in Figure 1.7 are self-evident.

1.4.2 Information Assurance Within the NCII

"Information assurance" is the term used in this report to describe how threats to and vulnerabilities of the NCII must be addressed to ensure the integrity of information and the timeliness and continuity of its flow for network-centric operations as a whole.

The NCII, like all information networks in the modern age, must routinely exhibit high reliability and must include safeguards against system failures due to overload, loss of critical nodes as a result of enemy action, and other operational factors. It will also face many threats to the quantity, quality, integrity, and continuous flow of the information it manages and provides, and it will have many vulnerabilities. Both the threats and the vulnerabilities are too numerous to elucidate in this summary discussion. They are noted briefly here and are described in detail, along with potential defenses and countermeasures, in Chapter 5.

Critical vulnerabilities for tactical networks are spoofing, jamming and other interruptions, interception, and ground terminal capture. Important sources of weakness in the NCII transport elements will derive from the use of commercial subsystems and from the outsourcing of important elements of the transport operations, and also from the need to connect with and share information with coalition partners. The key strength of the NCII in allowing the connection of disparate networks and functions is also, however, a source of risk. Among these connections is that linking the fleets' operational networks, in which a degree of secrecy and control can be maintained, with the naval force business networks that are essential for the logistic support of the fleet and that must be open to both the naval forces at sea and their shoreside commercial connections. A critical vulnerability in the nontransport part of the NCII derives from the threat posed by the potential malicious insider, who could, working alone or with outside adversaries, cause serious disruption to network-centric operations.

NCII information assurance must be achieved throughout the information infrastructure, including wireless links. In the design of the NCII, all components must be treated as vulnerable, and security vulnerabilities must be anticipated in any system component and even in any given protection mechanism. Overall, to meet the threats and mitigate the vulnerabilities, a defense in depth is required. It consists of three elements: prevention; detection of attack, assessment of the damage, and remedying of the effects of the attack; and robustness in its ability to tolerate penetrations.

Today, because of technology shortfalls at each level of the defense-in-depth strategy, it is not possible to completely implement such a strategy. However, in some cases steps that do not depend on technological remedies can be effective. For example, in a crisis certain functions may be considered so critical that any risk to their timely and correct functioning is intolerable. In such cases, the decision may be to not connect, and to use an air-gap defense (which inserts a deliberate break, to be connected by manual action, in a link of the network). Reducing the risk of damage by a malicious insider might be accomplished by reducing the scope of access and control available to any single individual, and by requiring two- (or more) person control of key functions. Monitoring user activities, coupled with exploring observed anomalies, is another risk-reduction technique.

Red teaming is often prescribed for exposing a system's vulnerabilities and weaknesses so that they can be remedied. However, it is important to understand and capitalize on red teaming's strengths while understanding the limitations of its use. Red teaming proves not to be a preferred way of discovering system vulnerabilities or learning how to mitigate threats, because the red teams come from the same culture that created the system. Red teaming's primary benefits are that it is the best tool for raising the level of security awareness within an organization and that it is useful as a method for ensuring that correct security configurations are maintained for the system. Red teaming for these purposes can be carried out by a system's security staff on a periodic basis.

In its review, the committee found that information assurance for the NCII is not receiving appropriate attention at high enough levels within the Department of the Navy to ensure that this critical problem area is managed in a manner consistent with its importance to successful network-centric operations. There is no single individual in the Department of the Navy charged with the responsibility for information assurance. Further, the Navy Department has no overall plan for information network security in its tactical networks. Mitigation of vulnerabilities will come from many measures in the defense in depth, with support from continual red teaming, but the organizational problems will have to be remedied as well.

In addition, because of the likelihood of attack on the NCII or its operational degradation, it is imperative that naval forces train for situations with impaired NCII function. Not only must the NCII system staff learn to quickly restore service, but the operational forces must also learn to deal with system failures. Beyond that, in recognition of the vulnerabilities the forces should be shaped such that they can fall back to operational modes that are at least as good as those that preceded network-centric operations. For example, the naval forces have a tradition of developing operational workarounds for loss or degradation of radio frequency communications in tactical operations. The same should be done for the NCII so that naval forces will be prepared to deal with these likely situations in practice.

In the spiral development process, especially in the experiments and proof-of-concept exercises that will attend the development of concepts of operation, the opportunity will exist to probe for the most logical vulnerabilities (e.g., jamming of tactical networks) and to design appropriate redundancies and fallback modes of operation.

Is it worth accepting all the vulnerabilities and the attending risks, as well as the cost and operational penalties of anticipating and remedying them? This is a question that cannot currently be quantified. However, in all recent military endeavors, including the Gulf War and operations in the Balkans, and in endeavors throughout the national and even the global economy, the gains are seen as being so great that the risks are accepted even while mitigation attempts are undertaken and their costs incurred. The trends in technology, force size and utilization, and U.S. global responsibilities are such that network-centric operations offer the only means of achieving the necessary mission effectiveness of U.S. naval forces.

1.4.3 NCII Functional Capabilities: What Exists and What Is Needed

Tables 1.1 and 1.2 summarize the status of the currently programmed baseline elements of the NCII and the challenges that must be met to give it the capabilities needed for network-centric operations to function as envisioned.

There are many naval, defense agency, and commercial endeavors that can contribute to the development of the NCII. These include the Navy's IT-21 strategy; the Navy/Marine intranet; the Global Command and Control System–Maritime; software radios that can emulate multiple legacy radios and also adaptively select appropriately robust waveforms; the design guidance in the Information Technology Standards Guidance; naval communications and software research at the Office of Naval Research, Naval Research Laboratory, and Space and Naval Warfare Systems Command (SPAWAR); and—in a broader sense—the DOD Global Information Grid as it becomes more specifically defined. In addition, there are valuable DARPA programs that can help advance NCII capabilities in the areas of challenge listed in Table 1.2, including work on information assurance and survivability, dynamic system resource management, agent technology, and data visualization, among others. However, these ongoing developments do not constitute a comprehensive approach to realizing the set of capabilities necessary for an NCII. An integrated overall plan, as well as changes in organizational focus, will be necessary to achieve the NCII.

Key problems include, but are by no means limited to, robust wireless communication networks for tactical environments, content-based system resource management, and scalable information dissemination management. Current conceptualizations of the operational and system architectures seem more suited to situations where requirements can be laid out fully in advance of development rather than to the flexible, iterative process necessary for construction of the

TABLE 1.1 Status of Programmed Baseline NCII

Capability	Assessment
Supporting Resource Base	
Communications and networking	Significantly increased in-theater satellite communications capacity planned, but stated Department of the Navy capacity requirements could be unrealistically low; only limited improvements in tactical communications planned.
Information assurance	Basic network security products being deployed; critical vulnerabilities remain to be considered.
System resource management	Communication channels can be assigned, but priorities cannot be assigned within Internet Protocol (IP) networks. IP advances offer quality-of-service enhancements.
Operational Function	
Collection management	Current capabilities are stovepiped by sensor; limited near-term enhancements are planned.
Information exploitation and integration	Automated extraction of individual targets is accomplished, but much manual work still required for overall battlespace picture.
Information request and dissemination management	Significant improvements in information location and access are promised by information dissemination management capabilities currently being deployed.
Information presentation and decision support	Dynamic two-dimensional, map-based displays of friendly and enemy platforms are in development; overall concept for information needed and means to display it still required.
Execution management	Dynamic mission planning for rapid direction and redirection of forces during operations is limited.

NCII. Sufficient information was not available to the committee to resolve the matter of communications capacity requirements, but it appears that stated future Navy communications requirements could be unrealistically low, even though the available military and commercial satellite communications (SATCOM) capacity is projected to increase significantly. The appropriate division between military and commercial communications will have to be a topic of continuing analysis, planning, and adaptation as the NCII is built and operated.

However the division between military and commercial communications is made, extensive use of commercial communications infrastructure will be inevitable. As pointed out in the Naval Studies Board's TFNF study,[9] this need will

---

[9]See Footnote 5.

TABLE 1.2 Some Remaining Challenges in Providing NCII Functional Capabilities

Capability	Challenges
Supporting Resource Base	
Communications and networking	Rapid configuration and reconfiguration of networks; flexible wireless networks; multifrequency, electronically steerable antennas.
Information assurance	Intrusion assessment; intrusion tolerant systems; preventing denial of service; hardening of legacy systems.
System resource management	Content-based priority management; dynamic allocation of resources.
Operational Function	
Collection management	Integration across sensors, with intelligent cross-cueing and dynamic tasking.
Information exploitation and integration	Automated integration of disparate information; increased automation of feature extraction from images.
Information request and dissemination management	Profile-based dissemination from large and heterogeneous collections of information sources; automated dissemination management policy.
Information presentation and decision support	Intuitive situational displays; comprehensive suite of necessary decision-support tools.
Execution management	Dynamic replanning; real-time simulation.

be most effectively and economically accommodated by direct use of commercial systems and technology. Such use will require the Navy and Marine Corps to adapt their system design and utilization practices to the demands of the commercial marketplace while ensuring security, priority, and uninterrupted access in times of emergency. Information assurance will be an essential factor in the NCII's evolution and adaptation for network-centric operations.

Finally, it must be noted that efforts to maintain the current distinction between the Joint Planning Network and the Joint Data Network, and likewise to maintain unique protocols for imagery data links, appear not only counterproductive in terms of such factors as interoperability, but also unnecessary in light of developing communications and network technology.

1.4.4 Recommendations Regarding the Design and Construction of the Naval Command and Information Infrastructure

1. The Department of the Navy should develop and enforce a uniform NCII architecture across the strategic, operational, and tactical levels of naval forces.

This means that, for all levels, (a) the same set of functions will apply (e.g., as defined in Figure 1.7),[10] (b) interfaces and standards associated with these functions will be the same, and (c) consistent definitions will be used for the data exchanged between the functions. Architectural concepts more advanced than the simple standards-based architectures currently being considered should be incorporated into the NCII to realize the flexible, rapidly configurable information support envisioned for network-centric operations. Standards should be imposed at a level that does not inhibit innovation in function or implementation; for example, radio standards should specify waveforms and transport protocols—not implementation details—to permit multiple generations of software radios to interoperate.

2. The Department of the Navy should develop a comprehensive and balanced transition plan for realizing the NCII. The functional architecture shown in Figure 1.7 provides a conceptual framework on which to base the transition plan, and the specific recommendations summarized at the end of Chapters 5 and 6 for each of the functional capabilities provide a starting point for the transition to use of the NCII.

3. The NCII should be developed in coordination and collaboration with the other Services, the joint community, and National agencies to promote interoperability and build on each other's efforts. It should also allow for incorporation of coalition capabilities, as appropriate, to missions involving coalition forces. One specific near-term opportunity for coordinating with other Services would be, for example, through participation in the joint expeditionary force experiments sponsored by the Air Force.

4. The Office of the Assistant Secretary of the Navy for Research, Development, and Acquisition (ASN (RDA)), in conjunction with other interested Navy and Marine Corps elements, should join with the other components of DOD to sponsor a vigorous, continuing R&D program aimed at meeting the challenges of creating an advanced NCII. As part of this effort, the Department of the Navy should give serious attention to the many DARPA and naval research programs that have the potential to meet the challenges.

5. The Department of the Navy should work with the ASD (C3I) and the other Services to make the operational and systems architecture products specified in the C4ISR architecture framework suitable for the flexible and rapidly evolving information support that the NCII must provide.

6. The Department of the Navy should conduct continuing comprehensive analysis of communication capacity requirements and projected availability, and should identify remedial actions if significant shortfalls exist. This analysis should include both long-haul communications and tactical data links, including direct links from in-theater sensors.

---

[10]The tactical domain will, in addition, have its own unique functions that are particular to warfighting mission areas. These are considered in Chapter 3.

OVERVIEW OF STUDY RESULTS 41

7. To the above recommendations that pertain to all applications of the NCII, including at the tactical level, the committee adds two particular recommendations concerning tactical communications:

 a. With few, if any, exceptions, new communications networks for tactical operations should conform strictly to the NCII goal architecture and should use appropriate gateways, firewalls, and encryption devices to ensure high quality of service.

 b. Terminals of the JTIDS and common data link families should be modified to use NCII standard protocols.

8. The committee also makes several particular recommendations in the information assurance area:

 a. Responsibility for information assurance should be assigned at a high enough level within the Navy Department and with sufficient emphasis to ensure that adequate attention is paid to all aspects of this problem in the design and operation of the NCII.

 b. A defense-in-depth strategy should be adopted, based on the premise that security vulnerabilities may always remain in any system components.

 c. Advances in security technology should be tracked and aggressively applied in the NCII, including its wireless, SATCOM, and land-based communication components.

 d. Procedural and physical security measures should be developed to further reduce the risk where the available technology is not adequate.

 e. Naval force information assurance efforts should include preparation and training for operations with impaired NCII functionality, including provisions for redundancy in appropriate places and fallback modes of operation.

 f. Research to address future critical NCII information assurance needs should be included as an explicit part of the R&D program that is the subject of recommendation 4 above.

1.5 ADJUSTING THE DEPARTMENT OF THE NAVY ORGANIZATION AND MANAGEMENT

1.5.1 Organizational and Management Needs

Four decision support processes are key to implementing the concept of network-centric naval forces for more effective operations:

1. *Requirements generation:* clearly stating operators' mission needs;

2. *Mission analyses (assessments) and resource allocation:* aligning program and budget resources to meet mission needs;

3. *System engineering, acquisition management, and program execution:* integrating, acquiring, and deploying for interoperability; and

4. *Personnel management:* acquiring personnel and managing careers to meet network-centric needs.

The entire decision-making process for definition, acquisition, and integration of forces to achieve network-centric operations is extremely complex and involves all parts of the Navy Department, as illustrated in Figure 1.8.

The committee reviewed the decision-support processes shown in Figure 1.8 and concluded that better integration was needed among them to attain significantly improved networked capabilities. Modifications to business practices in each of requirements generation; mission analysis and resource allocation; system acquisition and program execution; and personnel management, training, and education—as well as the integrated oversight of the entire complex—are needed to achieve the full benefits of network-centric operations.

The committee found that the information network and cross-platform interoperability are not as well represented in the fleet requirements generation process as are the platforms and weapons themselves. In addition, it found that the current requirements generation process is not sufficiently responsive to the demands imposed by the pace of information technology development to keep deploying naval forces at the leading edge of commercial practices. The committee also found that there is no one organization within the Navy operational community that has the credibility and authority to prepare requirements for the seams among subsystems and components supporting network-centric opera-

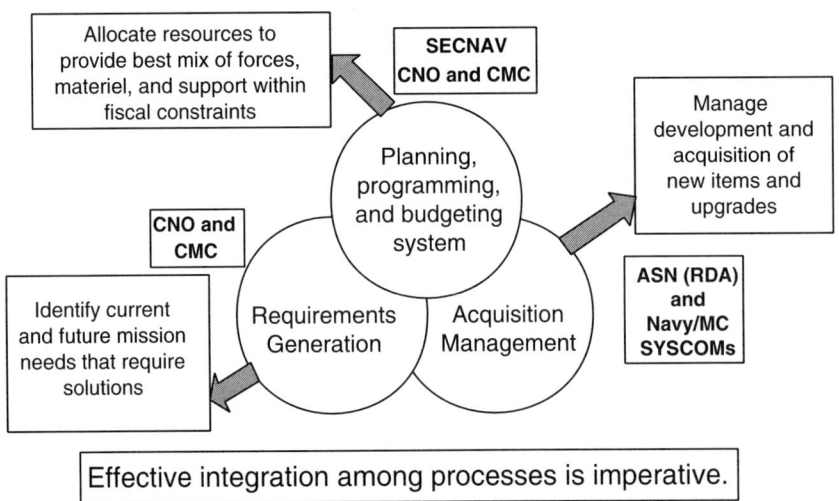

FIGURE 1.8 Major decision-making support processes in the Department of the Navy.

tions. In addition, joint efforts to improve interoperability need expansion. Thus, there is a need to augment the processes by which network-centric operations are internalized to become an integral part of the naval force system.

In the areas of mission analysis and resource allocation, the committee found that the naval forces, taken together, lack good measures of effectiveness (MOEs) and measures of performance (MOPs) for evaluating NCO systems as a whole and the contributions of their subsystems to the larger mission goals. And while the Navy, which has the ultimate responsibility for most naval force system acquisition, has recently taken some steps to enhance the system engineering process within the SYSCOMs (i.e., the NAVSEA 05, now NAVSEA 53, organization) and within the ASN (RDA) (i.e., the appointment of the Chief Engineer), there is insufficient system engineering discipline to ensure integration and interoperability of cross-platform and cross-SYSCOM subsystems of any overall NCO system. Possibly most important, in light of the demands of network-centric operations on force evolution and performance integrated across the naval forces and into the joint arena, is the need for more comprehensive review and oversight of the acquisition and program execution of the entire NCO complex of systems within the programming, budgeting, and implementation processes than the current business practices provide. Such review and oversight must include prioritization among the various subsystems.

Finally, some members of the committee believe that, due to the legacy of earlier maritime strategies, the Navy places insufficient emphasis on the power projection mission in the N8 organization and in the program executive office (PEO) structure. The N8 organization reflects submarine warfare, surface warfare, and air warfare, with power projection a part of each office but not the focus of any. Meanwhile, air dominance is well served by the focus of the office of surface warfare, and strategic deterrence by the office of submarine warfare. It appears that power projection lacks a true advocate in N8. The same may be true of sea dominance, although this issue was not examined in as much detail by the committee. In the PEO structure air dominance is the focus of the Program Executive Office for Theater Surface Combatants. At least five PEOs strongly relevant to power projection are primarily product oriented, the products being platforms and weapons in many cases. Therefore, management of end-to-end system designs and acquisitions as such is considered to be problematic. The same may be true for such system designs in other areas, although both the N8 and the PEO structures have been successfully adapted to the need in areas such as ASW and CEC and in the growing theater missile defense (TMD) effort. The ASN (RDA) has recently announced the redesignation of the Program Executive Office for DD-21 as PEO (Surface Strike), assigning it responsibility for NAVSEA Program Manager, PMS 429's Naval Surface Fire Support including the Advanced Land Attack Missile program, as well as the DD-21. This represents a major step in the direction of concentrating attention on power projection systems as a whole, in parallel with the concerns the committee expressed in this area. The committee's recommendations also pertain to making targeting an integral part of the strike system, to strike warfare from the air, and to the relationship between and coordination of naval surface warfare and air strike warfare. The committee commends the entire power projection area to further scrutiny of the kind that led to this most recent PEO reorganization, in both the PEO and the N8 contexts.

Within the context of this study, other members of the committee addressed and argued against making recommendations on these two issues; they favored what they regarded as more pragmatic recommendations to improve implementation of network-centric operations. Among other things they believe that recommendations on the two issues above will deflect Navy attention from recommendations made in more important network-centric challenge areas—i.e., the recommendations focused on (1) improving integration within and across all decision support processes and (2) developing improved output measures and mission/system component trade-off analyses and assessments. Given these divergent views and the uncertainty they reflect about the true management situation applicable to overall network-centric operations system planning and acquisition, the committee concluded that recommendations to the Navy Department and the CNO would be in order, to review the N8 and the PEO structures and adjust them *if necessary* and *as appropriate* to accommodate end-to-end system designs for NCO subsystems, including especially those relevant to the power projection mission. These recommendations are included with the others that follow.

1.5.2 Recommendations Regarding Department of the Navy Organization and Management

The committee believes that successful network-centric operations will require greater degrees of cooperation, trade-offs, and interaction than currently exist among the stakeholders responsible for the functions involved in NCO integration. It concluded that to best achieve this integration, the Department of the Navy should build on its existing organizations with some changes in emphasis, rather than attempt to totally restructure the department or create a new or additional "stovepipe" for all network-centric responsibilities. The difficulty with even attempting to create a new entity to be responsible for all, or a major portion of, network-centric operations is that such operations span almost the entire range of Navy and Marine Corps activities. Therefore the committee took a pragmatic approach respecting current laws and attempting to minimize organizational disruption.

In arriving at its recommendations, the committee recognized, of course, that internal and external considerations not known to the committee may lead the Navy Department to take other approaches to addressing the committee's findings. The recommended changes represent the committee's best judgments about the best means for the Navy Department to come to grips with the enormous

OVERVIEW OF STUDY RESULTS   45

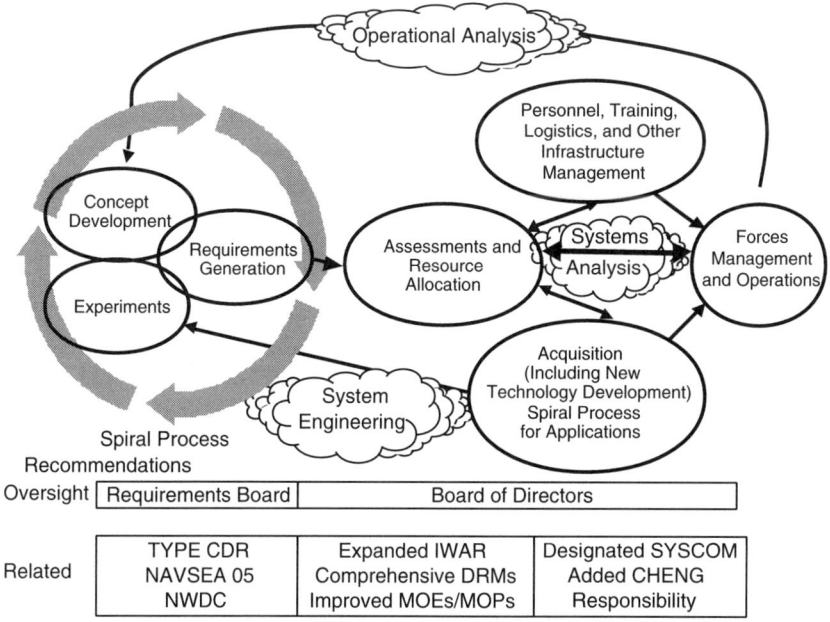

FIGURE 1.9 Functions for effective integration of network-centric operations shown in relation to major recommendations made in this report. CHENG, Chief Engineer of the Navy; DRM, design reference mission; IWAR, integrated warfare architecture; MOE, measure of effectiveness; MOP, measure of performance; NAVSEA, Naval Sea Systems Command; NWDC, Navy Warfare Development Command; SYSCOM, Systems Command; TYPE CDR, functional type commander.

complexities that will attend the evolution of the naval forces into the network-centric operations mode.

Figure 1.9 shows the processes specific to the Department of the Navy that are necessary for effective network-centric operations integration. The committee's major recommendations are indicated below the functions that would be most affected by the specific recommendations.

The major organizational and business process changes and recommendations are summarized in the following paragraphs. They are presented and discussed in full in Chapter 7.

1. The creation of one new position is recommended: a functional type commander,[11] the Commander for Operations Information and Space Command. This new functional type commander should report to only the three fleet commanders, in the same manner as the current platform type commanders report to individual fleet commanders. In addition to assigned operational responsibilities, including management of the fleet portions of the NCII and space assets, this new functional type commander should be the single point of information support to all the fleets, and should represent the fleet commanders' network-centric information operations needs and priorities in the program objective memorandum (POM) and budget processes. He or she would be involved in and support the fleet experimentation program and the recommended spiral development process for network-centric operations. The new functional type commander would also assume some of the functions now assigned to the Deputy Chief of Naval Operations (DCNO), Space, Information Warfare, Command, and Control (N6) (see Chapter 7).

In arriving at this recommendation, the committee considered various alternate approaches to carrying out the functions summarized above (and described in more detail in Chapter 7). The committee weighed the likely problems and benefits that would attend the creation of the new position. One alternative was leaving the organizational situation as it is now, with a lower-ranking officer functioning with each fleet to deal with its information network matters. This arrangement would not provide adequately for the broad and fundamental nature of the change needed to fully implement network-centric operations in the fleets. The committee also considered a recommendation for creating multiple flag positions for each fleet, but this approach did not appear to resolve the problems of achieving consistency of equipment, planning, and operational techniques in the operational forces throughout the Navy. Only a single individual could achieve that.

After considering the pros and cons of various alternatives, the committee concluded that the time is propitious for making information operations a warfighting mission with a fleet role comparable to that of current type commanders and that the need to achieve assured consistency and interoperability warrants having the functions be the responsibility of a single individual with a high enough rank.

2. A requirements board should be established to deal with operations information and to integrate various competing requirements as presented by the fleets for rapid improvement of complex at-sea operations. The proposed requirements board should be chaired by the VCNO and should have the N6 as the executive director (until the Operations Information and Space Command is established and is assigned that function). The membership of the requirements board should

---

[11] The flag officer responsible for all ships of a certain type in the fleet.

OVERVIEW OF STUDY RESULTS 47

consist of the deputy fleet commanders; the president of the Naval War College; the DCNO, Plans, Policy, and Operations (N3/5); and the DCNO, Resources, Warfare Requirements, and Assessments (N8). These members should have four broad functions: (a) develop policy and implement strategy for conducting operations based on the NCII, (b) advise the CNO on the strategy and doctrine, personnel, education, training, technology, and resource requirements for moving the Navy from platform-centric to network-centric warfare, (c) establish the linkage to the Navy of the future from this new level of warfare operations, and (d) prioritize emerging network-centric operations requirements based on fleet commanders' recommendations and the results of fleet experimentation.

3. Wherever NCO system needs involving both Navy and Marine Corps forces in joint operations intersect, the Navy and Marine Corps should arrange to coordinate their formulation of requirements.

4. A new board of directors consisting of individuals with the authority to make funding, scheduling, and program adjustments in relevant areas should be established for review, oversight, and prioritization of the acquisition, integrated installation, and program execution portions of network-centric operations. The Undersecretary of the Navy should be the chairman and the VCNO and the Assistant Commandant of the Marine Corps (ACMC) should be members of the proposed board of directors. Other members should be the ASN (RDA) (who should serve as the executive director); the Navy SYSCOMs; the Marine Corps Systems Commander; the DCNO, Plans, Policy, and Operations (N3/5); the DCNO, Resources, Warfare Requirements, and Assessments (N8); the Assistant Chief of Staff (ACOS), Plans, Policy, and Operations; and the ACOS, Programs and Resources of the Marine Corps staff. Requirements sponsors (N2, N4, N6, N85, N86, N87, and N88) should be advisory members to be consulted concerning operational impacts of potential program adjustments. The board's mission should be to provide a focus for network-centric operations and to ensure appropriate integration and interoperability for all acquisition and program execution (including installations in battle groups), for all cross-platform systems, including new subsystems, major subsystem components, and upgrades to existing subsystems and major subsystem components, of the overall system complex for network-centric operations.

5. The Department of the Navy should establish a three-star deputy to the ASN (RDA) for Navy NCO integration to carry out the acquisition and program execution directions of the proposed board of directors. The deputy should be a designated Navy SYSCOM commander and be double-hatted into this role. He or she should oversee all aspects of Navy system interoperability and integration and execution of NCO programs, including the NCII in Navy areas of responsibility. This also includes oversight of the activities of the Navy Chief Engineer and the NAVSEA battle force interoperability engineering function and working with the Commander, Marine Corps System Command, to ensure effective, coor-

dinated program execution in areas where the subsystems of both Services must operate together as part of an overall NCO system.

6. The Department of the Navy should define responsibilities, empower corresponding organizations, and provide adequate resources to (a) establish a comprehensive view of the capabilities and programs necessary to implement the NCII, and (b) see that these capabilities are realized. The assignments of responsibility for the NCII should be consistent with responsibilities for positions established in law and the other naval force organizational changes that are recommended herein. The assigned responsibilities should include interaction with other Services, the joint community, and defense agencies:

— Resource allocation and requirements sponsor: OPNAV N6;

— Operational NCII architecture: Commander, Operations Information and Space Command, with the support of OPNAV N6;

— Policy and standards: Department of the Navy Chief Information Officer;

— System and technical architectures (including enforcement): Navy Department Chief Engineer;[12]

— Acquisition and procurement: program management as designated by the ASN (RDA), and coordination of network-centric operations integration by the designated SYSCOM commander with functions described in 5, above; and

— Operational management of the NCII: Commander, Operations Information and Space Command.

7. Mission analysis and component trade-off evaluations should be strengthened by (a) providing staff and resources for the IWAR development process to enable continuous assessments from requirements generation through programming, budgeting, and execution; (b) developing output-oriented MOEs and MOPs for network-centric operations; and (c) developing a comprehensive set of design reference missions across all mission areas. Resource planning should be adjusted to support the spiral development process, including out-year funding to ensure that it is sustained.

8. The Chief of Naval Operations and the Commandant of the Marine Corps should review how system trade-offs and resource allocation balances are addressed in the Navy/Marine Corps staffs for all naval force missions, and particularly for the power projection mission, with a view toward orienting the process to the overall network-centric operations system concept.

9. Under the Deputy ASN (RDA) for Navy network-centric operations integration, the role of the Navy Chief Engineer should be strengthened to institutionalize the system engineering discipline for integration and interoperability of cross-platform and cross-SYSCOM subsystems and components of the overall network-centric operations system. The Navy Chief Engineer should oversee a system design and engineering cadre drawn from the three Navy SYSCOMs (and

---

[12]The operational, system, and technical architectures are defined in Chapter 4.

the Marine Corps SYSCOM when necessary, appropriate, and agreed to by the Services) for this purpose. The SYSCOMs should be provided with resources and staff to support this activity.

10. The ASN (RDA) should seek the best means to address the design and engineering of NCO systems, to eliminate as much as possible any distortion of the overall network-centric operations approach through undue emphasis on any single naval force mission or any one platform. In particular, the Navy Department PEO structure should be reviewed and provision made, as is found appropriate and necessary, for management of the acquisition and oversight of mission-oriented, networked major subsystems of the overall NCO systems. In doing this, special attention should be given to end-to-end (surveillance and targeting through effectiveness assessment) fleet-based land-attack (strike and fire support) subsystems for Navy, joint, and coalition missions.

11. The organization of the Navy's N8 office should be reviewed and adjusted as appropriate and necessary to increase emphasis on all aspects of the power projection mission, including strike and countermine warfare, amphibious and airborne assault, fire support, and logistics support of Marine forces from the sea.

12. The Navy and Marine Corps should recommend that J8 in the Joint Staff set up a joint organization for land attack, modeled on the Joint Theater Air and Missile Defense Organization (JTAMDO). Until such an office is set up, the Navy and Marine Corps should participate more actively in the "attack operations" pillar in JTAMDO that is looking at targeting of time-critical targets, such as mobile missile launchers.

Figure 1.10, reproduced from Chapter 7, summarizes the major organizational and business practice recommendations under the three major decision support processes affected most directly by the individual recommendations (including some additional recommendations at a greater level of detail that are included in Chapter 7). As noted on the bottom of Figure 1.10, NCO education and training are needed for all naval personnel.

1.5.3 Personnel Management, Training, and Education

Achieving gains potentially offered by modern technology for enabling force-wide network-centric operations is not likely with current DOD and Department of the Navy personnel management practices. Since information technology work in the military has been changing dramatically, it is not known exactly what skills will be needed for future efforts. It can be projected from the principles involved, however, that competent personnel will be required to address information and knowledge management (extraction, presentation, and application), technical design (architectures, network design) and sustainment (maintenance of

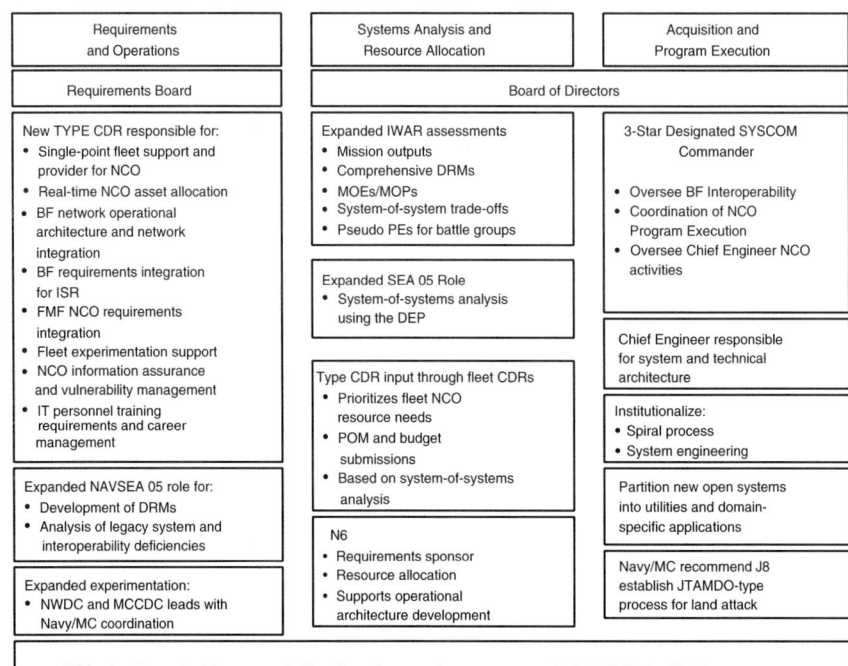

FIGURE 1.10 Key recommendations for managing network-centric operations. BF, battle force; DEP, distributed engineering plant; DRM, design reference mission; FMF, fleet Marine force; ISR, intelligence, surveillance, and reconnaissance; IWAR, integrated warfare architecture; MOE, measure of effectiveness; MOP, measure of performance; PE, program element; POM, program objective memorandum; TYPE CDR, functional type commander.

connectivity), and applications (for functional users). All future Department of the Navy personnel will need some level of information technology knowledge.

Current job skill codes do not provide the detail needed to fully define and manage the emerging workforce structure and skills pertinent to network-centric operations. While some progress is evident (e.g., SPAWAR initiated an analysis of the technical job codes used to identify information technology skills in the military), no systematic effort is under way to examine the job skills required for work involving use of information technology to convert data into knowledge. Within the Department of the Navy, career paths have been established for the newly named Information Technology Specialist rating. However, there are no established related career paths for civilian employees.

The national information technology worker shortage could become a serious problem for the naval forces. Workforce planning to meet information technology needs must begin now to take advantage of the important opportunity over the next 5 years to realign the workforce as large numbers of current employees retire. In addition, there is a need to analyze the content of the desired information technology work for both the military billet and civilian position structures.

Network-centric operations must be made pervasive in the education of Navy and Marine Corps officers, starting with the U.S. Naval Academy, the U.S. Naval War College, and the U.S. Naval Postgraduate School. Whereas in the past the basic education of naval officers, after leadership, has been focused on platforms—ships, aircraft, submarines—and then on weapons, combat units, and, finally, command, control, and related matters, that education will have to begin by conveying an understanding of the network-centric operations paradigm within which all the other naval force elements are embedded. Beyond that, network-centric operations will have to pervade all the training and education of naval force personnel and Department of the Navy civilian staff.

1.5.4 Recommendations Regarding Personnel Management

The following recommendations pertain specifically to personnel management:

1. The Department of the Navy and the naval forces should institute network-centric operations education and training at all levels across the Navy and the Marine Corps.

2. The Department of the Navy should develop a process for (a) identifying the qualifications for billets critical to network-centric operations (including both domain and infrastructure experts) and (b) identifying training and education needs for those billets. Military and civilian personnel should train together when the information technology learning requirements and facilities are shore-based.

3. The naval forces should develop career paths for both military and civilian personnel to retain and reward those with information technology expertise.

4. The Department of the Navy should analyze and describe the composition and qualities of the current and projected information technology workforce so that more informed decisions can be made about how to distribute specific elements of the work to active-duty or reserve military personnel, civilian employees, and contractor personnel.

5. The Department of the Navy should update information technology job codes to match the work that network-centric operations will require. This update should extend to both military billets and civil service positions.

2

Network-Centric Operations—Promise and Challenges

2.1 INTRODUCTION

2.1.1 Potential for Enhancing Mission Effectiveness

The promise of network-centric operations (NCO) for carrying out naval force combat and peacetime missions includes increased reaction speed and improved quality of decision making made possible by greatly improved situational awareness and access to widely dispersed forces and weapons. NCO are characterized by the rapid acquisition, processing, and exchange of mission-essential information among decision makers at all command levels, enabling them to operate from the same, verified, situational and targeting knowledge bases at the resolution and the decision cycle time required at each level. When coupled with a clear understanding of the higher commander's intent, this shared awareness will enable naval forces to reach joint action decisions more rapidly than would otherwise be possible and to focus the maneuvers and fire of widely dispersed forces to the greatest effect possible.

In NCO, all naval force elements will operate as a coherent whole in ways that were not possible with previous capabilities, with their actions synchronized in support of the commander's intent. The committee emphasizes, however, that network-centric operations must be conceived, designed, and implemented as systems consisting of sensors, human decision makers, forces and weapons, information repositories, and logistics. Every element of these systems must receive attention if the promised benefits of NCO—overwhelming naval warfighting superiority—are to be realized. It is envisioned that all levels of command, from the Chief of Naval Operations (CNO) and the Commandant of the Marine Corps (CMC) to individual sailors and marines, will engage in NCO over the complete spectrum of naval missions from humanitarian peacekeeping to full-scale war.

The Navy and Marines of the future have four fundamental missions: maritime dominance, power projection, deterrence, and air dominance. Increased effectiveness in these missions is the goal of network-centric operations. Because of changes in the geopolitical environment and a shift to continental U.S. (CONUS)-based forces, a premium is placed on forward presence and sea-based forces.

A major goal of NCO should be to have decision superiority, i.e., the ability to operate well inside an adversary's decision cycle so as to significantly reduce or lock out his options. When rapid decision making is coupled with access to a wider range of high-precision guided weapons delivered from more distributed locations on the network, the probability of achieving first-round-for-effect targeting with an accompanying reduction of collateral damage and logistic tail will be greatly increased.

2.1.2 Measuring Output

In NCO, combining sensors should enable naval forces to achieve results that surpass the sum of the results from individual sensor capabilities. For example, a single radar sensor can locate a target with great precision in range but with an angular uncertainty that can be orders of magnitude larger due to the width of the transmitted beam. (The resulting target location resembles a long, narrow ellipse, transverse to the target line of sight.) However, if a second radar sensor located at a different spatial position observes the same target at about the same time from a very different angle, the two regions of uncertainty intersect in a rather small overlap region. If both observations are combined to define the target position, uncertainty about its location is immediately refined in all directions to dimensions on the order of the range resolution (see Figure 1.4 in Chapter 1). Neither radar alone could provide the same overall location accuracy. Multiple-sensor cooperation in defining target location for precision-guided munitions will be a routine activity in NCO.

In a more revolutionary sense, NCO can enable the naval forces, as the first forces on the scene in many cases, to establish the command and control for an entire joint task force with responsibility for air and missile defense, initial land operations, and other support functions.

Benefits that derive from NCO include the greater flexibility of forces and support structure to conduct diverse operations faster than is possible today; the increased speed with which a commander in action can maneuver both forces and fire; the greater adaptiveness of pilots and controllers to shift en route aircraft to moving targets of opportunity; and the enhanced robustness of operations to the effects of uncontrollable events such as real-time enemy threats, tactics, and behavior, or the random events of nature and problems with technical systems.

Possibly the most important benefits for improved mission effectiveness are yet to be derived and will result from the development of new concepts of operations (CONOPS) made possible by a common information infrastructure (the Naval Command and Information Infrastructure (NCII)) and the development of highly integrated systems of human decision makers, sensors, forces, and weapons.

The potential for a substantial increase in mission effectiveness is the value proposition afforded by NCO. Realizing that potential will require that CONOPS be developed and doctrine changed with this top-level output metric in mind. Operations analyses, systems analysis, simulations, operations gaming, field experiments, and prototype forces must all be used to derive quantitative measures of improved, if not revolutionary, mission effectiveness as the output metric. Such measures might include target(s) destroyed, opposing forces turned back or defeated, success in completing a combined exercise plan, or other measures of mission accomplishment. Understanding this simple concept of output metrics is crucial before delving into the technical issues associated with networks, links, architectures, and other details of infrastructure. If, for example, NCO can make bomb damage assessment (BDA) more timely and accurate, then restrikes against destroyed targets can be avoided, thereby reducing risk to pilots and permitting a greater number of engaged targets. One study suggests that improving BDA may reduce the number of strikes by as much as 25 percent.[1]

Finding: While the Department of the Navy has a long tradition and in many cases leads the way in network-centric-like operations in such missions as air defense and antisubmarine warfare, it does not currently possess the metrics and measuring systems needed for the broad range of NCO mission areas envisioned. Department of the Navy efforts to implement NCO could be greatly improved by identifying output measures directly tied to mission effectiveness.

2.1.3 Evolving in a Changing Context

The naval forces—i.e., the Navy-Marine team—will continue to be a major forward-deployed arm of the United States around the world well into the foreseeable future. They are likely to be engaged in a wide range of operations from humanitarian relief to full-scale war. Engagements will occur at sea, sometimes far from friendly territories, and at times on land without the benefit of in-country support systems. The Navy-Marine team will sometimes have power projection ashore as a major mission, entailing many new challenges for which solutions do not currently exist. The Navy and Marines must develop an operational process

---

[1] Soules, CAPT Stephen, USN, *Joint C4ISR Decision Support Center [Norfolk Brief 99] (U)*, Office of the Assistant Secretary of Defense (C3I), Washington, D.C., February 16, 1999, briefing to the committee (classified).

for accomplishing this mission and must put in place the organization and structure to implement the process. This process includes preparing the battlefield through strikes, landing the Marines while dealing with mine warfare, and supporting the Marines once ashore with long-range fire, logistics from the sea, and control of the seas. Because of the dispersed nature of the likely engagement scenarios and the need for speed of action, and in some cases for new CONOPS, naval forces stand to benefit significantly if the move to global network-centric operations currently under way within the Department of the Navy can be planned, led, and executed cohesively.

2.1.3.1 Planning for Collaboration and Interoperability

Future naval force operations will require joint-Service collaboration and in most cases coalition involvement. Naval forces have a core set of equipment, doctrine, training, and responsibilities, but the other Services and agencies of the United States provide critically needed additional capabilities in almost all engagements. The Air Force provides bombers, in-flight aircraft refueling, specialized stealth bombers, long-duration manned and unmanned reconnaissance air vehicles, and other resources. The National Reconnaissance Office provides vitally needed overhead sensors of the battlespace. The Army provides large numbers of ground troops in any major land engagement and is much more richly endowed than the Marines in long-range weapons and support structure for sustained operations. The Navy and Marines cannot do the whole job by themselves. The naval forces alone do not have a complete system involving sufficient situational sensors, and forces and weapons, to successfully conduct many of the missions assigned to them. Moreover, the Department of Defense's (DOD's) vision of future operations is exceedingly joint and demands unprecedented integration, not mere defusing of conflict across the Services. In designing the NCII and planning for future network-centric operations, the Department of the Navy must accept the responsibility to provide the necessary interfaces to ensure effective interoperability with the sensors and assets from other Services and agencies because the Department of the Navy is the beneficiary of these resources. Joint force commanders of the future must be able to seamlessly integrate across the various Services. The design and implementation of the NCII and NCO planning must be fully compliant with the vision and intent of Joint Vision 2010.[2]

National interests will often dictate that the United States be part of a bilateral or multinational coalition force. Indeed, coalition operations will probably be—as they are today—the norm rather than the exception. The Department of the Navy and the DOD will need to develop and ensure effective methods of information interoperability with these coalition forces as new network-centric

---

[2]Shalikashvili, GEN John M., USA. 1997. *Joint Vision 2010*. Joint Chiefs of Staff, The Pentagon, Washington, D.C.

systems are developed and deployed. Coalition members can change from engagement to engagement and sometimes will not have procured the appropriate equipment or developed the appropriate doctrine. This presents many challenges—including the need to establish links and liaisons quickly in a crisis. Doing so can greatly leverage the capabilities of allied forces, which are often numerous and in place.

2.1.3.2 Providing Comprehensive Support for Decision Making and Action

To ensure smooth functioning across joint force operations, the NCII, the hardware and software that integrate seamlessly all the elements of NCO—namely, sensors, information and knowledge bases, logistics and support, commanders, and the forces and weapons and their subsystems (see Figure 1.1 in Chapter 1)—must be entirely consistent with DOD standards. However, investment in a common information structure alone is not sufficient to realize the significant potential benefits of NCO. In addition, investments must be made in sensors because the Department of the Navy lacks many of the sensor systems necessary to accomplish future missions. For example, naval aircraft are not equipped with appropriate sensors to track and destroy mobile and maneuvering land-based targets. The Marines need some form of a hovering observation and communications-relay platform over the battlespace to implement their land-attack plans. In the future, determining whether the desired effects of a military action have been achieved (the output metric) may require a collection of sensors that is not in place today from any U.S. resource.

Investments must also be made in supporting human decision makers so that they can reach more accurate decisions more quickly. Research in the cognitive sciences, in such areas as naturalistic decision making,[3] may provide answers regarding how humans make better decisions under stress and time pressures. The science of naturalistic decision making shows that, given time pressure, high stakes, and uncertainty, human intuition rather than analytic reasoning takes over. In stressful situations, experts recognize patterns and react immediately without building and evaluating multiple options. The Department of the Navy may need to train commanders in recognizing patterns in typical cases and anomalies encountered in operations to improve their mental simulation skills and enable quicker and better decisions.

Figure 2.1 illustrates the observe, orient, decide, and act (OODA) process in simple terms. At any point in time, Navy and Marine commanders at all levels are working in a context with specified objectives and constraints. This context

---

[3]Klein, Gary. 1997. *Sources of Power: How People Make Decisions.* MIT Press, Cambridge, Mass., November.

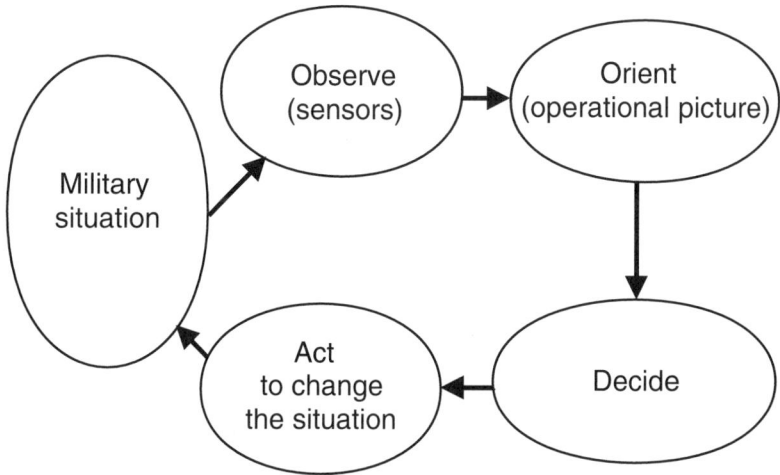

FIGURE 2.1 Steps in the observe, orient, decide, and act (OODA) loop.

is their military situation, which includes the strength, status, and location of friendly, coalition, neutral, and enemy forces; the political situation; environmental constraints; and any other factors, such as enemy tactics and morale, that can influence future actions and outcomes. The military situation is observed imperfectly by sensors of all types, ranging from satellite sensors to Aegis ships and E-2 aircraft, to Marine forward observers and even human spies. The information from all these sensors, some of which is erroneous and sometimes deliberately misleading or contradictory, must be collected and converted into a higher level of knowledge by staff personnel, or better yet by computers and software agents whenever possible, because of their speed. Validated information is presented to commanders so that they can make assessments, estimates, and judgments, i.e., orient themselves to the operational picture. Based on this situational awareness, the constraints presented by the military situation, and the time and resources available, commanders must decide what to do. Commanders can use a variety of instruments, the most potent of which are forces and weapons, to effect change in the military situation.

A commander who is planning what to do when tensions are rising may have enough time to seek additional input from sensors. A commander who observes that his ship is under missile attack may have only seconds to deploy defensive weapons. Time is a very important dynamic that overlays every OODA loop. Therefore, the NCII must be designed to reflect the time dynamic of most critical network-centric operations and to ensure that the OODA loop can be executed in the required time. Early in the NCII development process, requirements must be derived for response time and quality of information, based on analysis of likely future operations. When validated, these requirements must inform the overall NCII systems design. In some operations, the required time to complete the OODA loop may be so fast that it cannot be met by the response time of the NCII. In these cases, specialized closed-loop automated systems may have to be used.

The Navy and Marine decision makers who will affect military situations and outcomes range from the CNO and CMC to a ship commander, an aircraft commander, or a Marine platoon leader, and potentially to individual squadron leaders. This entire range of individuals could conceivably be operating simultaneously on the network, and the total number engaged at any time could be quite large. The average and peak numbers of users and their response-time requirements must be determined and analyzed as part of NCII system design. Each decision maker has a level of required information, with its associated level of granularity and specificity, as a basis for acting decisively in his or her own OODA loop time dimension. Special priority must be given to high-temporal-response OODA loops, such as in missile defense, for which traffic bottlenecks in the system could mean disaster and loss of a platform. The NCII must be designed to accommodate all these different requirements.

In addition, the type of operations being conducted by decision makers in their OODA loops at any given time will determine further requirements for the NCII. In operations ranging from operations other than war through major theater war, the tempo in each OODA loop and hence the demands on the NCII will increase significantly as tensions escalate. The NCII must be designed to respond dynamically to these changing requirements and to give each user confidence that the system will provide the necessary sensor information to permit deliberation, decision making, and execution that preclude the adversary's ability to respond.

2.1.4 Examples of Network-Centric Operations and Requirements for Success in Mission Objectives

As designers undertake the difficult job of designing the NCII to enable future NCO, it is useful to present brief examples or vignettes of missions or operations that occur in different parts of the four-dimensional space described above in terms of the OODA loop. In addition to indicating the range and characteristics of the information needed by the decision makers involved at various levels in resolving military situations, the scenarios also highlight technical requirements to be met by sensor systems and other sources of information in achieving mission success.

The committee points out here that its definition of NCO is quite general and does not prejudge important issues such as the form of command relationships, extent of delegation, dependence on automated systems, or globality of the networking. NCO encompass a broad range of activities over diverse circumstances. For example, the commander of a particular peacemaking operation might demand rigid control over even low-level actions, such as whether to engage a single enemy aircraft, because such actions could have strategic consequences. In another peacemaking operation, authority for on-the-spot decisions might be delegated down to a marine platoon. In large, intense wars against a highly competent enemy, operations might be driven by mission orders with extensive delegation and relatively little middle management; further, they might include—for certain periods of time—automated actions by air and missile defenses. In some instances, NCO might involve a fleet commander depending heavily on information provided from sensors and analysts many thousands of miles away (in an Internet-like fashion). In other instances, NCO might pertain only to the real-time sharing, within a much smaller region, of fire-control-quality information (in a cooperative engagement capability (CEC)-like fashion).

One of the distinguishing features of NCO is that mission objectives are achieved by coordinating functions across platform boundaries. NCO are thus a natural next step in warfighting that already includes multisensor cueing and networked defense systems. But network centricity is revolutionary, perhaps, in the sense that many critical mission components, including self-defense, targeting, and firing of weapons, will rely to an unprecedented extent on close multiplatform cooperation. In fact, the shift to NCO is driven in part by the inability of sensors on any single platform to provide the information necessary for force protection and power projection in the modern threat environment.

While traditional requirements are tied to platforms and platform subsystems, the technical requirements for NCO begin with the need to accomplish missions. Of the missions mentioned above in Section 2.1.1, the Navy has built considerable networking capability in deterrence, air power, and sea dominance, surface and undersea. The committee's judgment was, however, that the Navy's capability for the power projection mission, particularly the land-attack aspect, lags behind those of other mission areas. Hence in the examples below and in the remainder of the report, major emphasis is given to the land-attack aspect of the network-centric power projection mission.

2.1.4.1 Preparation for Major Theater War

When naval forces conduct strike planning for a major theater war during rising tensions and with a time frame of days or months, Navy and Marine commanders and staff are working with information at an intermediate level of detail on the numbers, location, and characteristics of targets. Because commanders in this situation must directly order and oversee execution of sensor and weapon missions, it is their responsibility to obtain the information needed to develop plans and a prioritized and synchronized target queue, including the type and number of forces and weapons to be used.

As tensions escalate, the effort and focus turn to indication and warning and a faster update of order-of-battle information through surveillance and reconnaissance, thus increasing the sensor tasking rate and the associated flow of information through the network. Given that many of the sensors to be tasked will not be organic to the Navy or Marines, the NCII must provide seamless connectivity to these joint assets so that the target queue can be updated continuously as targets are destroyed, as friendly weapons are no longer available, or as environmental conditions change. The position and mobility of the aim points must be understood at spatial and temporal resolution sufficient to ensure that any weapon or sensor will execute effectively. The full suite of sensors available on surface and air platforms within the sphere of influence must be accessible to commanders on the network so that they have the information required for flexibility and speed in adapting to changing requirements. The results of any attacks must be quickly ascertainable based on rapid input from appropriate sensors. For complex targets, such as military positions in urban environments, several different sources of data may have to be tasked, fused, and analyzed quickly. Upon firing, the weapons inventory will be decremented automatically and the information automatically presented to the commanders.

2.1.4.2 Long-range Targeting

The following scenario, focused on long-range targeting, illustrates the need for joint networked operations in many military situations and highlights the complexity of the technical requirements for success in this mission component.

> Satellite imagery shows enhanced activity at a terrorist base located 40 miles from friendly territory. The satellite imagery is presented through the NCII to the Navy battle group commander, who decides to monitor and attack if terrorist vehicles are directed toward the friendly territory. A Joint Surveillance and Target Attack Radar System (JSTARS) is deployed, and the synthetic aperture radar (SAR) imagery gathered in early flights is added via the NCII to the National Imagery and Mapping Agency's (NIMA's) point positional database (PPDB) (aboard JSTARS or located in CONUS) to determine the precise latitude, longitude, and elevation of fixed targets in the base. The data are entered into the automated planning system used by the battle group commander and his staff to preplan an F18 mission strike with joint standoff weapon-Global Positioning System (JSOW-GPS) missiles.
>
> On the fifth day of flight operations, the moving-target indicator (MTI) radar on JSTARS indicates significant movement in the angular sector that contains the base. Imagery from a Global Hawk unmanned aerial vehicle (UAV) confirms that the movement is due to terrorist vehicles leaving the base, and not to commercial traffic. The JSTARS data and the Global Hawk information are instantly provided to the

battle group commander, who decides to act by ordering an attack on the terrorist base and vehicles.

While on the carrier, the JSOW missiles on F18s are loaded with GPS coordinates for approved targets in the terrorist base. The F18s take off and head toward their target. Intelligence indicates that GPS jamming might be a problem, so the F18s ensure that GPS coordinates are accurate and release the JSOWs. As the JSOW missiles fly toward the base, each detects that its Inertial Navigation System (INS) and GPS coordinates differ by more than an acceptable margin, suggesting the effects of jamming. The INS in each missile now guides it to the selected target. The targets in the terrorist base are destroyed.

Because the terrorist vehicles are moving, they cannot be targeted with a GPS weapon. Based on the earlier alert status, special forces were landed and positioned to laser-designate any vehicular movement out of the terrorist base. The battle group commander decides to attack any moving targets with Maverick missiles fired from an F18. The F18 flies into enemy territory and releases its AGM-65C missiles. The missiles fly to the laser-designated targets and destroy them.

The technical keys to success in this mission scenario are as follows:

- Satellite intelligence;
- Precise localization of fixed targets by adding SAR data against NIMA's PPDB;
- Precise GPS localization of the aircraft before launch, and download of the data to the missile;
- Self-localization of the JSOW missile using inertial navigation when GPS is denied;
- MTI radar indications of movement;
- Imagery validation of potential moving targets using a UAV;
- Ground designation of moving targets; and
- Instant information on the situation provided by the NCII to the battle group commander.

The scenario illustrates the complex interplay between intelligence and tactical data that must be designed into the NCII. Satellite data are extremely valuable for identifying a potential target but often do not provide tactical targeting data. To provide the precision needed to target smart weapons, SAR and MTI data must be processed extensively, which works for fixed targets but not mobile targets. With the support of a network of sensors and platforms, GPS smart weapons are well suited for fixed targets. Currently, mobile targets can be detected by MTI but still require visual identification, which can be provided by imagery obtained from UAVs, and designation when targeted from the air, which involves the potential for significant risk to friendly assets. The critical capabilities are accurate identification to prevent kills of the wrong target and very timely localization to keep the target within range of the weapon. Reliance on National and joint assets for satellite imagery, the JSTARS SAR, and the Global Hawk information illustrates the importance of designing an NCII that has seamless interfaces to the valuable sensor assets enabling this kind of complex operation.

2.1.4.3 Individual Combat Missions

In a major theater war, individual sailors, marines, and aviators conduct combat in a time frame of seconds, minutes, or hours and are told the "what" of their commander's intent. With few exceptions, the "how" is left to these front-line operators, who work within the OODA operational model to plan and execute against the assigned target in a very stressful space-time dimension. They must have information about enemy defenses to outmaneuver them and must know or negate the target location (in four dimensions, including time) in the reference frame of the weapon or sensor to be used. Given that modern, high-speed, stealthy, and precision weapons are deployed by all combatants, decision times are short, and the effects of attacks must be determined dynamically with great precision and speed.

Because all the information for planning and execution must be timely and specific enough for mission completion, this situation represents the highest level of detail required and the most exacting time dimension. Combat in these circumstances will often place the greatest demands on the responsiveness of the NCII and movement of information through it and on the speed with which decisions can be made and acted on.

2.1.4.4 Network-Centric Expeditionary Operations

Expeditionary power projection operations include amphibious landing, fire and logistics support of forces ashore, and establishment of air superiority. At the same time, the task force commander must provide force protection, including theater, air, ballistic, and cruise missile defense, antisubmarine warfare (ASW), and mine countermeasures (MCM). Networking for each of these functions and/or missions will carry its own particular requirements for the NCII. Fully networking the overall expeditionary operations to provide and enable sharing of a comprehensive joint operational picture offers the potential of a very great improvement of efficiency and effectiveness in a joint system-like operation.

While the land-attack aspect of the power projection mission is emphasized in the ashore examples and throughout much of this report, it should be emphasized also that expeditionary power projection by the joint task force (JTF) will include littoral battlespace preparation involving ASW and MCM, as well as strike, amphibious landing, and fire and logistics support of the forces ashore. At the same time, to protect the forces afloat and ashore, local air dominance, and cruise and ballistic missile defense, and sea lane dominance including ASW and MCM, must be provided by the command JTF. Enabling these functional and mission areas brings its own requirements for networking in the NCII. Fully networked, the overall operation via the NCII will be very complex, as necessary to provide and enable sharing of a comprehensive common operational picture (COP), offering the potential for greatly improved efficiency and effectiveness of operations by the JTF.

2.2 BASIC CAPABILITIES REQUIRED IN A COMMON COMMAND AND INFORMATION INFRASTRUCTURE

As the critical core element that integrates the elements of commanders, sensors, information and knowledge bases, forces and weapons, and logistics enabling NCO, the NCII must be designed to meet the following basic requirements:

- Provide sufficient capacity, quality of service, and speed to meet operational needs as the level and tempo of conflict vary;
- Incorporate control mechanisms necessary to meet leaders' needs, e.g., for security, efficiency, and economy;
- Have costs of implementation and operation that are sufficiently low to ensure that all naval nodes needed to maintain operational effectiveness can be included in the network and that training needs can be satisfied; and
- Provide assurance regarding the overall security and reliability of the network and of the information it transports.

A revolution in commercial networking is now occurring that can be embraced to ensure that the NCII can be developed to meet these military needs. This revolution is increasingly converging on the Internet model of a single infrastructure that can accommodate all applications, with the characteristics of the network being determined by the requirements of the most demanding applications using it.

At the physical level, the NCII network will be made up of devices and media that physically connect nodes at which

- Data gathered by sensors can be injected or retrieved,
- Knowledge bases reside that were derived from previously collected data,
- Applications involving processing and fusion of data can be executed,
- Command can be exercised, and
- Actions can be implemented.

In addition, to enhance mission effectiveness, the elements of commanders, sensors, knowledge bases, forces and weapons, and logistics must all be integrated within the NCII (see Figure 1.1) to provide the following capabilities necessary for successful NCO in the 21st century:

- *Integrate.* Combine and present multiple elements of information.
- *Evaluate.* Analyze different courses of action, campaign plans, battle attack plans, and individual sorties and project the potential outcomes.
- *Predict.* Assess an enemy's view of the situation and forecast probable enemy behavior at all levels.
- *Cross-reference.* Express all relevant objects in a common space-time frame.
- *State.* Express and understand and/or estimate the time-referenced geolocation and movement vector of a relevant military object.
- *Catalog.* Know and keep current the details of all relevant military objects.
- *Associate.* Assign accurately and quickly the necessary information to relevant military objects so that they can be clearly understood. Automatic target detection, recognition, classification, identification, and fingerprinting are among the technologies that enable this capability.
- *Remain aware.* Maintain situational understanding in a relevant time frame. Enemy countermeasures, wartime reserve mode employment, and changes in enemy tactics are examples of activities that must be detected and monitored.
- *Provide assurance.* Maintain secure, uncorrupted, and timely delivery of information and knowledge.
- *Visualize.* Display an appropriate representation of the battlespace at all levels in all dimensions. This visual capability is appropriate when it is the best way to exploit the part of human cognizance associated with seeing.
- *Be dynamic.* Enable timely and decisive action that exceeds the enemy's capability by a large magnitude.
- *Assess.* Rapidly assess the effects of applying forces and weapons, including bomb damage as well as the effects of all military services' full range of weapons, from information operations to explosive devices.
- *Control.* Influence outcomes with the minimum expenditure of physical and human resources.

Chapter 4 discusses NCII concepts and architecture in some detail. The important issue of information assurance is addressed in Chapter 5, and Chapter 6 examines current capabilities and progress toward achieving the capabilities needed for effective network-centric operations.

2.3 THE NEED FOR SYSTEM ENGINEERING

The common command and information infrastructure required to support networked naval forces will be large and complex, with many different types of interfaces to external sensors, platforms, weapons, forces, knowledge bases, and human decision makers. Integrating all these resources in an efficient and effective way requires a disciplined approach—system engineering. The committee believes that the application of system engineering to the development of a successful NCII is mandatory. With few exceptions (Naval Sea Systems Command (NAVSEA) 05 was one[4]), it has not observed this methodology being applied in the network-centric effort under way now within the Department of the Navy.

Because system engineering is so important to success in developing a capable NCII, the six axioms of the methodology are outlined here, all to be applied with sound and creative engineering judgments as to where and how to allocate emphasis and resources:

1. *Set the requirements.* Develop a complete, consistent set of requirements. These requirements will relate partially to the NCII itself but also to the network-centric operations that the Department of the Navy wants to conduct. The importance of establishing the requirements for NCO early in concert with the developing new CONOPS cannot be overemphasized.
2. *Perform studies of the trade-offs.* Objectively and systematically select the best design concepts from among alternative solutions to satisfy the requirements within the available resources and schedule. Avoid point solutions; they are rarely optimum.
3. *Document the baseline.* Put the baseline design into a document for all to use.
4. *Manage the design.* Proceed from preliminary to final detailed design of the selected concept using accepted practices, components, and materials, and conduct major reviews with all the stakeholders present at the conceptual, preliminary, and final stages.
5. *Verify the design.* Continually verify that the design meets all the requirements under all expected environments and conditions.
6. *Document everything.* If it is not written down, it never happened!

Joint Vision 2010 presents an excellent conceptualization of future operations, but detailed plans are needed now for accomplishing the vision. Such plans need to be developed and prosecuted by those with large-system analysis and

---

[4]The committee was briefed by NAVSEA 05, Deputy Commander for Warfare Systems Directorate, to deal with the battle group interoperability problem such as the distributed engineering plant. With the establishment of the Chief Engineer, NAVSEA 05 has been designated as NAVSEA 53. In the remainder of this report, these steps and activities are attributed simply to NAVSEA.

engineering expertise. Only through the application of this disciplined approach, which has been used to design and develop many highly successful large systems (such as the fleet ballistic missile family, the Space Transportation System, and numerous military aircraft), can the naval forces have any confidence in the resulting design and implementation. Future critical missions must be defined, and operational analysts must determine the requirements to accomplish these operations. Designs must be developed to meet the firm requirements, and trade-offs should be studied to select the optimum design, given the various constraints. Only after critical design reviews should the hardware and software implementation begin.

The system engineering approach contrasts sharply with the approach currently under way in which viewgraphs paraphrase Joint Vision 2010 and the Defense Planning Guidance, and road-map charts merely identify the chronology of big events. Lists of miscellaneous desired operational capabilities that in many cases are ill-defined, open-ended, and more functional than operational will not result in an operational network. With few exceptions, the committee observed almost a total lack of system engineering rigor in the numerous presentations given to it for this study.

Finding: With few exceptions, a disciplined system engineering methodology is not currently being applied to the development of the NCII.

The hardest part of converting from platform centricity to network centricity will be changing the minds of those involved. Once begun, the momentum must not be seen to wane. This will require dedicated leadership, a constant and continuing reinforcement of the goals, and continuity of effort. This, in turn, calls for gathering a critical mass of formal and informal leaders throughout the Navy, carefully laying out a strategic plan and a campaign (operational/business) plan, anticipating where the weak points and/or potential failures lie, and developing contingency plans. One cannot tell people to believe in the concept of NCO and expect immediate acceptance. One can depict the desired outcome, define the desired behavior patterns associated with NCO, and reward the individuals who perform most effectively.

2.4 THE CRITICAL ROLE OF LEADERSHIP IN NETWORK-CENTRIC OPERATIONS

2.4.1 Technology and Doctrine for Supporting Decision Makers

A critical element in network-centric operations is the human commander. The human brain, although it remains limited in its ability to process the increasing amounts of information that networks and computers are capable of delivering, is still superb in making associations and recognizing patterns. Only human leaders can assimilate the information provided in NCO and convert it into the knowledge and understanding that lead to decisions and actions. Strong and effective decision makers therefore can be argued to be the most important element in network-centric operations. Better and more timely decision making—one of the significant challenges for improved mission effectiveness—requires high-quality information in a form that humans can rapidly recognize and understand. One example is graphical representations in which humans can easily recognize patterns and changes in patterns, as opposed to the textual representations used extensively today. Another challenge is to enable autonomous decision making for effective operations in local situations.

NCO must feature a mission style of command in which the commander's intent or the purpose of a task is explained and subordinates are given the freedom to accomplish that task in their own way within doctrinal guidelines. Senior commanders will need to hold a very loose rein, allowing for ingenuity and spontaneity in subordinates. Improvisation will often be the order of the day, and freedom of action the byword. Implicit understanding will reduce the need for detailed and lengthy instructions. In short, for NCO, restrictions on leaders must be minimized and their initiative and responsiveness maximized.

While the Navy command and operational decision structure has been evolving in this direction for some years, the succeeding steps needed to fully accommodate the needs and techniques of NCO could be wrenching for the Service. NCO could induce changes in the very meanings of the terms "command" and "leadership" and will also affect how coordination, cooperation, and teamwork are carried out. This goes beyond technological innovation to social revolution within the Service. The Navy's leadership will have to enter this new command and information world fully aware of its implications if the greatest advantage is going to be gained from the shift from platform- to network-centric operations.

Good leaders in the NCO mode want to have available the most up-to-date technology and will be well prepared to take full advantage of its sophisticated capabilities. However, experienced leaders are more restrained in what they expect technology to provide under the stress of combat than are many of the advocates of high-technology equipment, especially in the area of command and control. Mature leaders are realistic about the demands of battle, and they always anticipate the unanticipated. They realize that the "friction of war" will continue to haunt every corner of the battlespace.

Basic to NCO are the integration and interpretation of the reams of information streaming in from the many intelligence systems, sensors, and reconnaissance assets in order to present combat leaders a coherent "picture" that will provide situational awareness. Leaders must be able to discern meaningful patterns of enemy activity in conditions that appear, and in most cases are, disordered and confused. This knowledge, coupled with the experience, judgment, and intuition of well-trained leaders, will allow them to adapt to the situation at hand, identifying and exploiting enemy vulnerabilities while protecting their own.

Integral to NCO will be decision-centered command and control facilities designed by human-factors engineers and cognitive psychologists. The contributions from these experts will also be needed in the development of decision-centered staff organizations and decision-centered training programs. The strengths of computers will have to be balanced with the strengths of human minds—the application of intuition, improvisation, and creativity, especially in the face of new or unique problems. Disciplined application of ergonomics will be required to improve the interfaces between machines and humans, and across entire systems. Decision aids, including software agents and personal digital assistants, will have to become ubiquitous as they "mine" data, make comparisons, and otherwise draw on experience captured as lessons learned. Information that has been transformed to the knowledge level will be available in context and whenever possible in an image format. Anchor desks, common databases, and shared pictures will enable collaborative *thinking*, a more powerful and fundamental capability than collaborative planning.

Finding: The Department of the Navy needs to focus research and development (R&D) on methods to achieve improvement in human decision making because human decision makers are a key element in NCO, and their ability to make faster and better decisions is essential to mission effectiveness.

2.4.2 Leading the Transformation to Network-Centric Operations

To succeed, the planned transformation from a platform-centric to a network-centric naval force will require strong support from the top. This support must include a shared vision for NCO among the senior leaders of the Navy and Marine Corps, a set of strategic objectives, and a tactical plan for achieving the objectives. The plans must be supported by priorities, allocation of resources, appointments, recognition and reward of individuals and groups, and enthusiasm. Further, the top leaders are responsible for ensuring that those involved in change are meeting defined goals and objectives and persist in making progress over the long haul.

An important related aspect in transforming the naval forces is to develop concrete measures of output. The committee strongly recommends that the Navy and Marine Corps leadership use as a criterion whether proposed changes in operations will substantially enhance the capability of the joint and naval forces to accomplish critical military missions. This is in contrast, for example, to pursuing ill-defined and open-ended objectives such as "information superiority" without having any detailed measures for assessing achievement of the objective. Only by looking at operational objectives (missions) in a variety of circumstances can the naval forces develop the requirements to drive decisions about what is needed and how much is enough to support accomplishing the military objectives of the 21st century.

It is the responsibility of the top leadership to clarify the goals and the associated measures of success. This responsibility cannot be delegated to fleet commanders, ship captains, or systems commands. A consensus-building process that brings all the key stakeholders together to define the goals and requirements of network-centric operations is badly needed.

Finding: The naval force leadership needs to develop a shared vision of what network-centric operations can accomplish that includes concrete measures of improvements expected in mission effectiveness.

2.4.3 Creating the Environment for Transformation

Enlightened top-down planning to create an environment for the transition to network-centric operations should accomplish the following objectives:

- Set high-level operational (not functional) challenges to motivate and focus innovation.
- Identify crucial building-block capabilities in terms of forces, operations, and systems.
- Ensure development of integrative capabilities, i.e., command and control to operate adaptively by drawing on the building-block capabilities and providing the necessary tailoring, and doing so extremely quickly when necessary (inside the opponent's OODA loop and within the time scales of other critical events). These capabilities should be fully joint because, in many circumstances, the commander-in-chief (CINC) or JTF commander will be operating from a Navy ship and will be depending on naval forces for early critical operations.
- Establish a vigorous "marketplace" where innovations can be competed and rewarded.
- Support development of cross-cutting infrastructure (e.g., the information grid and standards driven by bottom-up considerations and commercial trends) with Department of the Navy funds.
- Encourage military science such that new operational concepts and operational phenomena are widely discussed, debated, and ultimately understood—not just in viewgraph terms or at the level of intuition, but in terms of system concepts and related methodologies.
- Establish mechanisms for ensuring that innovations move beyond a permanent test status and are implemented in the fighting force.

These objectives may seem straightforward, but it is revealing to contrast them with current practices. In the course of this study, few of the briefings received by the committee reflected an output-oriented approach. Instead, all too many repeated or rephrased general notions from Joint Vision 2010 or the Defense Planning Guidance rather than describing capabilities harnessed to accomplish missions. Discussion was typically quite abstract, whereas much of the real work in developing NCO will be at the level of defining and refining building-block forces, operations, and systems. The traditional U.S. approach to military planning, with its emphasis in peacetime on ponderous "deliberate planning" around a single operational concept and a myriad of assumptions, is almost the opposite of preparing for at-the-time adaptive plan development. To be sure, those engaged in deliberate planning develop many building-block operations and gain the detailed domain knowledge essential in crisis or conflict. To exploit NCO fully, however, the emphasis should be changed. Participants should practice developing plans rapidly from the building blocks rather than optimizing plans with ever-increasing levels of detail and refinement for postulated circumstances that probably will not apply—much as championship football teams play adaptively throughout a game rather than executing "the" operations plan. Such an approach would also make it easier to consider alternative concepts of operation. In the committee's view, this change in doctrine, which has great ramifications at the joint level, is critical to achieving the aims of NCO.

Similarly critical is the need to shift toward "system thinking" and to ensure that good ideas enter the operational force. Related is that it is essential for the Department of the Navy to ensure that senior leaders who are responsible for implementation of network-centric operations have appropriate technical education and experience; good system work is not a casually acquired capability.

With respect to moving ideas into the operational force, it is interesting to note that the Army's strategy in creating a strike force explicitly recognizes that real change requires translating ideas into provisional units operating in the fighting force (in the case of NCO, this could mean trying out a flexible command). Similarly, the influential Marine Corps Combat Development Command is working closely with the Commandant of the Marine Corps, who sees himself as the principal engine for change. The Navy, however, must use a different approach because of its very different organizational culture and balance of power. It is essential that the Navy's powerful fleet commanders play a key role in the Navy's transformation to network-centric operations—not just technically but also in terms of organization and doctrine. This effort will be challenging because of the fleets' continuing high operational tempo, but there are many examples of past innovation introduced in the fleets. Fortunately, NCO do not require setting aside scarce platforms and commands. Indeed, some of the important NCO concepts are potentially crucial to near-term challenges such as sea-based missile response to enemy artillery attacks on land (Korea), very fast strike and logistics resupply reaction against moving armies (e.g., the next Iraqi crisis), and sea-based defense against ballistic missile attack (e.g., the next Taiwan crisis).

Finding: The naval force leadership is not developing the type of rapid, adaptive, and innovative top-down planning required to realize the full benefits of NCO.

2.5 A PROPOSED PROCESS FOR DEVELOPING CONOPS FOR NETWORK-CENTRIC OPERATIONS

2.5.1 Overview and Recommendations

A new paradigm is needed to develop CONOPS that will enhance mission effectiveness in the network-centric world of the future. Because both Navy and Marine forces will be involved in future NCO, development of CONOPS should be implemented cooperatively by the Chief of Naval Operations (CNO) and the Commandant of the Marine Corps (CMC). A key challenge today for the Navy and the Marine Corps is learning how to migrate from their current information infrastructure architecture to the developing NCII. Each Service has designated responsible organizations to facilitate the transition.

The Navy established the Navy Warfare Development Command (NWDC) in 1998, to "focus and champion warfare concept development, design and lead the fleet battle experiment program and synchronize and standardize the Navy's doctrine."[5] The NWDC has three organizational components: a division for concept development, a doctrine division, and the maritime battle center, which is managing the fleet battle experiments. This new command is intended to produce new or alternative doctrine, insight into technologies in an operational context, identification of newly required operational capabilities, ideas for new warfare, and future experiments.

The U.S. Marine Corps established the Marine Corps Battle Laboratory (MCBL), an element of the Marine Corps Combat Development Command (MCCDC), as a focal point to expedite the evaluation and evolution of critical concepts through experimentation. The Sea Dragon process is being used to investigate future warfighting concepts, doctrine, tactics, techniques, and procedures (TTPs), organization, and advanced technologies. The Special Purpose Marine Air-Ground Task Force-Experimental was structured to function as a test organization.

The committee believes that the lead organization for the Navy portion of the CONOPS planning and development should be the NWDC and that the lead organization for the Marine portion should be the long-established MCBL. These two commands should work together closely, especially on operational missions such as power projection from the sea.

The recently established NWDC, however, has inadequate staffing in both number and qualifications to accomplish the envisioned NCO tasks. The NWDC should be supplemented with planning experts from the MCCDC and the other Services, operational analysis experts, systems engineering experts, and Navy

---

[5]Johnson, ADM Jay L., USN, Chief of Naval Operations. 1998. "The New Naval War College: Focusing on Forward Thinking," *Surface Warfare*, September/October, p. 2.

and Marine officers with broad operational experience, a system orientation, and an innovative spirit. Close cooperation with the proposed functional type commander for the recommended Information Operations and Space Command, described in Chapter 7, is mandatory. Indeed, the commander of the Information Operations and Space Command, who would become the single point for providing network-centric operations to the fleet, would have a major responsibility in providing the appropriate fleet-experienced officers to the NWDC for CONOPS development. These officers would then become the ambassadors for implementing new CONOPS into the fleet.

As emphasized above, the Navy and Marines (and the Department of Defense) need to develop and focus on *output* measures of effectiveness appropriate to the information era. These must include, among others, reduced decision cycle times, reduced engagement times for missile interception, improvements in BDA leading to reduced restrike missions, accuracy in predicting adversary actions, effectiveness of weapons in reaching targets based on improved location accuracy, and so on. Although traditional *input* measures of capability, such as numbers of divisions, battle groups, or wings, will still be of value, they fail utterly to capture the very capability-enhancing and outcome-improving features that NCO seek to strengthen.

Some of the measures required will deal with human capabilities (amidst suitable support systems) more than with raw measures of force. For example, the qualitative capability of officers to rapidly assemble and execute *good* plans involves more than merely shortening the cycle time for building them. Similarly, concepts of operations involving highly distributed operations (e.g., those of Marines operating in the rear area of enemy-occupied territory) must take into account the experience, morale, and comfort level of those involved. The feasibility of delegating authority to call in long-range fire will depend on the quality, training, and judgment of those young officers who have the authority.

Yet another class of measures relates to exploiting the potential of network-centric operations to affect the perceptions and resolve of both enemies and third-world countries. The ability to have major effects from long distances, without warning and with a high degree of precision and concentration, creates opportunities that are not yet well understood.

The committee recommends that the CONOPS planning group begin by selecting an initial set of operational concepts that meet the following criteria:

• Involve high-priority naval force missions that are difficult enough to demand new concepts of operations and/or capabilities (i.e., stressful *operational challenges*)[6] and can exploit the inherent advantages of a networked force engaged in NCO;

---

[6]The DOD has emphasized such operational challenges in the Defense Planning Guidance and elsewhere. The U.S. Atlantic Command has been increasingly emphasizing them in its planning of

- Have specific outcomes that can be measured for success; and
- Involve joint forces and perhaps coalition forces.

Some candidate near-term operational challenges that the committee believes meet these criteria are as follows:

- Rapid forced entry into land positions by the Marines supported by Navy fire to secure critical installations and defeat enemy forces early,[7]
- Attack operations against time-critical mobile targets,[8] and
- Rapid establishment of adaptive command and control centers at sea or on the land.

It is not enough to have broad challenges. Organizations also need more specific, even quantitative, goals if they are to get on with systematic problem solving and change. For example, quantitative methods are needed to characterize the ability to seize and secure some number of fixed facilities or positions against some specified level of opposition within some specified period of time in a range of operational circumstances. The Department of the Navy might wish to require the ability, assuming the presence in the region of a carrier battle group and an amphibious ready group, to seize and secure on the order of three lightly defended airbase or port-sized facilities or positions within 24 hours of an order to execute. More generally, the Department of the Navy should have a sense of what the emerging capabilities could accomplish in this regard. This understanding should reflect consideration of details such as warning time, threat level, terrain, whether the United States has information dominance (having the information it needs while denying the enemy the information it needs), and so on. Results should be characterized as "envelopes of capability in scenario space,"

---

experiment campaigns (U.S. Atlantic Command, 1998). These are quite distinct from such "functional challenges" as, for example, improving communications or improving collaborative en route planning. See also Davis, Gompert, Hillestad, and Johnson, 1998, *Transforming the Force: Suggestions for DoD Strategy*, RAND, Santa Monica, Calif.

[7]For discussion of how this challenge can be addressed systematically, see Davis, Bigelow, and McEver, 1999, *Analytical Methods for Studies and Experiments on "Transforming the Force,"* RAND, Santa Monica, Calif.; and Defense Science Board, 1998, *Joint Operations Superiority in the 21st Century: Integrating Capabilities Underwriting Joint Vision 2010 and Beyond*, Volume 1, Final Report, Office of the Under Secretary of Defense for Acquisition and Technology, Department of Defense, Washington, D.C.

[8]This operational challenge has been focused on by the U.S. Atlantic Command and its successor the Joint Forces Command in its joint experiments. The effort is supported by a large analysis group within the Institute for Defense Analyses. The work includes detailed human-in-the-loop simulation using synthetic-theater-of-war technology. It also includes gaming and more aggregate analysis more or less along the lines discussed in Davis et al., 1999, *Analytical Methods for Studies and Experiments on "Transforming the Force,"* RAND, Santa Monica, Calif.

not as ability to accomplish some point scenario. There are too many variables for any one scenario to be a good basis for planning.

The CONOPS group would proceed by conducting a detailed operational analysis of the selected mission, to include the following:

1. A systems approach in which all elements of the system are considered and traded off to lead to a balanced solution to the problem;
2. Identification of those elements of the operation necessary for its successful execution, to include numbers and types of sensors, information needs, forces, weapons, logistics, and decision aids;
3. Specification of the capabilities of and the detailed requirements levied on sensors, information, weapons, logistics, and other assets and elements of the military operational model described above;
4. Development of an initial operational, systems, and technical design to meet the mission objectives;
5. Studies of trade-offs intended to optimize the operational design and to avoid point solutions while managing risks (including security risks);
6. Introduction of new technologies when they can improve mission effectiveness;
7. Use of computer modeling and simulation and human gaming to develop insights into the operational design;
8. Testbed experiments conducted to verify critical design features, where appropriate;
9. Use of the proven "model-test-model" iterative or spiral development approach whereby incremental improvements are added to the design as a result of gaming, testing, simulation, new technology, and so on; and
10. Selection of the preferred operational approach as a result of the above effort.

The analytical approach that the committee suggests has a number of key features:

- A decision perspective supported by a decision-argument-hypothesis-analysis process, focusing research and experiments on issues central to potential decisions regarding capabilities and concepts addressing critical challenges;
- Hierarchical decomposition of the operational challenges into building-block challenges that can be studied more or less independently;
- A system perspective highlighting the need for well-understood building blocks that can be combined on short notice in integrated operations under diverse circumstances;
- For each building-block operation, an analytical architecture supported by a family of models that can be used for the following:
  —Exploratory analysis to understand issues associated with meeting the

challenges in a vast scenario space (including detailed circumstances) and to identify issues and context for in-depth study;

—In-depth, high-resolution analysis to understand underlying phenomenology—even down to the level of sensor logic, weapon times of flight, and command and control interoperability—and to use that understanding to help shape the higher-level, lower-resolution exploratory models; and

—Integrative analysis at the operational and strategic levels.

To implement this approach, the committee envisions a family of models ranging from analytic models that can be run and understood by a single analyst with a personal computer, to human games (which may be simulation-supported, as in synthetic theater-of-war work), to field experiments. The value of some models can be enhanced if human behaviors and decision making have been represented respectably (e.g., with so-called agent-based models). The point is that conducting research in a way that draws on the full range of analytical instruments is very different from what has traditionally occurred in Navy, Marine, or joint experimentation. Many opportunities have been lost.

The NWDC and the MCCDC next should subject the preferred operational plan to war games in which decision makers and adversaries will determine the plan's strengths and weaknesses. After the war game results are analyzed, any necessary modifications to the operational design should be incorporated.

The NWDC and the MCCDC should aim for an 18-month turnaround for the above spiral development process. The result would be a well-documented CONOPS that was ready for prototype implementation. The information required for conducting the operation would be captured and put into the form of adaptive templates. The templates would provide the initial set of information in an engagement, but as conditions changed the users could modify the templates easily.

At this point the recommended functional type commander, Information Operations and Space Command (see Chapter 7 for the organizational details), would introduce the CONOPS and related capabilities to the operational force on a provisional or prototype basis. For example, a single carrier battle group/ amphibious ready group in the Third Fleet (USS *Coronado*) or other elements of the operational fleet would be equipped to support the CONOPS. Such action will require changes in the acquisition cycle to expedite the procurement of new telecommunications and information equipment and software. Without this expedited procurement, realizing benefits from the new NCO in the near- or midterm will be impossible.

The committee recommends that large-scale fleet experiments involving the provisional force and other traditional forces be conducted. In this way fleet operators will develop experience with the new NCII and NCO, and the Navy and Marines can obtain a true comparative measure of the improvement in output effectiveness of the network-centric force. Depending on results, these CONOPS or revised versions would be taken up over time by other parts of the force. This approach to experimentation—after careful analytic development of CONOPS—is in sharp contrast to the current fleet approach that sets aside a portion of the naval forces for experiments, resulting in only incremental improvements without a plan for wide-scale implementation.

Assuming success in identifying and testing the new concepts and capabilities, the Navy and Marine Corps then would have to make plans for appropriate force-wide changes over time, develop and promulgate widely the relevant changes in doctrine, and make associated changes in the personnel system (including recruitment, education, and training). Some of these changes would begin early (e.g., developing initial doctrinal concepts before fielding even a provisional capability). They would co-evolve along with technology and concepts. The overall process of change from a platform-centric force to a network-centric force will take many years, especially in cases involving major acquisitions. Further discussion and details related to these recommendations are provided in Chapter 7.

Finding: There is no effective Navy and Marine Corps process for selecting, developing, and implementing CONOPS in the network-centric paradigm.

2.5.2 Transitioning Through Experimentation

To make the transition to network-centric operations as quickly as possible, a recommended strategy is to place key information technologies into the hands of naval warfighters at all echelons in a way that allows them to easily try out new ideas for using those technologies. Then, ideas that produced substantial warfighting value should be introduced quickly into the NWDC and MCCDC CONOPS development process and deployed more widely in an accelerated manner.

Experimentation with new technologies and processes holds the key to transitioning: "The purpose of an experiment is to explore alternative doctrine, operational concepts, and tactics that are enabled by new technologies or required by new situations. That is, new technologies or situations may call for different ways of conducting operations. But without actual operational experience in using those technologies or in those new situations, experiments are the next best thing, because they provide more of a basis for making informed doctrinal choices than does reliance only on analytical studies and/or simulations."[9]

Experimentation should occur at different scales, at different echelons, with

---

[9]Computer Science and Telecommunications Board, National Research Council. 1999. *Realizing the Potential of C4I: Fundamental Challenges*. National Academy Press, Washington, D.C., p. 210.

different mission types, and with different operational communities. Experiments should complement modeling and simulation activities and demonstrations such as the advanced concept technology demonstrations (ACTDs). They should be designed to provide insight into the ramifications of a new operational concept or innovative technologies. They should have hypotheses about and measures of effectiveness, and as such require rigorous analysis of results. They can fail in their ability to find the right solution but should always succeed in providing knowledge about the ramifications of new ideas and technologies.

Both the U.S. Army and the U.S. Air Force have incorporated experimentation in their own transition to network-centric architectures. Their programs have helped to refine not only system architectures but operational architectures as well. The spiral process they applied was essential to their transition strategy because it accelerated innovations into the field. Analogously, this core process is essential to the Navy's migration path for NCO, and it warrants further discussion.

2.5.3 The Spiral Process

2.5.3.1 Characteristics of the Spiral Process

The spiral process is also called evolutionary development because it "... is an innovative method to field a system quickly using commercial and government off-the-shelf equipment, with maximum user involvement throughout the process."[10] The first spiral is usually regarded as the first development cycle of a system. Subsequent spirals allow technology insertion, addition of new mission capabilities and upgrades, and enhancement of interoperability and integration, all in an environment of continuous user feedback.

The process characteristically partitions the more traditional development cycle into shorter, incremental cycles, during which operators get hands-on access to the evolving system in each cycle and provide their feedback and requirements to a development team that is prepared to respond with modifications. In so doing, the operators may modify their own operational processes and concepts based on use of the emerging capability. The spiral process is more than an acquisition process; it also supports reengineering the operational concepts. Each spiral has its own defined activities, performance objectives, schedule, and cost; each spiral concludes with a user decision to field the system, continue with evolution, or stop.

The spiral process has several distinguishing characteristics:

---

[10]Gilmartin, Kevin, Electronic Systems Command Public Affairs. 1998. *Spiral Development Key to EFX 98*. Department of the Air Force, Hanscom Air Force Base, Mass. Available online at <http://www.hanscom.af.mil/ESC-PA/news/1998/jul98/efx98.htm>.

- Continuous feedback is accepted from users throughout each spiral based on their *actual use* of the evolving capabilities. This is a preferred alternative to a paper-requirements process.
- It is *an acquisition process*—the operators' reactions are used to alter actual system capabilities during development.
- The operational concepts supported by the system capabilities are evolved as well, through a *reengineering of operational processes*, doctrine, tactics, and organizations.
- An experimentation *program* provides the framework for the spiral process to evolve new operational concepts and processes in addition to the new system capabilities.

2.5.3.2 Advantages of the Spiral Process

The spiral process is a powerful alternative to the traditional acquisition process. One of its advantages is that it offers a sound replacement in areas where technology is changing rapidly and cycle times in the commercial sector are short compared to the traditional DOD requirements and acquisition processes. It is difficult to specify requirements for revolutionary concepts in advance and equally difficult to anticipate how new and innovative capabilities will be used. Rather, such understanding matures over time. The spiral process as embedded in an experimentation framework enables a faster maturation of this understanding in incremental bursts and over discrete, short time periods.

The spiral process also accomplishes the following:

- It enables new capabilities to be developed based on *known* requirements (from actual use) rather than on *unknown* requirements (postulated many years in advance of deliveries into the field).
- It facilitates interoperability and integration of systems. Spiral development is effective at uncovering interoperability problems because the output of each cycle, though intermediate, is the result of a testing and integration process using operators with hands-on access. This is the best method for uncovering anomalies in interoperability.
- It reduces risk. It is possible to focus on higher-risk and unknown aspects of programs in early cycles of the process, rather than delaying until the final stages of a long requirements, design, and development process to detect problems and identify their solutions.
- It accelerates fielding of innovative operational processes and systems. The intermediate products of the spiral process can themselves be deliverables for operational use. Systems results can be fielded rapidly because there is a direct and immediate correlation between the product designed and developed and the operational process supported, which can be replicated in the field without another prolonged requirements-and-development phase.

2.5.4 Spiral Development in Army and Air Force Programs

2.5.4.1 Army Experimentation Program

The Army vision of battlefield digitization was articulated in the early 1990s. The goal is improved lethality and increased operational tempo through the application of information technology. Significantly enhanced situational awareness at all echelons is intrinsic. To evolve, the Army used a series of experiments to shape and equip its future force by evaluating networked forces equipped with information technologies. More specifically, the Army embarked on a series of experiments, simulations, and exercises, including several advanced warfighting experiments (AWEs), echelon by echelon. This process continues today with a view toward fielding an Army XXI over the next several years and evolving toward the Army After Next by FY 2020+.

Each experiment required changes to the then-current operational concepts and doctrine, supported by certain advanced information technology capabilities not fielded in the operational Army. The resulting systems architecture was a composite of experimental technologies integrated with legacy systems, designed and developed as an integrated product specifically for the experiment.

Because of continuing problems with interoperability, the Army evolved a technical architecture after soliciting responses from the commercial sector. At least two-thirds of this architecture was migrated into the first version of the Joint Technical Architecture (JTA). Today the Army's unique extension of the JTA is synopsized as JTA-Army. Compliance is addressed through acquisition oversight and certification testing conducted on systems before fielding. The Director for Information Systems, Command, Control, Communications, and Computers is the responsible architect and reports directly to the Army's top acquisition executive.

The Army's use of the spiral process in the migration strategy resulted from its experience with the Task Force XXI, an AWE that culminated in a force-on-force engagement at the National Training Center in March 1997. The preparation began with an operational architecture that described how a digitized brigade would conduct operations if equipped with all the information technology the Army had at the time. A spiral evolutionary process was used to deliver the systems architecture. This is discussed in an article by General Steven Boutelle, USA, and Alfred Grasso, in the *Army RD&A* magazine.[11] The Army has given much credit to the spiral process for the transformation. The process was used at the Central Technical Support Facility at Fort Hood, Texas, where operators

---

[11] Boutelle, BG Steven, USA, and Alfred Grasso. 1998. "A Case Study: The Central Technical Support Facility," *Army RD&A,* March-April, pp. 30-33. Available online at <ftp://204.151.48.250/docs/dacm/rda9802.pdf>.

trained with a series of operations-like drills on the systems architecture that evolved in increasingly robust stages. The evolutionary acquisition process allowed developers to adapt and/or correct while operators trained. The net result was an integrated "system of systems" that was used in the AWE.

For Task Force XXI, the architectural process began when an operational architecture was postulated. Legacy systems and digitization initiatives were evolved for the experiment to support that postulation and to conform to the then-current Army technical architecture. The actual conduct of the AWE was affected by some immaturity in certain advanced technologies used, but this is to be expected with an experimental process. The Army gained substantial knowledge from the event. The subsequent assessment of what actually happened during the AWE was used to accelerate certain key system acquisitions for subsequent fielding by the Army. The net result was that the Army moved to accelerate into the field operational concepts and a system architecture that incorporated key information technologies. This constituted an intermediate step toward a longer-term goal, one that will be achieved at a considerably accelerated pace in years over that allowed by the traditional acquisition process.

Today the Army is pursuing a migration strategy that incorporates the spiral process and experimentation as key components. Joint experimentation is being expanded, and an international coalition program for digitization is in the early stages, with specific international partners.

2.5.4.2 Air Force Experimentation Program

The vision of the battlespace infosphere proposed to the Air Force by the Air Force Scientific Advisory Board (AFSAB) is organized around information.[12] The architecture framework addresses not only the capabilities of network-connected command, control, communications, computing, and intelligence (C4I) components with database and communications services but also *all* forces and systems associated with conducting a military operation.

To move toward this vision, the AFSAB proposed jump-starting a prototype of the battlespace infosphere, starting with the colocation of elements of the Electronic Systems Command (ESC) and an aerospace command, control, intelligence, surveillance, and reconnaissance (C2ISR) center, and then moving rapidly to a major experiment applying many Defense Advanced Research Projects Agency initiatives, which, if successful, would result in "leave behinds" for operations. Locating this initiative near Norfolk, Virginia, was anticipated to improve "jointness." Use of the spiral development model initially developed at

---

[12]U.S. Air Force Scientific Advisory Board. 1998. *Report on Information Management to Support the Warrior,* SAB-TR-98-02. Department of the Air Force, Washington, D.C., December. Available online at <http://ecs.rams.com/afosr/download/sab98r1.pdf>.

ESC was intrinsic to the migration and was articulated as a specific recommendation: "... [T]he evolution model starts with a set of mature technologies plus an initial concept. The initial experiments will result in a revised concept and possibly a revised list of technologies. The art in using this spiral approach to concept and system evolution is to find the collection of mature technology that will support a meaningful test of the concept. If this spiral development approach is done correctly, this will simultaneously change the way people think about and deal with information while accelerating the development and maturation of enabling technologies."[13]

The migration process applied by the Air Force for its command and control (C2) architecture is illustrated by the expeditionary force experiments (EFXs) used to build the Expeditionary Aerospace Force. These are major and minor experiments conducted every year, alternating in scale every other year. EFX98 was a major experiment that used processes that align with the generic migration framework described above.

The EFX98 explored command and control using global networks for forces and information. The prototype operational organization was significantly reduced in footprint. A robust network linked shooters to C2 nodes to gain improved responsiveness. The objectives included reduced time lines and en route mission updates for changes in targeting based on an assessment of the situation more current than that available at the outset of the mission.

The operational architecture and systems architecture used in the actual experiment, conducted in September 1998, resulted from the "fourth spiral" of an evolutionary acquisition process begun at ESC many months earlier. The JTA-Air Force was applied as the standards and guidelines. Spirals occurred approximately every 3 months. Many operators exercised the evolving systems architecture that included many technology initiatives and continuously evolved until the time of the experiment. Their hands-on use stimulated many adaptations that eventually were stabilized in the architectures used for conducting EFX98. The result assessment is being used to establish an integrated C2 capability for the field.

The EFX98 was so successful[14] that the Air Force determined that the spiral process for evolutionary acquisition should be adopted Air Force-wide. The process is currently being documented in an Air Force instruction with the intent to mandate its application.

---

[13] United States Air Force Scientific Advisory Board. 1998. *Report on Information Management to Support the Warrior,* SAB-TR-98-02. Department of the Air Force, Washington, D.C., December, p. x. Available online at <http://ecs.rams.com/afosr/download/sab98r1.pdf>.

[14] As with Task Force XXI, "success" in an experiment does not imply that all innovations applied in the experiment are ready for operations. The knowledge derived from the experiment can be the most important product.

2.5.5 Navy and Marine Corps Experimentation

The Navy and Marine Corps have embraced large-scale field experimentation. The Navy used a war game, Global '97, to study ways that Joint Vision 2010 would be applied in the future for naval forces and also for joint task forces. A series of fleet battle experiments (FBEs) has been planned, and many experiments already have been executed to explore new concepts and systems. Among these are the maritime fire support demonstrator, the cooperative engagement capability, and new strategies for theater ballistic missile defense. ACTDs are also being used to explore emerging technologies with a view to earlier (than traditional) fielding.

FBEs Alpha, Bravo, Charlie, Delta, and Echo are completed. More are planned.[15] Alpha was linked with a prior Marine AWE called Hunter Warrior, conducted in March 1997. This experiment explored increases in lethality against time-critical targets with a robustly networked force of sea- and air-based shooters employing automated pairing of weapons to targets and allowing deconfliction (collision avoidance) of all objects in the integrated airspace.[16] Among the concepts tested were naval fire[17] coordination, C4I, the arsenal ship, and joint precision fire.

FBE Delta in September 1998 combined Navy and Army sensors and shooters, real and simulated, to combat a simulated attack by North Korea. Submarines, surface combatants, and aircraft were linked with a joint fire coordination network. The common operational picture enabled by Navy sensors was exploited by Army helicopters to react on time lines not previously demonstrated.[18] FBE Echo, in tandem with the Marine Corps' Urban Warrior experiment in the San Francisco Bay area, dealt with maritime asymmetrical threats in a littoral urban environment, using new concepts for undersea warfare. It also continued to explore naval fire, networked sensors, and strike/land-attack weaponry with command and control and theater air defense. FBE Foxtrot is currently in the

---

[15]FBE Foxtrot, Golf, and Hotel have been planned for December 1999, May 2000, and September 2000, respectively.

[16]Alberts, David S., John J. Garstka, and Frederick P. Stein. 1999. *Network Centric Warfare: Developing and Leveraging Information Superiority.* CCRP Publication Series, Department of Defense, Washington, D.C. Available online at <www.dodccrp.org>.

[17]Fire encompasses all ordnance deliveries and their required targeting, as well as integrating and coordinating mechanisms. See Soroka, Maj Thomas, USMC, 1997, *A Concept for Seabased Warfighting in the 21st Century,* Working Paper, Naval Doctrine Command, Norfolk, Va., October 31 (unpublished); and Maritime Battle Center, Navy Warfare Development Command, 1998, "The New Naval War College," in *Surface Warfare,* Vol. 23, No. 5, September/October, pp. 2-5.

[18]Alberts, David S., John J. Garstka, and Frederick P. Stein. 1999. *Network Centric Warfare: Developing and Leveraging Information Superiority.* CCRP Publication Series, Department of Defense; Washington, D.C. Available online at <www.dodccrp.org>.

planning stages to explore network-centric concepts for precision engagement, mine warfare, antisubmarine warfare, and counterweapons of mass destruction.

The FBEs alternate between U.S.-based and forward-deployed fleets. Each experiment is focused on a core mission, such as land attack. Results are assessed to establish how new technologies and tactics may enhance the capabilities of the fleet (and joint/allied forces).

As an example, a technology concept called Ring of Fire[19] has been tested and modified four times. The objective is to allow surface ships to respond quickly to a call for fire ashore using both existing and future weapons (simulated). To date, this experimentation has been used to demonstrate a significant increase in the speed with which targets can be identified and attacked. The concept is being evolved to include ground forces: Marine or Army, whichever unit is best positioned to engage. Ultimately the maturity of the concept will result in fielding a land or sea capability.[20]

The ACTD Extending the Littoral Battlespace, which had an initial demonstration in April 1999, provided new capabilities for theater-wide situation awareness, integration of sensors, and over-the-horizon connectivity. The objectives were to leverage C4I for improved precision targeting and mass remote firepower through integration and collaboration for use by dispersed units. Experimental capabilities included a central tactical information infrastructure for enhanced situational awareness and broadband communications networks.

In the U.S. Marine Corps' series of AWEs—Hunter Warrior, Urban Warrior, and Capable Warrior—each was preceded by its own series of limited-objective experiments; all are parts of a 5-year plan focused on an extended dispersed battlespace with varying terrain and including urban and near-urban littoral areas. Among the concepts being examined are unit enhancements that include long-range precision strike, urban operating capabilities for sea-based forces, and the effects of networking with weapons systems.

The Hunter Warrior experiment focused on tactical operations and equipped a Marine task force with a communications web over the theater of engagement, connecting all levels so that they could access the common digital picture of the battlefield. Enhancements were made to command and control, fire support, and targeting. Urban Warrior was conducted in conjunction with a CINCPAC-sponsored exercise, with FBE Echo, and with the first Littoral Battlespace ACTD. The objectives were to enhance the ability of naval forces to accomplish simulta-

---

[19]A joint fire coordination network that receives calls for fire, assigns a firing platform using the appropriate ammunition, keeps track of force ammunition inventories, and deconflicts fire in the joint operations arena, as described in *Surface Warfare*, September/October, 1998, p. 4.

[20]"Fleet Battle Experiments Set to Spearhead Future Technology," *Jane's Defence Weekly*, Vol. 31, No. 12, March 24, 1999, pp. 25-26.

neous noncontiguous operations throughout a littoral region. Capable Warrior will be used to integrate what was learned in the earlier series of experiments by using operational concepts, force structures, TTPs, and technologies that proved successful and modifying those that did not. It will be accomplished in conjunction with naval units operating at the level of a joint task force.

Broadly speaking, however, the committee believes that the Navy and Marine approach to experimentation has been inadequate. Among the problems have been the following:

1. A tendency to focus on a few critical "events" (e.g., major fleet experiments or short, intense Marine experiments) rather than a process of systematically studying a warfare mission and options for accomplishing it;
2. Extreme underutilization of analysis, modeling, and simulation (including virtual simulation with people in the loop); and
3. A failure to decompose the broad problems into components that can be studied in appropriate ways over time, whether with small-scale laboratory or operational experiments, analysis, systematic interviewing of experienced officers, or other methods.

In recent months the Department of Defense, the U.S. Joint Forces Command, and the Services have all received recommendations along the lines the committee urges here.[21] Sometimes this approach has been described as a recommendation to embrace the model-test-model paradigm (although "model" must be understood to include man-in-the-loop gaming).

2.5.6 Uniqueness of the Spiral Process

The spiral approach to designing network-centric naval forces—especially, the integration of major platforms into the information-based fleet network—will present many challenges to the current way of doing business. Methods of budgeting, planning, and allocating resources, congressional authorization and appropriation, enforcing accountability, and achieving standardization are needed to guide a rapidly evolving naval force configuration. Only in this way will the naval forces be able to evolve into their new configuration and modes of operation under the anticipated conditions of rapidly changing technologies and environment. The alternative is to remain with today's fragmented, stovepiped approaches that cannot keep up with changing technology and the demands of the

---

[21]Military Operations Research Society (MORS). 1999. *Proceedings of Joint Experimentation Mini-Symposium and Workshop* (Armed Forces Staff College, Norfolk, Va., March 8-11, 1999). Military Operations Research Society (MORS), Alexandria, Va.

"information economy" within which the naval forces are becoming embedded and will have to operate. This alternative is unacceptable, so that the naval forces will have no choice but to make the difficult and necessary adaptations to achieve the spiral process, including the negotiation of mutually acceptable approaches with the Office of the Secretary of Defense and the Congress.

2.6 SUMMARY OF FINDINGS AND RECOMMENDATIONS

In reviewing the naval forces development of network-centric operations to date, the committee arrived at a number of findings presented and discussed throughout the chapter and makes the following recommendations for improvement.

Finding: While the Department of the Navy has a long tradition and in many cases leads the way in network-centric-like operations in such missions as air defense and antisubmarine warfare, it does not currently possess the metrics and measuring systems needed for the broad range of NCO mission areas envisioned. Department of the Navy efforts at implementing NCO could be greatly improved by identifying output measures directly tied to mission effectiveness.

Recommendation: The Department of the Navy leadership should develop a set of strategic goals and expectations for NCO with accompanying measures of output performance. The current capability must be baselined, targets of improvement established, and progress verified as NCO become a reality.

Finding: With few exceptions, a disciplined system engineering methodology is not currently being applied to the development of the NCII.

Recommendation: The Department of the Navy should ensure that the NCII and the interfaces to external sensors, knowledge bases, human decision makers, forces, weapons, and logistics are treated as a system and that system engineering methodology is applied to all development aspects. Failure to implement this disciplined approach will have dire consequences.

Finding: The Department of the Navy needs to focus R&D on methods to achieve improvement in human decision making because human decision makers are a key element in NCO, and their ability to make faster and better decisions is essential to mission effectiveness.

Recommendation: The Department of the Navy should develop technology, techniques, and training for presenting information to human commanders in a way that increases the quality and speed of their decisions.

Finding: The naval force leadership needs to develop a shared vision of what network-centric operations can accomplish that includes concrete measures of improvements expected in mission effectiveness.

Recommendation: The naval force leadership should implement a consensus-building process that brings all of the key stakeholders together to define NCO goals and objectives based on expectations for improvement in the output measures of mission effectiveness.

Finding: The naval force leadership is not developing the type of rapid, adaptive, and innovative top-down planning required to realize the full benefits of NCO.

Recommendation: The naval force leadership needs to encourage and reward innovative system thinking to solve high-level operational challenges and ensure that the best concepts are moved into prototype and operational forces.

Finding: There is no effective Navy and Marine Corps process for selecting, developing, and implementing CONOPS in the network-centric paradigm.

Recommendation: The Navy Warfare Development Command and the Marine Corps Combat Development Command should work together on a few high-priority and challenging naval force operations that can be implemented more effectively using NCO. The committee believes that power projection from the sea involving the landing and engagement of Marines deep inland against an aggressor with long-range supporting fire from the Navy is one such operation. The NWDC, supplemented with the proper staffing, should analyze these missions as part of a spiral development process in which modeling and simulation, gaming, testing, experimentation, and new technologies are introduced to select a candidate CONOPS. The selected CONOPS should be implemented in a prototype fleet or in elements of the operational fleet. Fleet experimentation should be conducted, and measures of output effectiveness should be determined and used to evaluate performance. When finalized the CONOPS should be introduced into the fleet over time and the accompanying doctrine, equipment, training, and organizational structure co-evolved.

2.7 BIBLIOGRAPHY

Alberts, David S., John J. Garstka, and Frederick P. Stein. 1999. *Network Centric Warfare: Developing and Leveraging Information Superiority.* CCRP Publication Series, Department of Defense, Washington, D.C. Available online at <www.dodccrp.org>.

Beinhocker, Eric D. 1999. "Robust Adaptive Strategies," *Sloan Management Review*, Vol. 40, No. 3. Available online at <http://mitsloan.mit.edu/smr/past/1999/smr4039.html>.

Cohen, Secretary of Defense William S. 1999. *Annual Report to the President and Congress.* Department of Defense, Washington, D.C.

Computer Science and Telecommunications Board, National Research Council. 1999. *Realizing the Potential of C4I: Fundamental Challenges*. National Academy Press, Washington, D.C.

Davis, Paul K., David Gompert, and Richard Kugler. 1996. *Adaptiveness in National Defense: The Basis of a New Framework*, Issue Paper IP-155. RAND, Santa Monica, Calif.

Davis, Paul K., David Gompert, Richard Hillestad, and Stuart Johnson. 1998. *Transforming the Force: Suggestions for DoD Strategy*. RAND, Santa Monica, Calif.

Davis, Paul K., James Bigelow, and Jimmie McEver. 1999. *Analytical Methods for Studies and Experiments on "Transforming the Force,"* DB-278-OSD. RAND, Santa Monica, Calif.

Defense Science Board. 1996. *Summer Study Task Force on Tactics and Technology for 21st Century Military Superiority,* Vol. 1, Summary. Office of the Secretary of Defense, Washington, D.C.

Defense Science Board. 1996. *Summer Study Task Force on Tactics and Technology for 21st Century Military Superiority,* Vol. 2, Part 1, Supporting Materials. Office of the Secretary of Defense, Washington, D.C.

Defense Science Board. 1998. *Joint Operations Superiority in the 21st Century: Integrating Capabilities Underwriting Joint Vision 2010 and Beyond*, Vol. 1. Office of the Under Secretary of Defense for Acquisition and Technology, Department of Defense, Washington, D.C., October.

Defense Science Board. 1998. *Joint Operations Superiority in the 21st Century: Integrating Capabilities Underwriting Joint Vision 2010 and Beyond*, Vol. 2, *Supporting Analyses*. Office of the Under Secretary of Defense for Acquisition and Technology, Department of Defense, Washington, D.C., October.

Herman, Mark. 1999. *Measuring the Effects of Network-Centric Warfare*, Office of the Secretary of Defense (Net Assessment) and Booz-Allen Hamilton, draft, March.

Hundley, Richard. 1999. *Past Revolutions, Future Transformations: What Can the History of Revolutions in Military Affairs Tell Us About Transforming the U.S. Military?* RAND, Santa Monica, Calif.

Johnson, ADM Jay L., USN, Chief of Naval Operations. 1998. "The New Naval War College: Focusing on Forward Thinking," *Surface Warfare,* September/October, p. 2.

Joint Chiefs of Staff. 1997. *Concept for Future Joint Operations, Expanding Joint Vision 2010*. The Pentagon, Washington, D.C., May.

Klein, Gary. 1997. *Sources of Power: How People Make Decisions*. MIT Press, Cambridge, Mass., November.

Naval Studies Board, National Research Council. 1997. *Technology for the United States Navy and Marine Corps: 2000-2035: Becoming a 21st-Century Force*, Volume 9, *Modeling and Simulation*. National Academy Press, Washington, D.C.

Naval Studies Board, National Research Council. 1997. *Technology for the United States Navy and Marine Corps, 2000-2035: Becoming a 21st-Century Force*, 9 volumes. National Academy Press, Washington, D.C.

Shalikashvili, GEN John M., USA. 1997. *Joint Vision 2010.* Joint Chiefs of Staff, The Pentagon, Washington, D.C.

U.S Atlantic Command. 1998. *Joint Experimentation Plan 1999.56,57* Norfolk, Va., December. (This plan has now been superseded by: U.S. Joint Forces Command. 1999. *Joint Experimentation Campaign Plan 2000.* Norfolk, Va., September.)

3
Integrating Naval Force Elements for Network-Centric Operations— A Mission-Specific Study

3.1 INTRODUCTION

3.1.1 Scope and Approach

Network-centric operations (NCO) are performed by a set of networked assets the committee calls an NCO *system* (shown in Figure 1.1, Chapter 1). The committee has avoided the phrase *system of systems* because that phrase suggests a process whereby independently conceived and developed systems are somehow integrated. A useful approach to understanding requirements for effectively integrating these assets is to first postulate mission capabilities for the overall system and then allocate requirements among the various components.

In considering both the components of the system and the challenge of engineering and acquiring subsystems that will interoperate to perform a military mission effectively, the committee chose to focus on the Navy missions of air dominance and power projection, the first because examples of NCO exist, and the second because Navy leadership has given priority to capabilities that decisively influence events ashore.[1] (The four principal missions of the Navy,

---

[1]The committee did not study deterrence, and its examination of sea dominance was cursory. Although much of the surface portion of sea dominance is similar to power projection, current undersea warfare systems are often limited by the range of in situ sensors, and the function of remote sensors may be limited to cueing. In Appendix B, however, the committee acknowledges that there may be significant opportunities to employ networks of short-range sensors in a fully cooperative mode.

as viewed by the integrated warfare architecture (IWAR) assessment process, are maritime dominance, deterrence, air dominance, and power projection—see Figure 1.2 in Chapter 1.) Further, it focused on the naval forces' assets that interact over significant distances within rapid tactical time lines: the system of commanders and decision aids (tactical information processing); sensors and navigation; and forces and weapons. The committee believes that one or more coherent system designs are needed for NCO in each of these areas, although some systems may share components. Because distribution of components over space is central to NCO, the committee did not examine integration of assets located on a single platform.

3.1.2 Current and Potential Capabilities—What Is Possible

It is probably fair to say that the current broad interest in NCO was stimulated initially by the cooperative engagement capability (CEC) in air defense. The CEC (Figure 3.1) provides a robust information infrastructure, the data distribution system, that interconnects sensors at the radar return level. This information sharing permits a level of detection and tracking that can provide detailed engagement control. Weapons can be launched at targets the launcher cannot see, on the basis of shared tracking and target/weapon assignment algorithms. Because its embodiment is dispersed assets fighting as a coherent whole, the CEC network has been called a virtual capital ship by some.

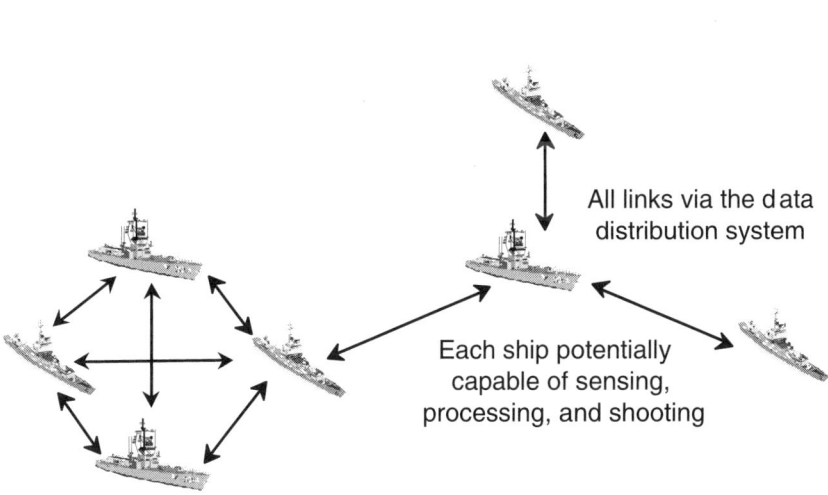

FIGURE 3.1 Cooperative engagement capability.

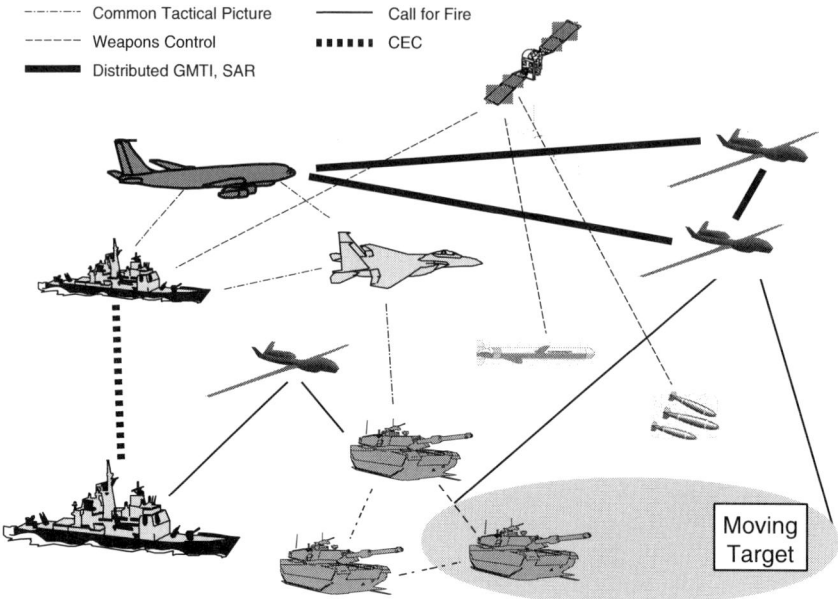

FIGURE 3.2 Potential future system for hitting moving ground targets.

An example drawn on throughout this chapter is the potential system, illustrated in Figure 3.2, that is intended to affect events ashore decisively. The fleet, standing offshore, protects itself via a CEC shield while projecting power ashore via the Marines, aircraft, and ship-based missiles. A number of unmanned aerial vehicles (UAVs) and a JSTARS aircraft provide continuous ground moving-target indicator (GMTI) coverage synthesized from all the distributed sensors as a single view, together with large volumes of synthetic aperture radar (SAR) imagery used for identifying tracks and responding to the moment's targeting needs. Theater and National signal intelligence (SIGINT) and image intelligence (IMINT) collectors provide data for context, cueing, and classification or identification. All forces (sea, air, land) contribute their geolocations and identity to a common tactical picture (CTP), which is augmented with information about enemy forces and neutral parties in the battlespace, derived in part from the real-time GMTI and SAR information. This CTP is distributed to all friendly forces to allow shared situational awareness.

Because, as both these examples suggest, naval planning and equipping are much more advanced for air defense than for land attack, the committee focuses below on discussing network-centric operations in the context of land attack.

3.1.3 Opportunities, Dangers, and Challenges— Need for a Total System Approach

Network-centric operations are more than just a good idea; they have already begun at the tactical level, and most observers deem further tactical use to be inevitable. The greatly extended range of current and planned weapons has already led to a tight, time-critical coupling between sensors, shooters, and the weapons themselves. Widespread use of the Global Positioning System (GPS) has already given rise to a battlespace in which all friendly elements are precisely geolocated in a real-time "map" that is shared among collaborating participants. There is every indication that such trends will continue and indeed accelerate in the Navy and other Services, even if no explicit action is taken to further this goal at the departmental level. The Department of the Navy's greatest challenge is that these efforts are currently diffuse and uncoordinated. A wide variety of tactical components are evolving independently toward participation in NCO. There are two principal dangers in the current state of affairs:

- *Incoherent components.* There will result a new set of "stovepiped" components that are optimized locally but do not properly internetwork, and an overall set of tactical capabilities that fails to match the Navy's needs. Such an outcome can be rectified, of course, but at the cost of time and money.
- *Dangerous new vulnerabilities.* Modern information networks can be interconnected fairly easily; without proper systems oversight, they may very well be connected in ways that lead to new, unforeseen, and dangerous vulnerabilities.

The need for planning of an entire integrated system is a recurrent theme in this chapter.[2]

3.1.4 Complexity of the Challenge

Enabling NCO requires the integration of existing components into a coherent system, and progress toward NCO will surely involve some evolutionary improvements that integrate legacy components planned and built independently. The committee believes, however, that the full power of NCO will be realized only if the sensors, weapons, and tactical information processing networked for NCO are planned and developed as coherent subsystems.

Building on the notional example of future power projection operations shown in Figure 3.2, one thread in this chapter's discussion is the complexity of

---

[2]Chapter 5 discusses controlling vulnerabilities in a common command and information infrastructure (the NCII), but some of the components discussed in this chapter have vulnerabilities of their own.

the interactions among these system components and the associated tight time lines. In the Figure 3.2 scenario, when the Marines encounter a moving enemy force, they issue a call for fire. This urgent request for aid leads to a high-priority revision of the current weapon-target pairing—a weapon in flight is diverted from its original target to a newly urgent target. More specifically, an already in-flight joint standoff weapon (JSOW) is issued GPS coordinates for the new target; these GPS coordinates are refreshed every few seconds (via a satellite link) to guide the weapon to the moving target, which is then destroyed.

This conceptual thread, which is easy enough to describe in general, poses enormous technical challenges for the various tactical subsystems and their linkages. For instance, how is the new target reconciled with the geolocated tracks provided by the UAV's GMTI system? How is this new information incorporated into the CTP? How does the call for fire interact with the weapon-targeting subsystem and give rise to a new weapon-target pairing? How is the enemy's ever-changing location continuously extracted from real-time GMTI information and relayed through a satellite to an in-flight weapon? And, most critically, how does all this happen within a few seconds?

In considering such questions, the committee found challenges in weapons, sensors and navigation,[3] and tactical information processing components of the NCO system on which it focused. Examples of these challenges are listed in Table 3.1.

In the committee's notional land-attack example, these platforms interoperate through a large number of linked components. Each component is complex in itself and involves processes and information flows that are distributed across a number of platforms. Table 3.2 indicates some of the capabilities required for success in this example and should give an idea of the complexity of the components.

3.1.5 Organization of This Chapter

The following sections explore some of the challenges to realizing the capabilities required in the four classes of components shown in Table 3.1. The discussions are condensed; fuller versions are found in referenced appendixes. In addition, the committee discusses the importance of system engineering and reiterates the need for a system of coherent components as basic to effective network-centric operations. The chapter ends with a review of the committee's findings and offers recommendations based on these findings.

---

[3]Navigation devices can be considered as sensors but are discussed separately here because of the crucial importance of gridlock in NCO. Support of commanders is discussed as part of tactical information processing.

TABLE 3.1 Examples of Leading Challenges in Developing Components of an Effective Network-Centric Operations System

Asset	Challenge
Weapons	Responsive, long-range, sustainable, affordable volume of fire for naval fire support; targeting for Global Positioning System-guided weapons
Sensors	Susceptibility to countermeasures; detection underground and under foliage; georegistration; target recognition
Navigation	Vulnerability of the Global Positioning System
Tactical information processing (decision making)	Extracting targeting-quality information from high-volume, multiplatform, multisensor data; coordinated, distributed weapon selection and support; flexible, adaptive software architectures; interoperable littoral operations

TABLE 3.2 Capabilities Involved in the Land-attack Example

Function	Capability
Common tactical picture	Provides shared situational awareness to all participants in the battlespace—Where am I? Where are my friends? Where is the enemy? This picture as a whole contains all objects in the battlespace, geolocated and annotated with other known information about the objects. Each participant, however, sees only those portions relevant to that observer's task.
Weapons control	Provides a prioritized list of targets, weapon-target pairing, authority to fire a weapon at a target, current target information, and means to update target locations for weapons in flight.
Distributed ground moving-target indicator (GMTI), synthetic aperture radar	Provides a more continuous, more extensive picture of the battlespace than can be obtained by isolated sensors. Linked unmanned aerial vehicles and a JSTARS could all contribute to a shared, real-time database for GMTI coverage; such a distributed system allows more continuous views in mountainous terrain and the like.
Call for fire	Provides a mechanism for time-critical requests from Marines or other land troops for weapons to be directed on enemy forces.
Cooperative engagement capability	Provides a highly effective defensive shield for forces afloat by tightly linking the radars and air defense missiles of multiple ships into one real-time system.

3.2 WEAPONS

This section presents as illustrative examples the use of weapons in three operations for which better connectivity and better use of networks to fuse sensor information seem desirable. Appendix D presents a broader view of current and near-term naval weapons and launch platforms, their uses, and targets and addresses the command and information support needed to employ them effectively.

3.2.1 Naval Fire Support: Targeting and Weapon Control

The Navy currently has little capability to provide prompt, long-range, surface-launched fire support for Marines or Army forces ashore but has a number of initiatives under way to develop longer-range naval fire support (NFS) weapons. The Navy is developing the extended-range guided missile (ERGM) by adding rocket power and combined inertial navigation and GPS guidance to a submunitions-dispensing artillery shell. ERGM will enable accurate fire to a range of 63 nautical miles. In a remanufacturing program, the Navy is adding GPS to convert existing, obsolete standard missiles (built originally for air defense) into the land-attack standard missile (LASM). LASM will enable accurate fire to ranges of over 100 nautical miles. ERGM and LASM will be retrofitted to Aegis ships and are projected to be used on the DD-21. The Navy is also beginning system studies for an advanced gun system that might be a 155-mm weapon and for an advanced land-attack missile (ALAM) intended for use on the DD-21.

ERGM's GPS receiver will have minimal protection against jamming during its range-dependent, 3- to 6-minute time of flight (TOF). The guidance component and the aerodynamic control authority of the weapon do not seem to support the accuracy of delivery that would be required for it to make effective use of a unitary warhead. Although foreseeable propellant upgrades may permit range extensions of this weapon to about 90 nautical miles, greater ranges will require a larger-diameter round. The weapon as currently designed will not support forces that are engaged in combat at ranges (~200 nautical miles) to which they can be delivered by the V-22 tilt wing aircraft.

The targeting concept for ERGM appears to be both ill-defined and inadequate. The targeting concept is that a forward observer or a sensor in an elevated platform will identify the GPS coordinates of the aim point. The data link that will be used by the forward observer has not been identified.

If the target moves during the weapon's extended TOF, there will be no means of correcting the weapon's trajectory. Even if a forward observer can call in corrected target coordinates, the weapon will not arrive at those coordinates for several minutes. Although the launcher is capable of rates of fire up to about six rounds per minute, high rates of fire may not be realized because of the time required for the targeting and aim-point correction processes.

To match the operational concepts that the Marine Corps is attempting to develop, NFS weapons will be driven inexorably to longer ranges. Inevitably, the problem of targeting rapid-fire, surface-launched weapons designed to attack targets at ranges beyond the line of sight will become more difficult. The solution will depend on development of closed-loop control to link a forward observer (or sensor) with the weapon and the launch platform.

The committee suggests that a robust targeting concept is needed to support the evolution of near-term and future NFS weapons. The concept should identify a doctrine for use of such weapons along with the links, sensors, and data fusion networks required for their employment in network-centric operations.

3.2.2 Air-to-Air Combat: Long-range Target Identification

In the area of air-to-air combat the United States has competent air surveillance radars on both the E-2C (airborne warning and control aircraft) and the Airborne Warning and Control System (AWACS), well-trained pilots, good tactical doctrine, high-performance aircraft, and good weapons (AIM-9X and AIM-120C). Evolutionary growth in aircraft performance, weapons range and agility, and airborne sensors is both feasible and programmed.

The problem of target identification (whether by cooperative or noncooperative means) has, for rules of engagement reasons, driven air-to-air engagements to ranges that are significantly shorter than the full kinematic range of available weapons. Although the AIM-9X is a world-class weapon, the outcome of a short-range air-to-air engagement depends on factors other than weapon performance. If the problem of identifying the target at long range can be solved, it will be desirable to engage the adversary at the longest feasible range even though the short-range weapons may be superior to those of potential adversaries.

In principle, the identification of targets at long range can be achieved by the fusion of data derived from theater and National sensors and from databases of commercial aircraft flight plans. These sensors and databases can be used to track hostile aircraft from takeoff. SIGINT may be used to deduce the mission objectives of hostile aircraft. If all available information can be fused together, the constraints imposed by restricted rules of engagement can be relaxed and engagement can be permitted at the maximum kinematic range of available weapons.

The committee believes that the expanded use of tactical networks to provide all available information to AWACS or the E-2C, and to the combat aircraft that they support, will enable air-to-air engagements to take place at the full kinematic range of current and future weapons. The advantages of future information networks that fuse all source data should be exploited to ensure the best possible outcome of future air-to-air engagements.

3.2.3 Attacking Low-signature Targets

Low radar cross section (RCS) targets, or targets that employ low and clutter-limited trajectories, are difficult to engage with existing or projected area-defense antiair warfare (AAW) weapons. Similarly, quiet submarines with reduced radiated acoustic signatures or submarines coated to reduce their effective acoustic (sonar) cross section (ACS) have become progressively more difficult to detect. Hostile submarines that are difficult to detect, classify, and localize are difficult to engage with even the best underwater weapons.

There is no simple counter to reduced-signature targets. In a general sense, the only way they can be detected is to exploit the fact that a target presenting a low RCS or ACS to a monostatic radar or sonar is likely to have large forward or specular scatter peaks. Also, a target that is buried in clutter when viewed from one aspect may not be obscured when viewed from another aspect. Thus a straightforward way to negate stealth technology is to illuminate a suspected target area with multiple illuminators and to use multiple independent sensors to detect forward and near-forward scatter peaks and specular glints. If the output of multiple sensors can be fused together, the probability of detecting low RCS and ACS targets will increase, along with the probability of successfully engaging them with current and projected AAW and antisubmarine warfare (ASW) weapons, in a network-centric operation.

3.2.4 Findings

Finding: Although new weapons are being developed for land attack, the range of surface-launched, short-time-of-flight weapons is currently too limited to support ship-to-objective maneuver at reasonable stand-off distances. Better targeting concepts are needed. (See Section 3.2.1.)

Finding: Target identification limitations inhibit the use of air-to-air weapons at their full kinematic range. (See Section 3.2.2.)

Finding: Weapons that attack low-signature targets will likely depend on guidance from networks of sensors and illuminators. (See Section 3.2.3.)

3.3 SENSORS

Effective network-centric operations require a wide variety of sensors ranging from distant sensors located in sanctuary that can provide precise target locations, to weapons sensors that can autonomously recognize targets. Current sensor capabilities and future growth possibilities are treated in detail in Appendix B. Here the committee summarizes general sensor technology trends, fundamental performance limitations, and prospects for both target detection and recognition.

3.3.1 Sensor Technology Trends and Limitations

Sensor capabilities are steadily improving through the use of modern electronic technology and the transition to all-digital and all-solid-state solutions. Distributed implementations are increasingly emphasized—both within individual sensors (e.g., radar phased arrays or optical focal plane arrays) and in the form of meta-sensors (e.g., multiple individual sensors operating cooperatively as a larger single equivalent sensor, as in CEC). Multidimensional signatures are collected to assist in classification and detection. Summarized in Table 3.3, these four trends in sensor technology are having an enormous impact on sensor capabilities.

These positive trends do not imply, however, that any sensor task or level of performance can be achieved. There are always engineering compromises to be made—trading performance for such practical aspects as cost, size, and weight—and the best possible performance is not always acquired.

Even when money and time are available, some sensing tasks are inhibited by the basic physical limitations listed in Table 3.4. Sensors are also susceptible to camouflage and deception, and to electronic countermeasures. All three sensor classes considered here—radar, electro-optics, and sonar—depend on the propagation of waves through various media and the interaction of these waves with material objects. Herein lie most of the basic physical obstacles.

For example, electromagnetic waves move at the speed of light, while sonar signals in the ocean move at about 1500 m/s. Sonar data inevitably take a much longer time to collect as compared with data from radar and optical sensors operating at similar distances.

The fundamental relationship between the angular spread or beam width of waves emitted by an electromagnetic or acoustic structure is that the beam width is of the order of the wavelength divided by the antenna diameter. Given the frequency band of the sensor, high angular resolution, which translates into small pixels on the target or background, requires a correspondingly large aperture. Optics, with the shortest wavelengths, can achieve very high angular image resolution (mrad to μrad) with millimeter- to centimeter-sized apertures; radar is characterized by much lower resolution ($\sim 1°$), with antennas measured in meters; and sonar, by even less ($\sim 3°$ to $10°$), with antenna sizes of meters to tens of meters.

Although it limits atmospheric propagation of radar and electro-optics to selected transmission wavelength "windows," media absorption is particularly troublesome for sonar because the absorption increases more or less quadrati-

TABLE 3.3 Trends in Sensor Technology

Trends	Implications
Digital technology	Stable, drift-free operation Compact, low-cost implementations Algorithm flexibility Increasing ability to exploit exponential growth of computing capabilities
Solid-state devices	High performance, e.g., sensitivity, power, and efficiency Miniaturization and low power requirements Low-cost integrated circuitry Compact integral packaging Novel microelectromechanical systems devices
Distributed components	Phased arrays for radar, electro-optics, and sonar Multiple sensor cooperation and networking, e.g., cooperative engagement capability Data fusion of multiple and diverse sensors for automatic target recognition (ATR) and geolocation Mobile sensors, e.g., unmanned aerial vehicles, unmanned underwater vehicles, and ground robots
Multidimensional signatures	Multispectral Hyperspectral Enhanced ATR and noncooperative target recognition

TABLE 3.4 Physics-based Limitations on Sensor Performance

Sensor Class	Fundamental Obstacles
Radar	Poor angular resolution with typical wavelengths and practical antenna sizes Absorption by and reflection from solid materials Frequency dilemma in foliage and ground penetration: low frequencies give poor resolution; high frequencies do not penetrate
Electro-optics	Serious weather scatter and absorption—electro-optic sensors require fair weather Resolution vs. coverage area dilemma Dimensional limits on electronic scan
Sonar	Slow, nonuniform oceanic sound propagation Interference from littoral noise and reflection Rapid increase of absorption with frequency Low frequencies imply need for very large antennas

cally with sound frequency. Only very low frequencies go long distances; given that achieving good angular resolution from practical antenna dimensions requires high frequencies, sonar imaging is thus limited by the physics to quite short ranges.

Attempts to use microwaves to penetrate solid objects (e.g., foliage, walls, the ground) suffer from the same conflict between penetration depth and resolution—low frequencies penetrate; high frequencies do not. Modern foliage penetration radars attempt to resolve this contradiction by combining ultrahigh frequencies (UHFs) that penetrate well with SAR techniques that do not require antennas with very large physical apertures.

Media scatter offers another persistent limit to the performance of radar and electro-optical (EO) sensors. The scattering from small particles (e.g., rain, fog, and dust) increases strongly with frequency such that most radars are little troubled by weather, but optical sensors fail in adverse weather and have limited atmospheric range even in good weather.

Sonar suffers much more severe media problems than do electromagnetic sensors because of the extreme inhomogeneity of the ocean and its effect on propagation. Unknown local variations in temperature and salinity deflect the acoustic beams into strongly curved unpredictable paths, and multiple nonuniform reflecting surfaces produce multiple confusing echoes. In addition, the ocean is full of natural and man-made acoustic noise sources, which seriously interfere with the detection and recognition of threats.

In addition, radar, EO, and sonar sensors are susceptible to deception techniques, such as the use of camouflage, decoys, or simply hiding, because these sensors collect reflections or emissions from objects. With the application of enough sophistication, the target may be so changed in appearance as to be undetectable or unrecognizable, whatever the capabilities of a given sensor. Technology alone may not be enough to solve this problem, but better sensor capability and the use of multiple sensing techniques will certainly increase the size of the investment the opponent must employ to be successful.

Finally, because only a very small amount of the radiation reflected or naturally emitted by the target typically reaches the remote sensor, the sensors are designed for high sensitivity and are therefore vulnerable to deliberately introduced radiation or jamming, which can saturate or even physically damage internal detectors. A trade-off of numbers, sensitivity, and distribution of sensors should be provided in a networked system.

3.3.2 Current Naval Organic and Joint Sensors

Today on its weapons' platforms the Navy employs many different local organic sensors—radar, EO, sonar, and electronic warfare (EW), as well as GPS and perhaps environmental and chemical and/or biological sensors (Appendix B lists representative naval organic and joint and National radar and EO sensors currently or soon to be available in the battlespace). Typically only a few of the many platforms get outfitted with any given version of a weapon or sensor, because when the next round is funded, the technology has changed such that better options are available. The newer ships get newer versions and combinations, while the previous generations of sensors remain in service. A recent exercise by the Office of the Chief of Naval Operations' (OPNAV's) Surface Warfare Division (N86) indicates that 22 different radars are currently deployed throughout the Navy, with plans to add 3 or 4 new, higher-performance radars over the next decade. The resulting eclectic collection of weapon and sensor components, which vary from ship class to class and sometimes from platform to platform, is mirrored in the variety of complex and somewhat personalized arrays of radio frequency (RF) antennas that top every Navy ship.

Up to now, many traditional Navy weapon-sensor suites have been designed primarily for platform self-defense in the open ocean or for attacking an air or ground target detected by sensors on the firing platform. The Navy boasts a large number of effective ship self-defense suites against attacks by aircraft, missiles, surface ships, submarines, and the like, but the location of primary sensors and associated weapons on the same platform often forecloses the possibility of attack beyond the horizon.

Below the surface of the water, there are a variety of both active and passive acoustic sonars capable of detecting submarines and other ships, frequently at considerable distance but with limited ability to localize the targets because of the uncertain nature of sound propagation in the ocean. Unfortunately none of these acoustic sensors perform well in the critical littoral environments that characterize one of the Navy's primary interfaces with the land for force projection.

3.3.3 Using Sensors in Network-Centric Operations

3.3.3.1 Targeting Ground-Attack Weapons

The current vision of decisively influencing events ashore includes a strong emphasis on force projection onto the land and to the purchase of the many land-attack, largely GPS-guided, long-range weapons. However, most of the high-performance radar and EO sensors deployed today throughout the surface Navy provide little or no capability to detect and localize targets on the land, even at short distances inland. Striking land targets at the long ranges permitted by modern missiles is a primary objective, but weapons' ranges exceed the horizon of surface-based sensors.

Airborne, mobile, and/or long-range sensors must identify and precisely locate targets of interest and must communicate this information to the shooters on the weapons' platforms. To implement its future vision, the Navy must have access to capable sensors, including not only its own organic sensors, but also joint and National sensors. A rational design for an NCO system would utilize all three of these resources—naval organic, joint, and National.

Several joint and National airborne and spaceborne sensors (e.g., JSTARS, Global Hawk, and U-2), expected to be present in the battlespace, can provide great capability in SAR ground imaging and the GMTI detection, location, and classification of slow-moving vehicles. Space-based equivalents, for example, Discover II, are being considered for acquisition.

Although these joint and National sensors could play an essential role in completing an effective system for power projection from ships using the Navy's precision GPS-guided long-range weapons, some in the Navy fear that they cannot rely on sensors they do not control and are reluctant to include them in the design of a power projection system. The Navy does not now possess any organic airborne sensors capable of providing targets for naval GPS-guided land-attack weapons, and although initiatives to provide this capability—for example, SAR options for a vertical-takeoff UAV platform to be developed—are commendable, the Navy can greatly improve its capability by investing in connectivity to the joint and National sensors.

3.3.3.2 Sensor Synergy

In spite of the current limited availability of appropriate land-targeting sensors, the Navy still possesses a large inventory of deployed, highly capable sensors, which have been persistently underutilized. Cooperating sensors can produce results that are much more than the sum of the individual capabilities. For example, in a single observation a radar can locate a target with great precision in range, but with an angular uncertainty that can be orders of magnitude larger due to the width of the transmitted beam. The resulting uncertainty about target location resembles a long, narrow ellipse, transverse to the target line of sight. However, as illustrated in Figure 3.3, if a second radar at a different spatial location observes the same target at about the same time from a very different angle, the two regions of uncertainty intersect in a rather small overlap region; if both observations can be combined, the two-radar target location is immediately refined in all directions to dimensions on the order of the range resolution. Neither radar alone could provide the same overall location accuracy.

Similar benefits are gained by fusing data to combine near-simultaneous observations from different classes of sensors that measure different physical characteristics of a scene. There are also benefits in using similar sensors in different physical locations where, for example, some sensors suffer from terrain masking or low monostatic cross section and others do not. Historically, however, naval organic self-defense sensors have been optimized for a particular weapon or suite of weapons on a single platform. This stovepiping has long

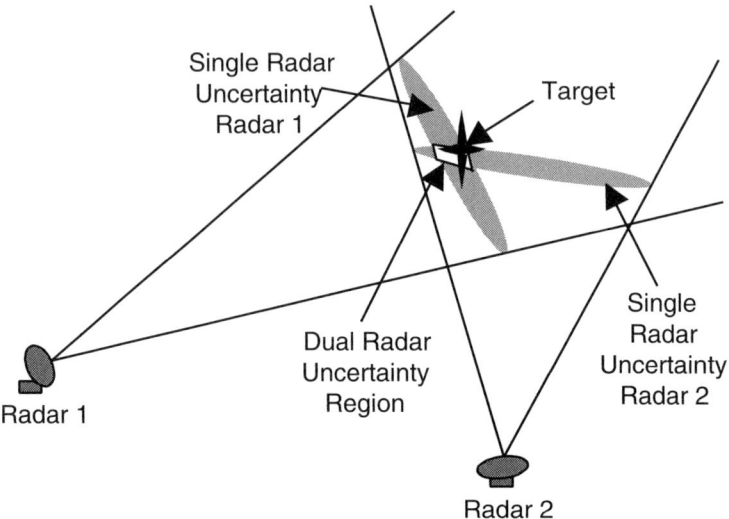

FIGURE 3.3 Multiple sensor cooperation can increase the precision with which a target is located.

characterized Navy practice, from system concept through acquisition, and has led to the current profusion of parallel, relatively independent capabilities.

Partly through organizational resistance (reluctance to rely on offboard data) and partly through technical difficulty, it is only recently that the benefits of cooperative, multiplatform sensor coordination have been strikingly demonstrated through the CEC. By means of a sophisticated, point-to-point, high-bandwidth, phased-array communication capability, CEC provides to each Aegis platform in the battle group the dwell-by-dwell detections observed by every other operating air-search radar in the group (see Figure 3.1). On each platform, all the observations from all the radars are combined into a single radar picture. The resulting view of the battlespace includes highly precise locations of objects that are simultaneously in the radar field of view of several platforms, as well as information on objects that are beyond the range of the particular platform's radar but within that of others in the group—a synergistic effect well beyond the sum of individual capabilities. Through CEC, all the Aegis radars in the fleet act together as a single, very large distributed meta-radar.

On the other hand, it has been common practice throughout the Navy to operate a subset of complementary sensors cooperatively on a single platform through a weapons control center where the information is fused and the appropriate weapon selected, targeted, and launched. For example, the Aegis system uses a long-range search radar (e.g., SPS-49) and perhaps a shorter-range surface search and navigation radar (e.g., SPS-55) for initial target detection and tracking, the SPY-1 four-faced phased array radar for precision tracking and the generation of command guidance information, and several simple continuous wave RF illuminators (e.g., SPG-62) that are aimed by the SPY-1 and provide the terminal-phase RF illumination of the target needed by the semiactive seeker of the standard missile. There are a number of integrated weapons control suites in the fleet, with different combinations of weapons and sensors.

Cooperative engagement on a single platform has been a long-standing practice in the Navy; the challenge now is to extend cooperation over many locations. CEC does this for air defense, and NCO should be extended to other missions as well. Achieving sensor synergy requires that sensors are suitably designed—for example, reporting confidence data to aid the fusion process.

3.3.3.3 Sensor-Shooter-Weapons Teams

The concept of NCO is far broader than CEC, which represents only one particular implementation of a system of sensors and weapons. NCO involve a multiplicity of individual taskable sensors of different kinds distributed throughout the battlespace and interconnected to fusion nodes, decision makers, and weapons via the NCII. Some sensors will provide surveillance contributing to general situational awareness, while others will be tasked opportunistically to form temporary, tightly coupled sensor-shooter-weapons teams involving dispersed sensors and shooters.

In this cooperative environment, it is envisioned that networked platform sensors will flexibly share and fuse data across sites to create the desired synergistic effects, while supplying data to and responding to requests from remote or co-located decision makers and weapons as needed. Additional mobile sensors, available for temporary arrangements, will no doubt become more prevalent in the future. They will permit adaptive situation awareness sensing to provide additional information or to compensate for deficiencies in other sensors' field of view or for limitations arising from environmental obscuration.

It will be tempting to create temporary sensor-shooter-weapons teams that will act in a tightly coupled manner to accomplish some immediate tactical goal and then to disband them and assign the sensors to other functions or teams as needed. However, providing the flexible, guaranteed, close-coupled communications required by an effective sensor-shooter-weapons team is an important challenge discussed in some detail in Appendix E.

3.3.3.4 Sensor and Target Geolocation

Tacit in this discussion is the requirement for highly accurate determination of the geoposition of each sensor and the local sensor-to-target orientation, so that a consistent overall map of the battlespace—the common operational picture (COP)—can be guaranteed.[4] Here, as in the guidance systems of most U.S. long-range, land-attack precision weapons, accurate, high-precision GPS geoposition measurement capabilities are generally assumed. Since GPS capabilities can be negated by simple techniques, it is important either to defend GPS or to find an alternative method of position location. This matter is so crucial that the committee discusses it in detail in Section 3.4.1.

3.3.3.5 Imaging Sensors and Automatic Extraction of Information

Whether radar, EO, or sonar, most battlefield sensors used for generating situation awareness and targeting information produce images. Some sensors, such as visible or infrared (IR) optical cameras or microwave SARs, often report something for every pixel, placing a heavy bandwidth or transmission time burden on the communication component used to interconnect this sensor into the system. Other sensors heavily process the raw data and report information only about candidate targets, greatly reducing the communication bandwidth required. Preprocessing reduces requirements on external communications. Given the current relentless exponential growth of computational capabilities, this can be an effective trade-off—if appropriate algorithms can be devised for automatic information extraction. However, constructing algorithms that can approximate human abilities to recognize poor-quality, partially obscured images has proved to be difficult.

Automatic target recognition (ATR), often sought in imaging applications for target detection, classification, and aim-point selection as well as for terminal weapon guidance, is a familiar example of automatic information extraction. The contribution of ATR in attacking moving targets is discussed in Section 3.6.1, and its role as a hedge against GPS jamming is discussed in Section 3.3.4. Without ATR, the large amount of data produced by sensors can overload communications channels and human analysts. Image compression can help overcome only the communications overload. Evolving in small steps over the years, ATR has proved effective in many applications, although powerful general solutions continue to be elusive.

Template matching may be the most direct technique if target dimensions and geometry are precisely known—knowledge that could be derived from SAR and three-dimensional imaging ladar—but an enormous number of possible templates must be scanned when dealing with images characterized by variable illumination and unknown target size and orientation. Extraction of features can sometimes simplify this process. Model-based vision techniques can extend feature- and template-based techniques by providing robustness to variations in target configuration and sensing conditions.

---

[4]Difficulty in achieving this has restrained COP development.

The state of the art in ATR employing SAR imagery is represented by the ongoing Defense Advanced Research Projects Agency (DARPA) Moving and Stationary Target Acquisition and Recognition (MSTAR) program. The MSTAR program takes a model-based vision (MBV) approach to ATR based on high-resolution SAR imagery. In this approach, targets are detected and initial classification and hypotheses are developed using a conventional template-based ATR approach (the MSTAR "front end"). MBV techniques are then used to reason about target component articulation, obscuration, and other real-world effects that cannot be handled using template-based approaches.

Comparing the results of raw, single-look ATR performance as indicated by the operating characteristic for the MSTAR Version 7.1 (March 1999) and the MSTAR Version 6.2 (September 1998) shows rapidly improving performance (Figure 3.4). The results do not reflect use of techniques such as object level change detection and target context analysis, which can further significantly reduce the false-alarm rate. The crucial importance of the false-alarm rate is demonstrated in Section 3.5 and in Appendix C.

Table 3.5 presents the single-look classification performance of the MSTAR software and several other ATR components. The target set includes a number of similar targets (e.g., XM-1 and M60 tanks). The current baseline is template based. Note that in the laboratory it performs very similarly to the MSTAR front end, which is also template based. The full MSTAR, including the MBV back

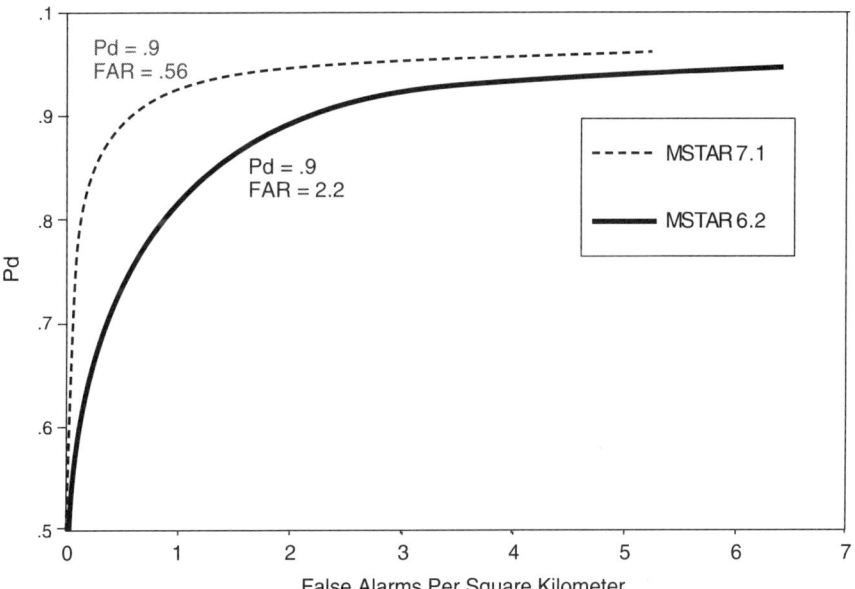

FIGURE 3.4 MSTAR operating characteristic.

TABLE 3.5 Probability of Correct Identification by Various Automatic Target Recognition (ATR) Components

Targets in Scene	ATR Component	Probability
Few	Best laboratory prototype	0.94
18	Full MSTAR, 1999	0.87
15	Full MSTAR, 1997	0.78
18	MSTAR front end	0.68
20	Baseline ATR (in laboratory)	0.64
20	Baseline ATR (in field)	0.35

end, achieves very high levels of recognition, currently approaching 90 percent on an 18-class problem. For comparison, the performance of a laboratory prototype in classifying a limited data set is shown; this level of performance (about 94 percent) is a practical upper bound on MSTAR performance. Also shown is the field test performance of the current baseline. The relatively poor field test results are due in part to variations in target configuration, target component articulation, and imaging geometry. MSTAR is designed to reason about these variations with the goal of achieving performance in the field similar to that achieved in the laboratory.

As promising as the new algorithms are, none yet approach the information-extraction abilities of humans. ATR is a well-defined challenge in that the objects being sought and their characteristics are well known in advance. Detecting "anything unusual" in a surveillance scene, without knowing just what to expect, can be a much greater challenge and probably requires algorithms incorporating completely new insights and concepts—a topic for future research.

As the inevitable data and communication overloads materialize with the proliferation of even more sensors viewing the battlespace, mastery of automatic information-extraction techniques must be diligently pursued. Investment in concepts and algorithm development is relatively inexpensive, but the payoff can be very large.

3.3.4 Findings

Finding: Sensor capabilities are improving through exploitation of digital and solid-state technology. (See Section 3.3.1.)

Finding: Adversaries can exploit fundamental physical laws and make detection by sensors difficult in certain situations. (See Section 3.3.1.)

Finding: Deployed Navy sensors span a ranges of types, but most were designed for platform defense, are stovepiped, and exhibit a mix of old and new technologies due to the budget-limited practice of incremental upgrades over a long period. (See Section 3.3.2.)

Finding: The Navy has no organic sensors capable of guiding its precision, long-range weapons to ground targets. Emerging doctrine assumes access to joint or National resources in the battlespace, but the Navy is only beginning to invest in such connectivity. (See Section 3.3.3.1.)

Finding: Multisensor cooperation offers significant performance advantages. (See Section 3.3.3.2.)

Finding: Temporary sensor-shooter-weapons teams are natural in network-centric operations but offer flexibility and quality-of-service challenges for the communication infrastructure. (See Section 3.3.3.3.)

Finding: Geolocation in the same absolute or relative coordinate system of the sensors and targets in the battlespace is mandatory. Use of the Global Positioning System is often assumed to be the sole technique employed but may not always be available. (See Section 3.3.3.4.)

Finding: Automatic target recognition avoids overload of communications and of image analysts, may be necessary for remote attack of moving targets, and provides a hedge against GPS jamming. Model-based vision may overcome the limitations of template matching. However, more general capabilities for automatic information extraction continue to be elusive and must remain the subjects of continuing research and development (R&D). (See Section 3.3.3.5.)

3.4 NAVIGATION

In this section the committee compares means for the navigation of weapons to a remotely selected target, considering dispersed assets that require a coordinate method of designating locations. It does not consider here simple closed-loop systems such as fire control and guidance radars in which the target and the weapon are visible from the same sensor on the launch platform. Table 3.6 identifies some options for navigation.

3.4.1 Evolution to GPS Guidance

In the past, precision land-attack weapons were directed precisely to a target either by closed-loop guidance from a platform that fired and controlled the weapon while it saw the target, or by locating absolute target coordinates and

TABLE 3.6 Options for Navigation of Weapons

Technique	Strengths	Limitations	Examples[a]
Use of inertial measurement units	Small, light	Very expensive for low drift rates	
Multilateration	High potential accuracy	Need for multiple references	GPS, CEC, JTIDS, LORAN
Automatic target and landmark recognition	Independence from absolute coordinates	Expense, time to build templates	TERCOM, DSMAC
Satellite Doppler	Simple receiver	Intermittent availability	TRANSIT
Range and bearing	Simple apparatus	Horizon limitation, limited accuracy	VOR/TAC

[a]GPS, Global Positioning System; CEC, cooperative engagement capability; JTIDS, Joint Tactical Information Distribution System; LORAN, long-range navigation; TERCOM, terrain-contour matching; DSMAC, digital scene matching area correlation; TRANSIT, Navy Satellite Navigation System; VOR/TAC, very high frequency omni-range/tactical air control.

using inertial measurement units (IMUs) to guide the weapon to those coordinates.

IMUs measure accelerations to deduce position relative to the beginning of the measurement period, usually the time of weapon launch. If the position is precisely known at this time, then an IMU can provide an estimate of absolute position and velocity throughout a weapon's flight.

Many weapons navigate primarily by using IMUs. However, the cost and complexity of IMUs capable of navigating long distances with low error are daunting. Intercontinental ballistic missiles used very expensive IMUs, despite the fact that the destructive range of their thermonuclear warheads reduced the requirement for accuracy of delivery. Area ammunition can be used to overcome navigational imprecision, but the focus here is on precision weapons that reduce collateral damage.

The major source of error in an IMU is a displacement of the computed from the actual track that accumulates during the flight through the integration of errors of acceleration measurement. This error is cumulative; for the same IMU, longer flights lead to larger position errors. Today, although integrated optics promises lower-cost precision IMUs in the future, most practical systems use some other method of geolocation to update their IMUs in flight and negate the IMU error accumulated up to that point.

Prior to the availability of GPS, IMUs in cruise missiles were updated by terrain following and scene matching. Both techniques require time-consuming preparation of reference scenes and relatively expensive sensors. Now, the preferred method for attacking precisely located targets from over the horizon is GPS guidance updating a relatively low-cost IMU.

3.4.2 Robust Navigation for Precision Attack

Except for one fact, GPS would be in universal use to update low- and moderate-cost IMUs in precision weapons. However, the unfortunate fact is that jamming the GPS signal is not difficult. The signal is transmitted with powers of tens of watts from distances of thousands of miles; jammers are likely to be of higher power and much closer. In the basic civilian GPS mode, a jammer with a few hundred watts of power is effective to its horizon, and a jammer of only a few watts of power can jam GPS at significant ranges. The frequencies that GPS uses are comparable to those in a microwave oven. Oven power sources of hundreds of watts are manufactured for $25 or less.

There are four principal strategies for dealing with this threat:

- Strengthening the resistance to jamming of a platform or weapon using GPS-aided navigation,
- Attacking GPS jammers,
- Substituting other reference sources for the GPS satellites, and
- Abandoning entirely the use of GPS-like multilateration.

These four strategies are discussed below.

3.4.2.1 Strengthening Resistance to Jamming

The Joint Requirements Oversight Council is considering an operational requirements document (ORD) to make GPS more robust in the face of jamming. At the time of this writing the committee knew neither the contents nor the probable fate of that ORD, so it examined jam resistance from physics and engineering standpoints.

Table 3.7 compares various methods of increasing the resistance of GPS-guided weapons and platforms to GPS jamming. It distinguishes between en route jamming, which may involve a few high-power jammers, and target-area jamming, which may involve larger numbers of low-power jammers.

If a weapon's GPS receiver is not jammed over its entire flight, it is possible that an IMU alone could guide it through the jamming to the target. Against short-range point-defense jammers this is feasible.[5] However, IMUs capable of

---

[5] Without the use of onboard GPS, the joint attack direct munition (JDAM) can maintain its specified accuracy over a large fraction of its kinematic range. Therefore, some may think of GPS as the "icing on the cake." However, JDAM accuracy in this mode depends on high-accuracy navigation by the releasing aircraft, which faces the same GPS jamming environment as the short-range weapon.

TABLE 3.7 Means of Increasing the Resistance of Global Positioning System-guided Weapons to Jamming

Means	Antijam Feasibility and Performance		Issues
	En Route Jamming	Target-area Jamming	
Inertial measurement unit	May be too expensive	Feasible at short ranges	Cost vs. accuracy
Automatic target recognition and automatic landmark recognition	Low productivity	Unproved	Reliability, field of view
Signal processing	Limited	Limited	Spectrum, compatibility
Spatial processing	Good	Probably good	Number of jammers

precision navigation over the scores (or even hundreds) of miles of effective en route GPS jamming may be too expensive to include in weapons.

Resistance to the effects of GPS jamming may possibly be increased by tightening the generally loose coupling between IMU and GPS outputs. Each has a separate Kalman or equivalent filter to estimate position and velocity from that subsystem's data alone. These estimates are fed to a separate application that produces a combined estimate and feedback for the filters. It has been demonstrated that performance could be improved by unifying the filters so that the raw observations from both sources would be tracked together. This technique may reduce the accuracy requirement for IMUs and permit the use of more affordable ones. Even deeper integration that permits IMU data to optimize the signal processing within the GPS receiver in real time offers the possibility of substantial additional jam resistance.

Despite the low productivity of automatic landmark recognition (ALR) and the immaturity of most ATR, alternatives to ATR for hitting moving targets are hard to attain.

Signal processing alone cannot overcome serious GPS jamming but can contribute to jam resistance. The GPS signal occupies a spread spectrum, and the military-only codes are spread further by a cryptographically secure pseudo-random sequence. Receivers in these military modes have a substantial advantage over their civilian counterparts in resisting jamming. Although this advantage is not large enough to overcome serious GPS jamming—leading to the rating of "limited" in Table 3.7—it increases the power that the adversary must use to achieve jamming at a given range.

Once military receivers acquire the civilian code, they can synchronize to the military codes, which allow more precise measurements and which are unavailable to parties not possessing the proper cryptographic key. The secure codes also prevent spoofing of the GPS signal, that is, introducing deceptive signals that cause false readings at the receiver. In principle, the civilian code can be spoofed under some circumstances, although measurements of received power greater than that expected from the GPS satellites would be a good indication that spoofing was being attempted.

Most military GPS receivers first synchronize to the less robust civilian code before acquiring the military codes. This mode of operation assumes that initial synchronization takes place outside the view of the jammer. Dependence on this assumption can be avoided by providing a correlator that acquires the military codes directly. Such correlators are now within the state of the art but remove a vulnerability only in initial acquisition; they do not increase the jam resistance of the military codes. Alternatively, it may be possible for the launch platform to initialize the weapon so that it can acquire the military code promptly.

Raising the power in the GPS transmitters would help overcome jamming, but because generating power in space is expensive and because power increases that affect the basic spacecraft design would lead to high nonrecurring engineering costs, increases in power cannot alone overcome the jamming threat. Nevertheless, it may be economical to bear high costs in a few dozen satellites to avoid even moderate costs in tens of thousands of military GPS receivers. An intermediate strategy is to use a high-gain antenna to increase the effective radiated power seen in areas where jamming is expected, without greatly increasing the required power generation capabilities of the satellite.

It is possible that alternative or additional GPS waveforms would yield more spread-spectrum processing gain and easier direct synchronization from the present military signals. However, it would be difficult to obtain additional spectrum for purely military GPS; indeed, current frequency assignments are threatened. The plans of the Department of Defense (DOD) need to be harmonized with the civilian navigation community's desire for an additional civil frequency.

Simply changing existing waveforms would make existing military GPS receivers obsolete. The GPS program office is investigating the possibility of adding new codes on a new dual-use frequency that would facilitate direct synchronization and improve processing gain, and therefore the jam resistance, of these new signals. Existing GPS receivers would continue to operate normally but would not take advantage of the new signal.

Spatial processing distinguishes real GPS signals from jamming signals by exploiting their different directions of arrival. Two techniques can be considered.

One technique is to take advantage of the fact that the true GPS transmitters will be above the horizon, while most jammers will be at or below the horizon.

An antenna that looks only above the horizon with very low side lobes mounted on a body that prevents leakage into the antenna of signals transmitted from below may, in some scenarios, suppress the jammers sufficiently so that signal processing can reject the jamming signals.

Another technique is to build an adaptive antenna array that analyzes all incoming signals within its field of view and that builds an antenna pattern that delivers to the receiver the same power from all sources by providing the lowest gain to the strongest signals. Once the true and the jamming signals are made comparable in power, signal processing can reject the jamming signals. This technique is sometimes called null steering.

The first technique is vulnerable to airborne jammers, while the second, null steering, is vulnerable to large numbers of low- or moderate-power jammers. Because of their long range, airborne jammers may be the more serious en route threat, while multiple low-power ground-based jammers may be found in the vicinity of high-value targets. Because only a limited number of signals can be equalized in power by the antenna array, a combination of the two techniques may be required in the vicinity of the target.

Although spatial processing is not inexpensive, in combination with existing signal processing it can often overcome plausible jamming threats. Recent news reports have appeared of European trials of spatial processing to protect airline GPS receivers from interference by high-power UHF television stations. If spatial processing gains wide use in the commercial sector, competition for a larger market could reduce prices. However, many weapons have little real estate on which to place highly directive arrays.

3.4.2.2 Attacking GPS Jammers

An often-discussed option is to attack GPS jammers, and a variant of the high-speed antiradiation missile (HARM) is being developed to provide this capability. The limitation of this approach is the poor cost-exchange ratio.

A HARM costs in the vicinity of $500,000; a moderate-power GPS jammer can be made for $100. Thus, the use of antiradiation weapons would be restricted to high-value targets, such as high-power jammers on aircraft, or to those jammers whose effects cannot be overcome by other means. Some have argued that a 10-kW jammer might be worth attacking, but spatial processing can be used against small numbers of jammers.

The challenging case is that of dozens of low- or medium-power jammers. The large number would overcome spatial processing, and their low individual cost would make them unappealing targets for HARMs.

It is the committee's impression that although a few jammers might be attacked by HARMs, and although the existence of this weapon might demoralize crews that operate or maintain GPS jammers, antiradiation missiles cannot be substituted for a proper mix of IMU performance and signal and spatial processing.

Because even moderately high-powered GPS jammers are easily relocatable but not easily recognizable on sight, antiradiation weapons are the least implausible method of attacking GPS jammers. Consideration should be given to developing low-cost antiradiation weapons for this function; they need not be high speed, and their homing range need be no better than the accuracy to which the jammer can be located by electronic support measures.

3.4.2.3 Substituting Reference Sources

Three alternative reference sources for multilateration are considered here: the Global Navigation Satellite System (GLONASS), surrogate satellites, and incidental reference satellites.

GLONASS provides navigation service comparable to that of GPS in its civilian mode; to the best of the committee's knowledge, it has no military codes comparable to those of GPS. Many civilian GPS receivers also receive GLONASS. Although GLONASS uses frequency division multiplexing instead of the code-division multiplexing used by GPS, it is easy to build a jammer that will work against both.

Many possible schemes exist for using high-power airborne antijam transmissions to overcome GPS jamming in the vicinity of a navigating platform or weapon. In some scenarios, surrogate satellites (sometimes called pseudolites) range and provide robust reference sources for multilateration.

Among the disadvantages of airborne pseudolites are their lack of global coverage and the need to attain air superiority before their deployment. Development costs for the pseudolites and new navigation receivers would be expected to be high. The original Link 16 of the Joint Tactical Information Distribution System (JTIDS) provided relative navigation through multilateration. If one member of the network could determine its absolute location, then all members could deduce theirs. However, the accuracy of Link-16 navigation is significantly lower than that of GPS. Link-16 terminals are too expensive for weapons, and some terminals lack the navigation feature.

Terrestrial pseudolites have the disadvantage that a weapon approaching its target may be far forward of most of the pseudolites, causing isochrones to intersect at small angles, leading to imprecision in locating their intersection. This phenomenon is known as geometric dilution of precision (GDOP) and is avoided by having the satellites surround the receiver.

GPS satellites provide a code that facilitates accurate time measurements as well as broadcasts of data that determine the satellite's position. In an attempt to limit the utility of GPS to foreign powers, a selective availability (SA) mode was included wherein errors of the order of 100 meters would be introduced in the civilian signal.

Numerous experimenters found that SA could be defeated for relative navigation by ignoring the code and counting the cycles of the stable carriers. Relative navigation can be converted to absolute navigation provided that the initial position is known in absolute coordinates and that no interruption of signal occurs thereafter.

Techniques of this sort raise the possibility of navigation through monitoring signals from satellites that are launched for other purposes. The European Union desires an alternative to GPS and is considering the use of these techniques as an alternative to developing and deploying its own network of navigation satellites. Although no decisions have been made, some satellite producers are planning for high-stability carrier transmissions to keep the alternative viable.

The GPS and GLONASS orbits, chosen to minimize GDOP by keeping always in view a large number of satellites at different azimuths and elevations, lie in a high-radiation region that would not be selected for any other purpose. Most other satellites are used to relay communications. With a few exceptions, communications satellites are found either at a synchronous equatorial altitude or in dense low-orbit constellations. The use of synchronous-altitude satellites could lead to a severe GDOP problem. In the Northern hemisphere, all the satellites will be seen as being in the southern sky and will not meet the goal of surrounding the receiver.

Low-altitude constellations have many satellites and might initially seem to offer a solution. However, these systems are designed to conserve precious spectrum. To permit frequency reuse, they are designed so that as few satellites as possible are transmitting to the same point on Earth. The more distant satellites are shielded from the receiver by the horizon. It is unlikely, therefore, that a single low-altitude constellation would provide enough simultaneous reference points for accurate navigation. The committee suggests investigating the possibility of using whatever low-altitude satellites are in view to reduce the GDOP associated with the use of synchronous equatorial satellites.

The cycle-counting apparatus used to defeat SA is complex and may not work well from moving platforms. The use of incidental satellite transmissions might involve multiple receivers. However, European interest in reducing dependence on GPS may lead to refinements of these techniques.

3.4.2.4 Abandoning GPS-like Multilateration

Table 3.6 notes the limitations of navigation components such as the VHF omni-range (VOR) navigation system and of Doppler navigation techniques such as were used with the Navy satellite navigation system TRANSIT.[6] Ab-

---

[6]TRANSIT, the world's first operational satellite navigation system, was conceived in the early 1960s to support the precise navigation requirements of the Navy's fleet ballistic missile submarines.

sent an unexpected breakthrough in affordable very-high-precision IMUs, abandoning multilateration may imply operation in relative coordinates or dependence on some form of ATR or ALR.

In connection with describing terrain-guided Tomahawk operations, the committee noted that route planning was slow and tedious. GPS was adopted as an upgrade to Tomahawk to avoid the necessity of terrain guidance. A guess is that the right approach for fixed targets may be to improve IMUs such that they could guide a missile to a recognizable point and then attack a target whose position is known relative to that point. In the best case limit, the digital scene matching and correlation would include the target, simplifying the endgame.

It is also possible that a target might be newly discovered subsequent to the initial depiction of the scene. In this case, the observation that detected the target need only be registered to the scene, and not in absolute coordinates.

Planned National Aeronautics and Space Administration missions that will produce high-resolution surface elevation maps raise the possibility of greater reliance on terrain navigation, provided that the route planning can be automated.

For moving targets, decoupling of observation and attack time is not possible. The utility of absolute geolocation is in the coupling of sensor and weapon coordinate systems. But there are other ways to synchronize sensors and weapons without resorting to absolute coordinates.

Suppose the sensor is an all-weather device, such as a radar or an intercept receiver, and is at some distance from the weapon launch point, as occurs when the sensor is in the space sanctuary or when it is in an aircraft kept out of harm's way. In these cases, an attack aircraft can be vectored to the target's vicinity in relative coordinates, provided that the sensor can see the weapon in that coordinate system.

The aircraft attack scenario just described does not achieve the goal of over-the-horizon fire. Such fire could be achieved if the observation platform could guide the weapon in flight to converge with the target in its relative coordinate system. The launch platform would hand off control of the weapon to the observation platform in what has come to be called a forward pass. Once the pass has been accomplished, the endgame becomes simple, although some measure of ATR would be needed unless observations were very precise and frequent. If observations were intermittent, as from sparse constellations of low-altitude satellites, the attacks would have to be launched soon after detection to ensure observation of the target throughout the flight of the weapon.

3.4.3 Findings

Finding: No single technique will make GPS-aided weapon navigation invulnerable to GPS jamming. Practical solutions are likely to involve a combination of cheaper, precise IMUs, better ALR and ATR, improved satellite signals and receiver signal processing, and the use of spatial processing. (See Section 3.4.2.1.)

Finding: Available antiradiation weapons do not solve the GPS jamming problem because the jammers can be easily replicated and the weapons cost many times more than the jammer. Suitably modified HARMs could be used to attack aircraft carrying high-power jammers, and the presence of such HARMs in inventory might demoralize crews operating GPS jammers. (See Section 3.4.2.2.)

Finding: Although navigation through the use of satellites not designed for that purpose is possible, the difficulties of using these techniques in weapons are formidable. Nevertheless, European interest in these techniques will cause the difficulties to be assessed and perhaps overcome. (See Section 3.4.2.3.)

Finding: Passing control of a weapon forward to a sensor that holds the target in view is a plausible means of reducing or eliminating dependence on GPS and similar systems. (See Section 3.4.2.4.)

3.5 TACTICAL INFORMATION PROCESSING

Network-centric operations require that component weapons, sensors, and platforms be combined into a coherent warfighting subsystem. The information connectivity and functional capability discussed in Chapters 4 and 6 are necessary, but not sufficient, for realizing NCO. Algorithms, software, and human-computer interfaces are needed to process the data exchanged across the NCII and to enable interactions with commanders. This section discusses the information-processing functions required for tactical-level NCO, especially strike warfare for ground targets, an area of increasing importance to the Navy and one that presents significant challenges to making NCO a reality. Similar issues arise also regarding effective NCO for theater air and missile defense and for undersea warfare.

3.5.1 Generic Tactical Processing Functions for Network-Centric Operations

Figure 3.5 depicts a generic tactical processing functional architecture. Sensors and friendly position- and status-reporting systems provide the external data that drives tactical-level NCO. (Organic Navy sensing assets are described in Section 3.3.3.2.) In addition, the Navy currently relies heavily on National sensors and will in the future rely increasingly on the sensors of other Services.

In the discussion that follows, attention is restricted to real-time control of combat operations and to kinetic energy weapons. However, as indicated in

Figure 3.5, non-real-time planning is important to position platforms, determine sensor coverage assignments, choose priorities for target classes, provide rules of engagement, and so on.

Tracking, determining the kinematic states of hostile, neutral, and friendly platforms and weapons, is central to tactical-level NCO. Links 16 and 11 are currently employed to distribute platform-derived tracks to develop a common tactical picture. The Navy is currently deploying CEC, which distributes radar returns to provide a common, low-latency track that can be used for fire control in air defense NCO. Studies and experiments are under way to extend the CEC concept to theater ballistic missile defense. The extension of the concept to land targets as well can be considered, although the committee believes there will be some significant differences, driven in part by the nature of the targets and their environment, and in part by the use of several different kinds of platforms in the notional strike system described in Figure 3.2. There will also be some strong similarities, as the strike system takes advantage of CEC techniques for creating tracks from measurements provided by distributed sensors. The undersea warfare mission presents new challenges given the current emphasis on littoral rather than blue water operations.

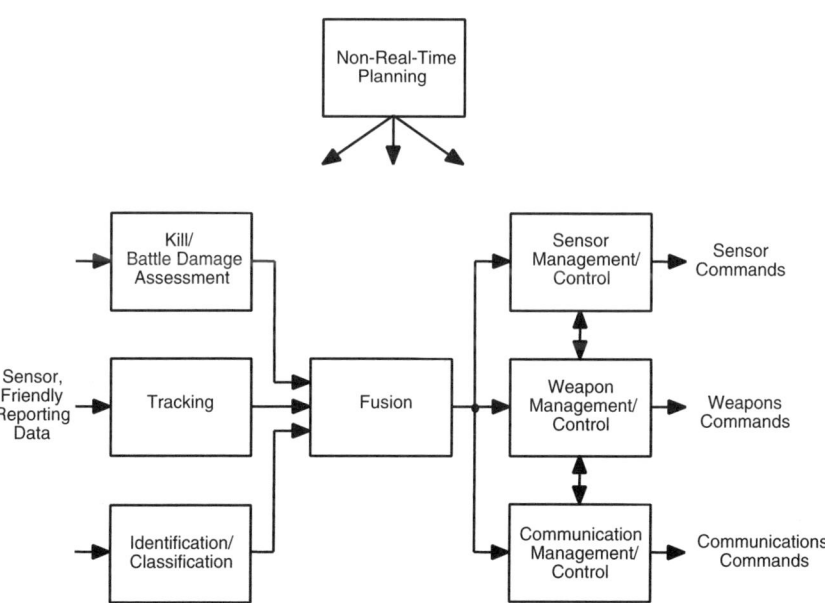

FIGURE 3.5 Generic tactical processing functions.

Classification, determining the type of a particular platform or weapon, and *identification*, determining its organizational relationships (at the simplest level, hostile, neutral, or friendly), are required for situation awareness and targeting. Classification and identification can be direct, based on sensor signature or transponder information, or indirect, based on inferences from processed data (e.g., an object with a track originating from a hostile platform or base might be assumed to be hostile). Classification and identification continue to be problematic, as evidenced by engagement of friends, neutrals, and decoys in recent conflicts and peacekeeping missions. The role of ATR in this function is discussed in Section 3.3.3.5.

Kill/battle damage assessment (BDA), determining the status of targets, also presents problems. For example, in kill assessment for theater missile defense, it is not enough to determine that an interceptor has hit an incoming tactical ballistic missile (TBM); it is also necessary to determine that the TBM warhead is no longer functional. BDA for land targets is not being accomplished with the timeliness and accuracy required, as illustrated in Desert Storm and the recently concluded air campaign in Kosovo.

Fusion is the combining of kinematic, classification, and status data into a CTP. Fusion requires that data from disparate sources (e.g., radar, imagery intelligence, signals intelligence) be combined with higher-level information. In practice, fusion depends heavily on having humans in the loop. However, as the number of sensor sources and the data rates for each source continue to multiply while staffing is being reduced, automation is becoming increasingly important. As noted in Section 3.3.1, there are difficult problems (targets in foliage, underground targets, low observable targets) for which single sensor solutions are currently unavailable. Although not a total solution, fusion of data from multiple networked sources may be the only viable approach in such cases. Fusion is also important for developing a COP in support of NCO planning. Accurate sensor geolocation is a key fusion enabler.

Weapon management and control includes assigning weapons to targets, selecting the optimal engagement time, planning for weapon delivery (including making the airspace free of conflict), and arranging for sensor and communication support to weapons. With the increasing range of naval weapons, coordination of fire with other Services is becoming an increasingly critical requirement. Thus NCO require combining sensors and weapons into coherent systems that cross Service boundaries, increasing the complexity of an already difficult task.

Sensor management and control is the allocation of limited sensor resources for target detection, tracking, classification, and identification; kill and battle damage assessment; and weapons support. With the development of longer-range, multimode sensors capable of supporting multiple missions and functions, sensor management and control is becoming an increasingly complex and important function.

Communications network management and control is the allocation of limited communication resources to support sensor, weapon, and processing functions. These resources can include frequencies, JTIDS slots, and antenna time lines. Most communications management today is done in a non-real-time planning mode, but there is an increasing need for a more flexible real-time allocation. Recently, the Joint Theater and Missile Development Organization has developed the concept of a joint interface control officer (JICO) to perform communications network planning, management, and control for the tactical data links associated with developing the CTP.

The generic tactical processing functions depicted in Figure 3.5 are equally applicable to a platform-centric system, such as the combat direction system of a single ship, or to a network-centric system such as CEC. What is unique about NCO is that sensors, weapons, and processing functions are distributed across multiple platforms, connected by tactical networks. Figure 3.6 contrasts platform-centric and network-centric architectures.

The sensing, weapons, and processing functions depicted in Figure 3.5 must be allocated across platforms to optimize the overall combat effectiveness of the collection of platforms as a whole rather than that of any individual platform. The functional allocation must take into account the communications loads imposed on tactical networks by the functional allocation. For example, a direct downlink of imagery data from a sensing platform to a shooting platform typically requires a very-high-capacity data link. However, if processing is performed on the imagery to extract target parameters prior to communication, data link requirements may be greatly reduced.

Although not explicitly represented in Figure 3.5, human operators can play a major role in carrying out any of the functions shown. When mission time lines permit, humans may be involved in searching imagery for targets, extracting aim points, selecting weapons, and performing mission planning in a manual

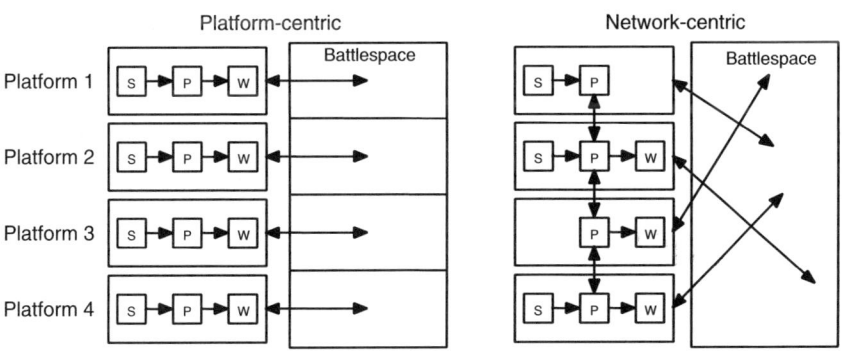

FIGURE 3.6 Platform-centric versus network-centric architecture.

or semiautomated mode (e.g., Tomahawk strike planning). In other cases, the human role may be to monitor and supervise a highly automated system (e.g., Aegis air defense). A key issue for NCO, in which humans distributed across multiple platforms must cooperate to conduct tactical operations, is design of components, procedures, and training to facilitate effective, real-time, distributed team decision making.

3.5.2 Example: Capabilities Required for Littoral Warfare

3.5.2.1 What Is Needed

Maritime forces operating in the littorals require the following capabilities that are not currently available. The need for many of them was highlighted in the extended littoral battlespace experiments,[7] including:

• *Flattened, rapid, webbed, distributed command and control (C2) processes.* Echelons of command should not be a hindrance in exploiting information technology. Real-time or near-real-time C2 is critical to success. Networks based on World Wide Web technology seem to fit the informational and procedural needs of the force.

• *Common situational understanding.* To execute the complex activities required in a modern littoral environment, a CTP is critical. To synchronize execution, coordinate planning, and ensure the massed effects dictated by modern warfare, the decision makers require a common framework, view, and understanding of the battlespace.

• *Fully coupled decision, planning, and execution components (sea/land) on a shared battlespace network.* Owing to the pace of a littoral campaign, planning and execution are concurrent. The components that support these functions must be integrated. In addition, they should "ride" on the same backbone, the NCII.

• *Intelligent networks.* Information must be provided not only to the decision makers but also in many cases to the executors. Tailored intelligence must be provided to various echelons of the littoral force and presented in a manner that can be readily assimilated by combat forces at the tactical, operational, and strategic levels.

• *Improved combined fire response time.* Calls for fire must be responded to in a timely manner. Targeting must be accurate. The appropriate munitions must be scheduled and the priorities of the fire set in a dynamic manner to ensure

---

[7]Cole, Ray. 2000. *Office of Naval Research Demonstration Manager's Campaign Plan: Extending the Littoral Battlespace (ELB) Advanced Concept Technology Demonstration (ACTD).* Office of Naval Research, Arlington, Va., forthcoming.

that they are in concert with the concept of operations, rules of engagements, and emergency pop-up threats.

• *Interoperability with joint, combined, and coalition forces.* The systems must enable U.S. maritime forces to interoperate with and share information with joint, combined, and coalition forces that may be interspersed with the littoral force, supporting, or adjacent to the force.

3.5.2.2 Current State

Current capabilities do not match what is needed for effective littoral warfare operations. For example, force track coordination is a key ingredient of a robust CTP, but current doctrine does not provide a strategy for land operations. Traditional force track coordination is based on AAW activities. The JICO concept, established to overcome joint and combined interoperability deficiencies related to management of the joint force multi-tactical digital information link (TADIL) networks, was successfully demonstrated at numerous joint exercises and has been effective in managing the complexity of the electronic battlefield, thereby improving the joint force commander's ability to engage hostile forces and prevent fratricide. However, there does not appear to be a similar capability for land or maritime surface combat forces.

The Navy and Marine Corps use different radios and radio frequencies. The Navy is converging its data link activities to Link 16, whereas the Marine Corps plans little procurement of Link 16. Marine Corps and Navy forces will require a gateway to the U.S. Army. Since most current military satellite terminals do not have the support of "C2 on the Move," new generations of terminals and relays of line-of-sight radio frequency systems are essential, but, to the committee's knowledge, none are programmed for acquisition.

Position location information (PLI) is a key aspect of developing the land CTP. PLI components could provide two-way targeting information, track generation, supporting arms coordination, and other activities. PLI is important not only to the forces operating on land but also to the air and sea forces supporting the land forces. Several PLI components exist within the Navy and Marine Corps, and the Army has others. These components currently do not fully interoperate. If mixed components were to be operating in a combined area, it would be difficult for them to share information to form a COP and near impossible to form the CTP.

Even CTP data must have some filtering mechanism to optimize it for mission, component, function, and echelon use. Such optimization is particularly important as this information is moved to battlefield users disadvantaged in communication connectivity. Otherwise, inappropriate information will clog tactical networks and end-user devices.

3.5.3 Example:
Capabilities Required for Attacking Mobile Land Targets

Before mobile and moving targets can be selected for attack, it is necessary to find such objects, identify them unambiguously, and track them to maintain knowledge of their identity and position. Absent ATR, frequent and precise positional updates are required in the weapon endgame. Although SIGINT (for cueing), electro-optics (for unambiguous target identification), and other sensing modalities can play important roles, only radar has the combination of high area rate, all-weather coverage required to provide surveillance, and fire control support for long-range targeting of mobile and moving targets.

Although the Navy has a very limited organic capability for long-range, stand-off tracking, classification, and kill assessment of land targets, DOD is investing significant resources in the development of manned airborne radar platforms (e.g., JSTARS, U-2), UAVs (e.g., Global Hawk, Predator), and space-based radar platforms (e.g., Discoverer II). Current capabilities include high-resolution SAR (<1 m) for stopped targets and low- to medium-resolution MTI (> 10 m) for moving targets. Capabilities in the R&D stage include interferometric SAR, high-resolution MTI (HRMTI), and moving-target imaging (MTIm). The Navy needs to obtain assured two-way connectivity to these platforms and the capability to utilize effectively the data they provide for targeting and fire control.

The volume of data produced by these sensors requires automation to assist analysts in sifting through the data to find high-value targets. As discussed previously in Section 3.3.3.5, ATR capabilities, applied to SAR, HRMTI, and MTI, have been developed to the point that they can play a significant role in reducing operator workload and system response time.

3.5.3.1 Tracking

Maintaining mobile target identity, whether obtained from SAR imagery or by other means (e.g., SIGINT or EO imagery), requires high-quality tracking of ground targets. Tracking the ground target is very difficult owing to terrain obscuration, minimum detectable velocity thresholds, the extreme maneuverability of ground targets (including stopping), and other factors. Multiple radar tracking algorithms are under development that utilize both SAR and MTI data to track high-value targets through multiple move-stop-move cycles, using features derived from high-resolution radar modes to maintain vehicle identity through coverage gaps and in the presence of "confuser" vehicles.

3.5.3.2 Detection and Classification

Consider the mission depicted in Figure 3.7 of finding and classifying high-

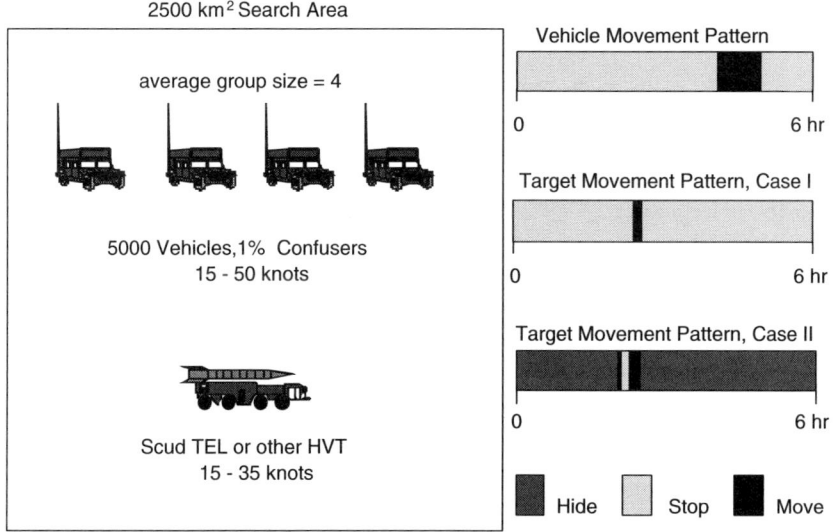

FIGURE 3.7 High-value-target tracking and classification mission.

value targets (HVTs) using the Global Hawk radar. The HVTs are assumed to be mixed in among 5,000 vehicles in a 2,500 km² area. Most of these vehicles are relatively easy to distinguish from the HVTs, but some 1 percent (the confusers) are not. Two cases are considered. In the first case, characteristic of ephemeral targets such as theater ballistic missile transporter-erector-launchers, the HVTs are hidden from SAR except for a brief period when the HVT emerges from hiding to conduct a mission. In the second case, characteristic of relocatable targets such as the elements of a mobile surface-to-air missile (SAM) unit, the HVTs are visible to SAR except when they are moving (when they are then visible to MTI).

Assumed sensor and processing specifications are listed in Figure 3.8. The assumed classification and tracking performance is aggressive but is potentially attainable with advanced processing technology. Note that the current Global Hawk radar does not include an HRMTI mode, but the development of such a mode for both the Global Hawk and U-2 radars is planned by the Air Force under the Advanced Synthetic Aperture Radar System (ASARS) Improvement Program. The concept of operations (CONOPS) for SAR is to search in strip mode and to check detections (which may be false alarms due to clutter or non-HVT vehicles) by using spot mode. The CONOPS for MTI is to classify vehicles using one-dimensional ATR based on HRMTI. Because of the poorer classification performance obtained using HRMTI as compared to spot SAR, three looks are used.

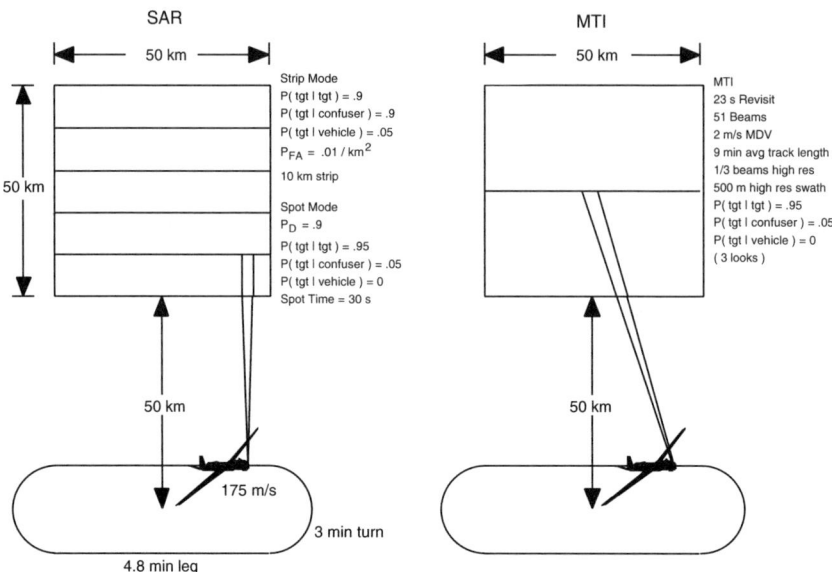

FIGURE 3.8 Synthetic aperture radar (SAR) and moving-target indicator (MTI) sensor, tracking, and processing performance specifications.

Table 3.8 gives the expected time to detect an HVT and the expected number of false alarms for a relocatable stationary target when the HVT is exposed to detection by SAR. SAR is the preferred sensor mode for this case. With the deployment of Global Hawk and the AIP-equipped U-2, and with the installation of the common high-bandwidth data link (CHBDL)—(the Navy's version of the common data link (CDL))—on Navy carriers and large-deck amphibious ships (general purpose and assault), the Navy will have the connectivity to sensors that can detect, classify, and provide targeting-quality data against relocatable HVTs. As discussed in Section 3.3.2, the Navy is developing precision weapons capable of engaging these targets but needs to develop the capability for timely processing of the sensor data that comes down the CHBDL.

Also indicated in Table 3.8 is the expected time to detect an HVT and the expected number of false alarms for the ephemeral stationary target that is hiding and not conducting its mission. Neither sensor mode is entirely satisfactory for this mission. SAR has an unacceptably long search time. The MTI search time is acceptable, but the number of false target nominations is not. Thus improved sensor and processing technology, multiple sensors, and/or fusion of additional sensor types are needed. For example, the expected number of false target nominations for MTI can be reduced using MTIm and two-dimensional ATR, albeit at the expense of an increase in the expected time to detect a

TABLE 3.8 Radar Search for Exposed and Hiding Stationary, High-Value Targets

	Exposed		Hiding	
	SAR[a]	MTI[b]	SAR[a]	MTI[b]
Expected time to detect (seconds)	2	7	39	6
Expected number of false alarms	1	34	14	34

[a]SAR, synthetic aperture radar.
[b]MTI, moving-target indicator.

target. The time to detect a target can be reduced by increased power aperture and electronic scanning for the radar. Combined use of SAR and MTI to track targets through multiple move-stop-move cycles can increase track length and reduce classification requirements. Multiple radar platforms can greatly extend track length and hence reduce the classification load if coverage is coordinated to avoid data gaps during turns. Radar classification can be supplemented by unattended ground sensors. Although sensing and processing technology is being developed along these lines, developing a capability to attack ephemeral targets is a longer-term effort than developing a capability for attacking relocatable targets.

3.5.3.3 Sensor and Weapon Management and Control

Sensor management and control algorithms are needed to assist operators in coordinating the use of sensor resources. For example, when a track is lost on a high-value target, a SAR image should be requested to see if the target has stopped. Likewise, when a new track is initiated in the vicinity of a stopped high-value target, a SAR image should be requested to see if the target has moved. When a target is selected for engagement, sensor resources must be applied to reduce track errors to limits acceptable for the weapon employed.

When targets have been identified and located, weapon management and control decisions must be made. Algorithms and decision aids are needed to assist operators in selecting the optimal weapon, and to ensure that adequate sensor coverage is available during weapon fly-out and that communications links are available to provide in-flight target updates to the weapon, if required. Successfully engaging mobile and moving targets requires that decisions be made in seconds to minutes rather than the hours to days acceptable for fixed targets.

The processing functions described can be performed by a single node on the network (centralized) or by multiple nodes (distributed). In a joint environment with platforms that are supporting multiple missions simultaneously, some form of distributed implementation is perhaps inevitable.

The above discussion focuses on the real-time aspects of attacking distant, mobile, high-value land targets. However, there are planning issues also. It is necessary to obtain the sensor resources required for targeting and fire control, to have shooter platforms in position to take advantage of opportunities that arise, and to deconflict fire with other missions.

3.5.4 Needed Research and Development

The data rates of emerging sensors, the time lines required to address fleeting targets, and the complexity of resource allocation and scheduling decisions make apparent the need for semiautomated algorithms and decision aids (e.g., ATR algorithms) for NCO tactical information processing. Deploying this technology as it continually evolves and integrating it with legacy information processing systems will require flexible, adaptive, distributed tactical information processing architectures, to include human-machine interfaces. The state of the art in these areas is such that a continuing research effort is required.

Fortunately, DARPA has an active program of research in information processing directly relevant to NCO. The committee recommends that the Navy increase its level of participation in these efforts. Naval officer and civilian personnel should be encouraged to serve as DARPA program managers (PMs), an approach that may require changes in personnel policies so that assignment as a DARPA PM is viewed as career enhancing. The Office of Naval Research (ONR) and supporting naval organizations should serve as agents for DARPA programs. ONR should establish appropriate 6.2 programs in NCO tactical information processing and in human-machine interfaces and interactions. Also, a continuing 6.3 program to develop and evaluate prototype NCO tactical information processing capabilities is needed.

In addition to technology, tactics, techniques, and procedures (TTPs) are needed for tactical NCO. Information processing functions must be allocated to humans and computers and across platforms locations. Effective human-computer interfaces (HCIs) for distributed tactical NCO information processing must be developed and must be evaluated in both normal and abnormal situations.

To satisfy these needs, the Navy should develop standard measures of effectiveness (MOEs) and a strong analytical capability focused on NCO and should couple this capability tightly to research, experimentation, and development for NCO. The Navy should continue its fleet battle experiments and its strong participation in joint warfighting experiments with a focus on developing and refining TTPs for NCO. The Navy should also conduct continuing experimental evaluation of the tactical NCO information processing prototypes developed under the 6.3 program recommended above. As in an advanced concept technology demonstration (ACTD), such evaluation should be ongoing, matched to an evolutionary development process, rather than a single evaluation of military utility. Mature versions of tactical NCO information processing prototypes should be used in sensor and weapon system operational evaluations.

For deployment of NCO tactical information processing components, the most pressing need is the ability to exploit current and emerging nonorganic sensors to support land-attack missions. In the near term, the focus should be on relocatable targets, current sensors (e.g., JSTARS, Predator), limited automation, and to extension of the CTP to land targets. In the mid-term, the focus should be on ephemeral targets, preplanned product improvement (P3I) sensors and sensors being deployed currently (e.g., Global Hawk, U-2 AIP), and decision aids. In the long term, the focus should be on moving targets, new sensors (e.g., Discoverer II), and automated systems with human monitoring and override.

3.5.5 Findings

Finding: There is no mechanism to coordinate the development of Navy and Marine Corps doctrine and apparatus for littoral operations, or to coordinate such functions as tracking and network control. (See Section 3.5.2.2.)

Finding: There is no mechanism for coupling NCO research, experimentation, and development with the refinement of doctrine and then assessing the military value of the proposed improvements. (See Section 3.5.2.2.)

Finding: To achieve NCO, research and technology development, experimentation, and development and deployment of tactical information processing capabilities are required. (See Section 3.5.4.)

Finding: The Navy needs to position itself to exploit the fruits of DARPA investment in technology that can provide tactical information processing capabilities. (See Section 3.5.4.)

Finding: To project power at long ranges ashore, the Navy must be able to use nonorganic sensors and so should pursue connectivity to some of these sensors as vigorously as possible. (See Section 3.5.4.)

3.6 SYSTEM ENGINEERING

In this section, the committee uses the challenges of the notional land-attack system shown in Figure 3.2 to illustrate the need for analysis and engineering of the total system of complexly interacting components performing network-centric operations. It then comments on what it perceives as a lack of a unified approach in the specification and development of components.

3.6.1 System Requirements to Hit Moving Targets

Presented here is an example of the recommended system engineering approach that focuses on solving the war-fighter's problems and thereby derives the characteristics of the component systems instead of starting with these characteristics as a "requirement." An acute problem at present is that of hitting moving targets on Earth's surface. Surveys show that moving targets normally constitute a high percentage of the targets in theater; tanks, armored personnel carriers, and patrol boats are examples. An important specific case is a high-value target such as a missile transporter-erector-launcher that is usually in hiding when stationary and therefore vulnerable to attack only when on the move. The committee conducted an example analysis to do the following:

• Quantify requirements of various concepts for end-to-end systems to hit moving surface targets, considering a range of realistic environments and target behavior;
• Explore trade-offs in how to balance the burden of performance among system elements; and
• Examine how networking concepts can be employed to achieve system requirements.

The specific problem to be solved is that of hitting a moving surface target among randomly distributed false contacts (real physical objects that can be confused with the intended target). The intended target deliberately maneuvers to avoid engagement.

The committee considered three weapon system concepts, from simple to complex. With important exceptions (such as main battle tanks and SAM radars), moving targets are often numerous and individually of low value, so simple, inexpensive weapons are often desirable. However, targeting system complexity must increase to meet the demands of a simpler weapon. This was one of the key trade-offs examined.

Outlined very briefly here is the committee's analysis approach; Appendix C describes the approach and findings in more detail and presents the mathematical model, which builds on one used for a previous Naval Studies Board report,[8] which showed that the targeting system should provide a steady stream of reports to the weapon, as opposed to a single report. The targeting system

---

[8]Naval Studies Board, National Research Council. 1993. *Space Support to Naval Tactical Operations (U)*, 93-NSB-494. National Academy Press, Washington, D.C. (classified).

must be able to classify a target and associate multiple reports with a single track. With these capabilities a targeting system can then provide a steady stream of reports that enable a tracking filter to estimate speed and heading. The targeting component is characterized by three parameters: the position accuracy, report interval, and data time delay. It can be assumed that a weapon or launch platform that attempts to reacquire the target is successful if (1) the target is inside the sensor or seeker area of regard and (2) the search finds the intended target before a false contact is misclassified as the target. The probability of satisfying these two conditions depends critically on accurately predicting the target's location.

Summarized very briefly here are the results of the analysis. Requirements to target a weapon of intermediate complexity (and cost) are not onerous compared with those to target a complex weapon (e.g., a manned aircraft with capable sensor suite). For the simplest weapon, one that does not reacquire the target, the targeting requirements are difficult to achieve.

The analysis showed that system requirements are driven by the environment, principally the density of false contacts. How can one design a system for all likely environments? Design for very dense environments would be overdesign by large margins for less stressing cases and appears to be prohibitively expensive for widespread deployment. The answer may be to provide the commander with the tools to control assets flexibly in order to focus assets and tighten the targeting-system-to-weapon-system loop when necessary.

Can networking enable the requirements to be met? The committee believes several networking concepts may help. First, fusion of data from multiple sensors at different geometries can greatly improve the accuracy of the target position measurement; the radars' precise range estimates provide the accuracy refinement. Second, targeting data can be put into a common navigational coordinate system by communicating among all targeting and weapon system platforms to control the specific GPS satellites they all track.

To summarize, hitting moving targets will require a tight network of distributed sensors, processing facilities, command and control facilities, weapon launch platforms, and weapons. In many circumstances, weapons with simple, inexpensive seekers and links for in-flight targeting updates may provide the best balance in distributing the burden of performance between targeting and weapon components. In the more distant future, networking concepts may permit the use of low-cost weapons without seekers. A network-centric operations system that is both affordable and yet effective in all likely situations will have to be flexible and adaptable to the commander's tasking, and it will have to make available for use in the most challenging high-density traffic scenarios some means of target recognition on the weapon or on the platform controlling it.

3.6.2 Coordination of Component Development

Figure 3.9 diagrams some of the interactions among components involved in hitting land targets by indirect fire for the case in which the weapon receives no external guidance after launch. The required ATR false-alarm rate is a function of the area to be searched. That area is a function of target location error and navigation error. Target location error is a function of targeting sensor accuracy and latency, target motion, and weapon time of flight. Navigation error is a function of resistance to GPS jamming and the performance of the IMU that guides the weapon, after GPS guidance has been lost, to the vicinity of the target.

Although a system analysis can be performed to allocate requirements among the components, opportunities and challenges arise during the course of component development. An increased GPS jamming threat could be addressed by investing in some combinations of better IMUs and ATR. A breakthrough in ATR could ease requirements on weapon time of flight. The +20 dB spot beam proposed for future generations of GPS satellites would reduce the effective range of a terminal jammer by a factor of 10, easing the requirements on IMU drift rate or ATR coverage by a similar factor. For an open-loop attack on a fixed target, the probability of hit is determined by target location error, navigation error, and ATR performance; time delay is not an issue. However, for an ephemeral target, that is, one that is detectable and stationary for only a limited time, the weapon must arrive before the target moves. The sum of the delays in sensing, decision making, and weapon time of flight must be smaller than the

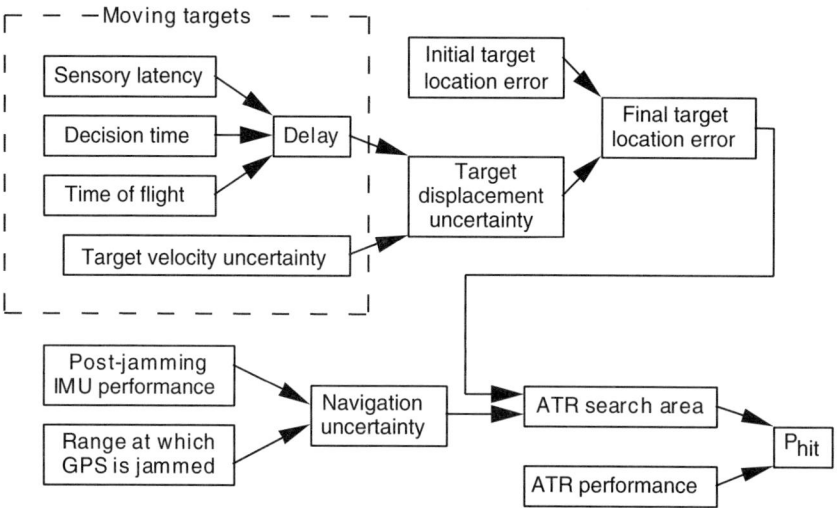

FIGURE 3.9 Component performance interactions (no external guidance after launch).

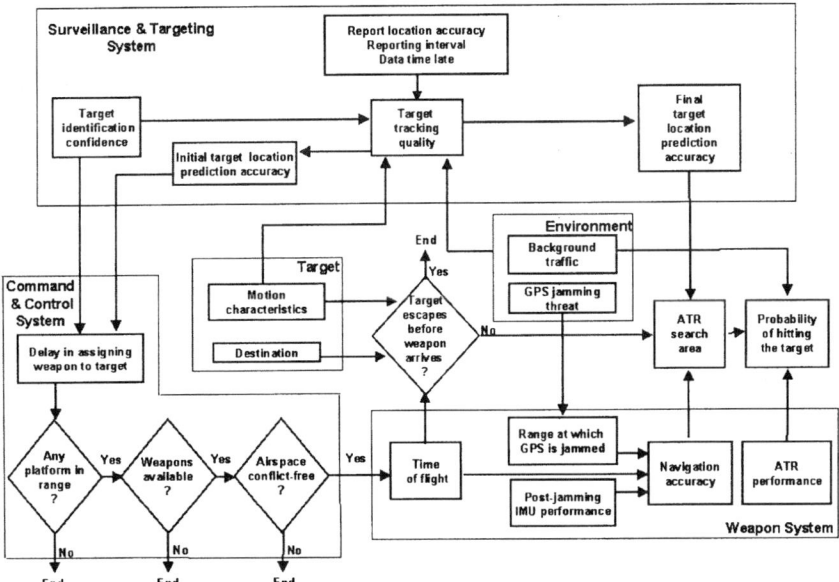

FIGURE 3.10 System factors in delivering firepower ashore against moving targets (in-flight updates from an external sensor).

period the target will remain stationary. For a moving target, the case shown within the dashed lines in Figure 3.9, there will always be uncertainty about the target's location, and short times of flight and excellent ATR will be needed to hit it.

Complex as these interactions are, the situation analysis becomes even more complex when the weapon receives in-flight updates from an external sensor, the case that is analyzed in Appendix C. Figure 3.10 displays these interactions. Absent a breakthrough in ATR, the committee believes that closed-loop control will usually be required to hit moving targets. This belief motivated the recommendations to provide control links to weapons and to consider developing and deploying organic sensors that could provide near-staring control of the weapon's endgame.

The committee had the opportunity to hear from many officials responsible for the development of components that will be used to constitute the NCO systems. These officials were uniformly knowledgeable about the challenges implicit in meeting the specifications laid down for their components, but, as focused program managers, were less interested in the derivation of these specifications or the possibility of network-wide trade-offs.

The committee found that coherent analysis and development were best exhibited in antiair warfare (AAW), perhaps because a single organization performs the system engineering and program execution, or perhaps because AAW has occupied the Navy's attention for many years. The least coherence was found in strike and over-the-horizon naval fire support—perhaps because diverse, independent program offices are developing the component subsystems; perhaps because the Navy and the Marine Corps do not have common doctrine for naval fire support; and perhaps because the Navy's focus on decisively influencing events ashore is relatively new.

The committee is aware of ongoing work in land-attack targeting, for example, the activities of the land-attack targeting integrated process team and of the DD-21 program office. Its comments are not intended to be critical of these activities, but rather to indicate that additional resources, scope (e.g., involvement of the air community), and authority are needed.

Among the problems the committee found in strike and over-the-horizon naval fire support were the following:

- Need for responsive, long-range, low-cost, high-volume weapons for compatibility with stand-off distances imposed on naval platforms by antiship missile or other threats, and for Marine Corps plans for ship-to-objective maneuver;
- Inadequate targeting for naval surface fire, including lack of an agreed-upon method, backed by program actions, for transmitting target coordinates from a deep inland forward observer to an over-the-horizon firing ship; and
- Inadequate capability to detect, identify, track, and engage moving targets.

One reason for this lack of overall system engineering is clear: the Navy has undertaken a new mission—to influence events ashore decisively—and has not fully adapted itself to execute that mission. Of course, organizing to perform end-to-end system engineering over a sphere of activity as large as naval strike and surface fire is a daunting challenge. But the Navy has done exactly that, twice in past decades.

In the 1960s and 1970s, the Navy faced a formidable submarine threat posed by the Soviet Union. Meeting the antisubmarine warfare (ASW) challenge required system improvements on aircraft, surface ships, and submarines and in surveillance systems. In response, the Navy established an office, PM-4, in the Naval Materiel Command, and gave it responsibility and authority for development of the Navy's ASW capabilities. PM-4 performed end-to-end system analysis, trading among ship, submarine, aircraft, and surveillance system components, and enabled communication among programs so as to accomplish the end-to-end system engineering needed to develop an effective ASW capability. Another important factor in the Navy's success was an OPNAV sponsor responsible for the entire ASW capability. The OPNAV sponsor directed operational and system analyses to support funding allocations.

In the 1990s, the Navy faced another challenge—a spectrum of air threats, ranging from low-flying, stealthy cruise missiles to theater ballistic missiles, being acquired by a large number of potential adversaries. The Navy responded by forming the Program Executive Office (PEO) for Theater Air Defense, then consolidated that office with the PEO for Surface Combatants to form the current PEO for Surface Combatants and Theater Air Defense. This consolidation allows the Navy to conduct end-to-end system analysis, trading among the multiple layers of air defense, and, most relevant to the topic at hand, develop systems that cross platforms, including the CEC system that is the exemplar of NCO. Here again, the Navy is well served by an OPNAV sponsor responsible for the entire capability. In the first decade of the new century, the Navy's challenge will be to build the capability to influence events ashore decisively, particularly by projecting power ashore.

The Navy's two successful examples demonstrate what will be required. Future naval strike and surface fire will encompass naval air, surface, and subsurface platforms, air- and sea-launched weapons, and associated command, control, and communications components. Even if development of components is decentralized, someone must be responsible for the development of the overall system and must have the status and resources to manage interfaces with other Services and with National sensor systems. The CNO must clarify responsibility in OPNAV for the power projection mission. The development of new warfighting concepts and doctrine and the rebalancing of the materiel components must coordinate throughout the evolution of the system.

3.6.3 Finding

Finding: Hitting ephemeral, relocatable, and moving targets is a vital capability that will require improvements in sensors (e.g., platforms for surveillance in high-threat areas), processing (identifying targets and maintaining tracks on targets moving through high-density traffic), command systems (capability for frequent and rapid decisions on weapon-target pairings), and launch platforms and weapons (e.g., affordable communication links and simple seekers). Many trade-offs can be made among system components, and many network concepts can be brought to bear to improve performance and reduce overall system cost. (See Section 3.6.1.)

3.7 SUMMARY AND RECOMMENDATIONS

A network-centric operations system comprises a number of subsystems, each designed and engineered to accomplish a military function. The subsystems are networks of components—such as sensors, weapons, command elements, and mission-specific information processing—tied together by the NCII that is described in Chapter 4.

The sections above describe the characteristics of the components and illustrate the interdependencies among element performance and their effects on subsystem performance as required for power projection, the Navy mission chosen by the committee for study because this mission has only recently been emphasized by Navy Department leadership and because much work will be needed to realize the potential of NCO in this mission. In particular, concepts are needed for the targeting of short-time-of-flight weapons from adequate stand-off.

Consideration of what is needed for effective power projection—in terms of weapons, sensors and navigation, and tactical information processing—revealed a number of potential trade-offs across elements for effective operations, for example, GPS jam resistance against ATR performance, guidance accuracy against warhead lethality, and sensor latency against weapon time of flight. The complexity of the interactions led to the committee's conclusion that the design and development of new subsystem components must be coherently managed so that the trade-offs can be continually reexamined to account for developmental difficulties and breakthroughs.

Attacking moving targets with an in-flight link from the targeting sensor would require either warheads that are lethal over large areas or excellent ATR performance. While recommending further development of ATR, the committee also recommends that sensors, weapons, and the NCII should be designed to support the use of such a link.

Sensors have physical limitations and are subject to camouflage, deception, and information operations. Diversity in location and phenomenology, together with the ability to form ad hoc networks, can overcome some of these challenges.

The committee's consideration of sensors showed some promising ATR work to which the Department of the Navy's technical community is not strongly coupled. The high potential value of theater and National sensors able to interface with Navy platforms is not receiving high Navy Department priority.

3.7.1 Principal Recommendations

Based on the findings presented throughout the chapter, the committee's principal recommendations are as follows:

Recommendation: The Naval Warfare Development Command and the Marine Corps Combat Development Command should formalize their relationship and ensure joint development of littoral NCO concepts. In particular, they should reach agreement on the need for a family of short-time-of-flight over-the-horizon weapons from adequate stand-off distances and concepts for their targeting.

Recommendation: The Department of the Navy should design and engineer, as a coherent whole, the mission-oriented subsystems of the NCO system, trading off performance goals across components to achieve required mission performance. Some reform of the acquisition community from platform-centric to mission-centric should be considered, especially for the power projection mission.

Recommendation: The Department of the Navy should facilitate the power of networks of sensors at disparate locations and employ disparate phenomenologies by moving more smartly to connect to National and theater sensors and by designing new sensors to permit cooperative behavior in ad hoc networks.

Recommendation: The Department of the Navy should seek the capability of in-flight guidance of new weapons designed to be fired from over the horizon against ephemeral, relocatable, and moving ground targets. In addition, the Department of the Navy should work to enhance connectivity to joint moving-target indicator (MTI), synthetic aperture radar, and electro-optics sensors and consider the acquisition of organic airborne near-staring MTI sensors to provide closed-loop endgame weapon control.

Recommendation: While participating in endeavors to increase the jam resistance of Global Positioning System receivers in naval platforms, the Department of the Navy should continue to seek technology for better long-range target identification (including ATR) and should interact more strongly with the relevant DARPA programs.

3.7.2 Summary of Findings and Associated Recommendations

The following subsections repeat the findings presented in the text of this chapter and offer, in addition, individual recommendations based on those findings.

3.7.2.1 Weapons

Finding: Although new weapons are being developed for land attack, the range of surface-launched, short-time-of-flight weapons is currently too limited to support ship-to-objective maneuver at reasonable stand-off distances. Better targeting concepts are needed. (See Section 3.2.1.)

Recommendation: Examine targeting concepts before specifying weapons.

Finding: Target identification limitations inhibit the use of air-to-air weapons at their full kinematic range. (See Section 3.2.2.)

Recommendation: Pursue technology for reliable, long-range identification.

Finding: Weapons that attack low-signature targets will likely depend on guidance from networks of sensors and illuminators. (See Section 3.2.3.)

Recommendation: Provide capability to accept in-flight guidance.

3.7.2.1 Sensors

Finding: Sensor capabilities are improving through exploitation of digital and solid-state technology. (See Section 3.3.1.)

Recommendation: Continue basic technology and advanced sensor development.

Finding: Adversaries can exploit fundamental physical laws and make detection by sensors difficult in certain situations. (See Section 3.3.1.)

Recommendation: Investigate new physical phenomena that exhibit different physical limitations while continuing to explore the existing technology for design concepts that can extend performance limits.

Finding: Deployed Navy sensors span a ranges of types, but most were designed for platform defense, are stovepiped, and exhibit a mix of old and new technologies due to the budget-limited practice of incremental upgrades over a long period. (See Section 3.3.2.)

Recommendation: Develop and acquire all new sensors as a consequence of NCO top-down systems engineering. Build in enablers for cooperative behavior of dissimilar sensors, accommodation of new technology, and participation in ad hoc networks.

Finding: The Navy has no organic sensors capable of guiding its precision, long-range weapons to ground targets. Emerging doctrine assumes access to joint or National resources in the battlespace, but the Navy is only beginning to invest in such connectivity. (See Section 3.3.3.1.)

Recommendation: Address the nature of the Navy's mix of organic and joint or National sensors. Consider the acquisition of a Navy synthetic aperture radar/ground moving-target indicator sensor for unmanned aerial vehicles.

Finding: Multisensor cooperation offers significant performance advantages. (See Section 3.3.3.2.)

Recommendation: Design all future sensors to accommodate flexible data exchange and cooperative behavior.

Finding: Temporary sensor-shooter-weapons teams are natural in network-centric operations but offer flexibility and quality-of-service challenges for the communication infrastructure. (See Section 3.3.3.3.)

Recommendation: Impose flexibility requirements on sensors and their information links. Factor this requirement into the initial design and engineering of the Naval Command and Information Infrastructure.

Finding: Geolocation in the same absolute or relative coordinate system of the sensors and targets in the battlespace is mandatory. Use of the Global Positioning System is often assumed to be the sole technique employed but may not always be available. (See Section 3.3.3.4.)

Recommendation: Develop protection for and alternatives to the Global Positioning System.

Finding: Automatic target recognition avoids overload of communications and of image analysts, may be necessary for remote attack of moving targets, and provides a hedge against GPS jamming. Model-based vision may overcome the limitations of template matching. However, more general capabilities for automatic information extraction continue to be elusive and must remain the subjects of continuing R&D. (See Section 3.3.3.5.)

Recommendation: Support R&D on automatic target recognition and related information extraction approaches as well as image-compression algorithms.

3.7.2.3 Navigation

Finding: No single technique will make GPS-aided weapon navigation invulnerable to GPS jamming. Practical solutions are likely to involve a combination of cheaper, precise IMUs, better ALR and ATR, improved satellite signals and receiver signal processing, and the use of spatial processing. (See Section 3.4.2.1.)

Recommendation: Perform analysis to determine what combinations of improvements would be required to overcome foreseeable Global Positioning System jamming. Fund technology base work to determine whether these improvements are attainable.

Finding: Available antiradiation weapons do not solve the GPS jamming problem because the jammers can be easily replicated and the weapons cost many times more than the jammer. Suitably modified HARMs could be used to attack aircraft carrying high-power jammers, and the presence of such HARMs in inventory might demoralize crews operating GPS jammers. (See Section 3.4.2.2.)

Recommendation: Do not depend on physical attacks against jammers as a general solution to Global Positioning System vulnerability.

Finding: Although navigation through the use of satellites not designed for that purpose is possible, the difficulties of using these techniques in weapons are formidable. Nevertheless, European interest in these techniques will cause the difficulties to be assessed and perhaps overcome. (See Section 3.4.2.3.)

Recommendation: Monitor European and commercial progress in navigation through incidental satellite transmissions.

Finding: Passing control of a weapon forward to a sensor that holds the target in view is a plausible means of reducing or eliminating dependence on GPS and similar systems. (See Section 3.4.2.4.)

Recommendation: Design weapons and sensor platforms so as not to foreclose the possibility of endgame control of the weapon directly from the sensor.

3.7.2.4 Tactical Information Processing

Finding: There is no mechanism to coordinate the development of Navy and Marine Corps doctrine and apparatus for littoral operations, or to coordinate such functions as tracking and network control. (See Section 3.5.2.2.)

Recommendation: Formalize and institutionalize the relationship between the Marine Corps Combat Development Command and the Navy Warfare Development Command with regard to NCO innovation, tactics, techniques, and procedures, and doctrine in the littorals.

Finding: There is no mechanism for coupling NCO research, experimentation, and development with the refinement of doctrine and then assessing the military value of the proposed improvements. (See Section 3.5.2.2.)

Recommendation: Develop an analytic capability and measures of effectiveness to support the evolutionary improvement of NCO tactics, techniques, and procedures and tactical information processing. Continue experimenting; emphasize experimental design and measurement.

Finding: To achieve NCO, research and technology development, experimentation, and development and deployment of tactical information processing capabilities are required. (See Section 3.5.4.)

Recommendation: Maintain Navy Department technology programs underlying tactical information processing.

Finding: The Navy needs to position itself to exploit the fruits of DARPA investment in technology that can provide tactical information processing capabilities. (See Section 3.5.4.)

Recommendation: Interact more strongly with DARPA and offer strong candidates for leadership of appropriate DARPA program offices.

Finding: To project power at long ranges ashore, the Navy must be able to use nonorganic sensors and so should pursue connectivity to some of these sensors as vigorously as possible. (See Section 3.5.4.)

Recommendation: Establish a continuing 6.3 nonacquisition program for prototyping and experimentation.

Recommendation: Move smartly to ensure connectivity from nonorganic sensors to Navy control and firing platforms and to ensure the ability to process data from these sensors.

3.7.2.5 System Engineering

Finding: Hitting ephemeral, relocatable, and moving targets is a vital capability that will require improvements in sensors (e.g., platforms for surveillance in high-threat areas), processing (identifying targets and maintaining tracks on targets moving through high-density traffic), command systems (capability for frequent and rapid decisions on weapon-target pairings), and launch platforms and weapons (e.g., affordable communication links and simple seekers). Many trade-offs can be made among system components, and many network concepts can be brought to bear to improve performance and reduce overall system cost. (See Section 3.6.1.)

Recommendation: The Department of the Navy should engineer the capability to hit ephemeral, relocatable, and moving targets as an end-to-end system.

4

Designing a Common Command and Information Infrastructure

This chapter begins by articulating the concept of a common command and information infrastructure for the naval forces, the Naval Command and Information Infrastructure (NCII). In particular, Section 4.1 notes the general attributes that the NCII should possess, including adaptability in the face of changing needs and new technologies, and the functional capabilities it should provide to support users. The NCII supports all echelons—strategic, operational, and tactical—with a uniform architecture that uses commercial network protocols, a concept that, for tactical networks, runs contrary to the current situation. Section 4.2 develops this important point, and it is further elaborated on in Appendix E. Since an architecture is key to realization of the NCII, Section 4.3 discusses the products and processes the Department of Defense (DOD) and the Department of the Navy currently use for developing architectures and comments on their suitability for developing an NCII architecture. Section 4.4 gives the committee's recommendations based on the material presented in the preceding three sections.

4.1 THE NAVAL COMMAND AND INFORMATION INFRASTRUCTURE CONCEPT

4.1.1 Definition and Properties

The broad and rapid exchange of information and the ready assimilation and use of this information are at the heart of network-centric operations (NCO). In NCO, the individuals involved have access to information from a wide variety of sources and can operate in an effective, coordinated manner by exchanging information, even if the force elements are widely dispersed. As a consequence, decision making should be more informed than is now the case, collaborative planning among dispersed elements should be more timely and complete, and distributed engagements involving sensors, fire control authority, and weapons at separate locations should be more readily executable.

Underlying this exchange and use of information is the Naval Command and Information Infrastructure, so named to indicate an infrastructure that supports not just the manipulation of information but also the actual functions of command. Such an infrastructure should possess a number of attributes:

- It should integrate and support operations at all levels of command;
- It should be responsive and assured, providing a continuously available, secure, high-integrity resource to support all information needs;
- It should facilitate information management by offering consistent, tailored operational information to specified recipients;
- It should be dynamic and self-organizing, automatically healing breaches and forming and automatically maintaining high-priority, low-latency broadcast or normal communication channels;
- It should be independent of location, providing great operational flexibility in the geographical positioning of component units; and
- It should be easily scaled and evolved, adaptable in size to meet changing needs, and capable of being modernized easily through the use of common, open interface standards and functionally modular design.

The NCII (see Figure 1.6 in Chapter 1) comprises the communication and computing assets necessary to accomplish two things: (1) effect the exchange of information among information repositories, sensors, command elements, forces and weapons, and logistic and support elements and (2) allow this information to be used for both human decision making and automated processes pertaining to command and execution.[1] The communications and computing components embedded with sensors, platforms, weapons, and support systems are not considered part of the NCII, but the effective operation of the NCII requires that their interfaces to the NCII satisfy standards established in the overall NCII design. Information repositories are part of the NCII if they are naval assets directly supporting command, but other naval information sources that may be called upon (e.g., personnel databases) are not part of it, although their inter-

---

[1]The word "infrastructure" as used in this report includes both the underlying communications base and the common support applications that ride atop the base; specific mission applications are not included.

faces to the NCII must be compatible.[2] Joint and intelligence information sources are regarded similarly.

An information infrastructure like the NCII has natural analogs at the operational and strategic levels of warfare (e.g., the Global Command and Control System). The infrastructure also applies at the tactical level since it refers to general means for manipulating and transporting information, although one needs to define its scope at the tactical level carefully. Two cases in point illustrate the issue—tactical digital information links (TADILs) and the cooperative engagement capability (CEC). Because TADILs represent a general information exchange capability, they would fall within the scope of the NCII. The issue here is that the current TADIL implementations would not comply with the architectural standards that will probably be chosen for the NCII (e.g., Internet Protocol (IP) networks). However, over time (see Section 4.2 below), the TADILs could migrate into compliance. The CEC could be a different matter. It is a specialized implementation necessary to meet particularly demanding performance requirements. It is not clear, at least at this time, if the general standards chosen for the NCII would satisfy CEC performance requirements. If they do not, then the CEC would remain outside the NCII, but its interface to the NCII would have to satisfy standards established in the overall NCII design.

4.1.2 Information Use and Design Considerations

The next step in developing the NCII is to characterize more precisely the functions it will perform. That, in turn, requires that the concept for information use be considered.

4.1.2.1 Internet Paradigm

Network-centric operations embody the idea of rapid, ready, and flexible access to information, with the Internet in some sense serving as a model or paradigm. The Internet, as commonly referred to in popular discussion today, has two components: (1) a robust, underlying networked communications base and (2) the applications that make use of the communications base to provide information to a widely dispersed user population. The communications base derives from the ARPANET begun in the early 1960s, while the applications have appeared mostly within the last decade. The ease with which these applications have been able to make use of the underlying communications base has been a critical factor in the rapid growth of the Internet and the effect of the applications on society. The point for NCII design is to view the infrastructure as having two layers, a supporting resource base (e.g., communication) and

---

[2]The close association of logistics with command and control means, however, that the logistics databases should be part of the NCII.

applications, which should be easily able to make use of the supporting resource base.

The Internet model has two aspects, top-down and bottom-up, both of them important. On the one hand, some top-down principles and standards are necessary so the applications can easily use the communications base and so users can interact with applications. On the other hand, the applications were developed from the bottom up, i.e., by a diverse developer population. This diversity means there is a broad base for innovation, which has contributed greatly to the widespread popularity and utility of the Internet. The point for the NCII is that it should use a framework of standards that would permit its applications to come from a diverse set of sources.

Further insight can be gained by focusing on the user. The Internet today (and even more so in the future) allows access to an "information marketplace." Users' needs are not satisfied with a predefined set of information; rather, they seek widely for the information they need. This behavior has a direct analogy in military operations in the current and anticipated future world environments. In the more prescribed scenarios of the Cold War, one could define the information requirements quite well, but now, uncertainty as to the type and location of future military operations precludes that. Furthermore, different operators may vary in their approach to a situation and hence in their information needs. Certain information requirements can be predicted—e.g., a unit will obviously want to know when it is under attack—but overall, detailed information requirements cannot be predicted in advance and will vary from user to user. From this it follows that the NCII should provide users ready and flexible access to information.

While the NCII should allow widespread dissemination of information, it must also be able to accommodate the need of commanders for some degree of control over dissemination (e.g., for security purposes and bandwidth management). Furthermore, this information dissemination enables greater decentralization of command, but at the same time it allows for the centralized collection of information and hence for greater centralization of authority. There is no single appropriate point on this centralization-decentralization spectrum; it will depend on the nature of the military operation. The NCII must able to support these varying modes of command.

Finally, it is critical to recognize that the manner in which information is used in the NCII will change continually as operational concepts are refined and new technologies introduced. This is the lesson of personal and business use of information technology: One cannot fully anticipate all the myriad ways that it can be used. Rather, one has to work with the technology to explore its uses, and these uses will suggest new technologies to be explored, which will lead to further new uses. Thus, the NCII must be designed to allow this continual evolution in information use and the introduction of new technology.

4.1.2.2 Decision Focus

The ultimate end use of information in military operations is to support decision making. Since the final measure of any information system is the quality and timeliness of the decisions reached, the decisions should be central in assessing the functionality provided by the NCII.

Figure 4.1 is a highly simplified schematic of the information flow in the decision-making process at all levels of command. It makes clear that there are two aspects to the information support of decisions. First is the information gathering and generation stage, to prepare for and support making a decision. As discussed above, decision makers (or their staffs) should be able to easily find information and draw it to themselves. Second is the command dissemination stage—that is, the decision itself is information that must be conveyed to appropriate elements. The NCII applications should be specifically designed to support this decision-making process.

Not shown in Figure 4.1 but important to realize are the different time scales that can be involved. Generally speaking, there are two such scales. At the operational-strategic time scale, information is gathered, decisions made, and results disseminated in a time span ranging from minutes to hours (or even days). In the tactical time scale, the same process takes place in a matter of seconds or fractions thereof. In highly compressed tactical situations the decision-making function can be short-circuited by passing the information directly from the sensors to the weapons or possibly with automated processes replacing human decision making. Similarly, very rapid iterations in the decision process can be made in response to the changing tactical situation.

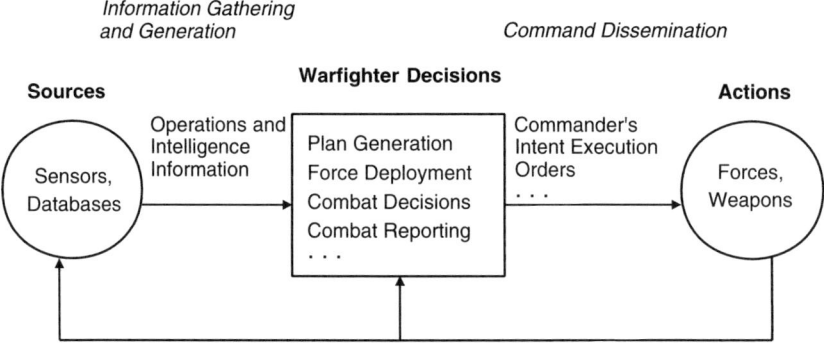

FIGURE 4.1 Information flow in decision making.

4.1.2.3 Summary

The NCII must provide a set of functional capabilities, i.e., a set of system functions to support the user. These capabilities would be partitioned between applications and a supporting resource base. The application layer would contain specific functional capabilities to support the decision-making process. The supporting resource base provides more generic functional capabilities (e.g., communications), although generally speaking, these capabilities support decision making, too.

Specification of functional capabilities is not sufficient for designing an NCII, however. For example, the NCII must be easily configured to meet specific mission needs. This requirement is not satisfied by a functional capability; rather, it refers to a property of the system as a whole and is achieved by applying certain design principles to the overall system. Thus, both the functional capabilities and the system properties must be specified.

There are thus three general classes of requirements for the NCII:

- Functional capabilities that directly support decision making both in the information gathering and generation stage and in the command dissemination stage,
- Functional capabilities for the supporting resource base (e.g., communications), and
- Design practices to achieve system properties (e.g., configurability).

4.1.3 Functional Architecture

The functional architecture shown in Figure 4.2 describes the capabilities that the NCII must provide and shows the interrelationships among these functions. Figure 4.2 follows from Figure 4.1 by inserting specific functions (e.g., collection management) to support the decision-making process, recognizing that these functions ride on top of a supporting resource base. The four functions to the left of the decision box in Figure 4.2 enable information gathering and generation, and the single function to the right of the box supports command dissemination.

The functions for the supporting resource base may be described as follows:

- *Communications and networking.* These are the basic services that provide the communication links and form networks from them. Some communications could be point to point and not necessarily part of a network.
- *Information assurance.* This function protects the information content from unauthorized disclosure or modification and ensures its delivery to intended users. Protection against both malicious threats and system failure would be considered in realizing this function.

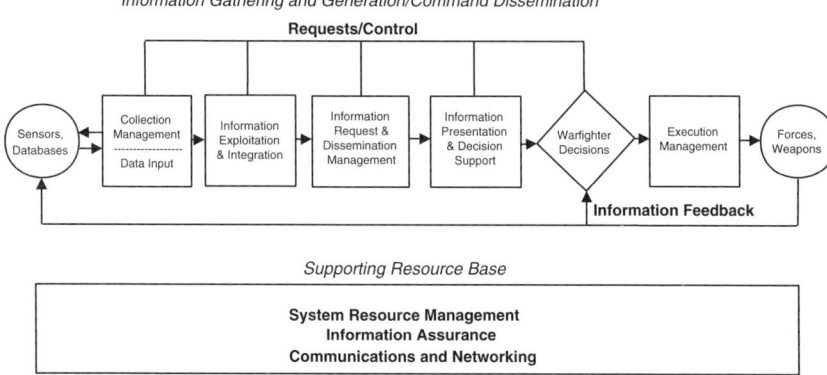

FIGURE 4.2 NCII functional architecture for information gathering and generation and command dissemination.

- *System resource management.* This function furnishes services so that applications use the supporting resource base in an efficient, coordinated manner, in keeping with established priorities. Thus, the function would include services to manage and allocate bandwidth and to provide end-to-end quality of service.

There is more to the supporting resource base than the three functions given. There are also processing and storage functions. However, for the scope of this report, the three functions described were considered to be the ones requiring the most emphasis and assessment.

The functions supporting information gathering and generation and command dissemination may be described as follows:

- *Collection management.* This function determines the tasking of sensors to collect data. It should task the sensors based on an integrated view of the sensor assets available and should support cross-cueing between sensors.[3]
- *Information exploitation and integration.* This function extracts basic information from the initial input data and further refines that output by correlating, fusing, and aggregating it.
- *Information request and dissemination management.* This function provides information based on user-specified requests for a given type of informa-

---

[3]Data input parallels collection management by drawing data from stored databases. Standard database retrieval methods are involved. Given the scope of this study, this function does not receive further treatment.

tion. Its operation is transparent in that users do not have to know the details of where the information is located. This function will also provide information to users based on the directions of any other authorized party.

- *Information presentation and decision support.* Information presentation is the graphical display of information to users. Decision support is a set of automated tools that allows users to manipulate information for the purposes of making a decision.
- *Execution management.* This function supports delivery of decisions to the intended recipients and allows for dynamic adaptation of those decisions in the light of rapidly changing events. It could have been included under decision support but was believed to be important enough to be singled out.

The logical flow among the functions is easily seen. For example, if a user wanted a certain type of information, he or she would go to a particular display (information presentation) or request it in some generic manner (information request and dissemination management). Either of these functions would then draw on the base of exploited and integrated information, and, if necessary, sensors could be tasked to gather further data.

The set of functions given here seems complete, apart from the omission of the processing and storage functions noted above. The next step in development would be to specify the subfunctions or services that make up each of these functions and then to implement them in software or hardware. Particular attention should be paid to defining the interfaces to each of these functions and subfunctions so that they can interact appropriately. There are existing programs for implementing some of the functionality, although not generally within the context of the integrated view expressed in Figure 4.2. Chapter 6 discusses the status of implementing these functions.

4.1.4 System Properties

As discussed in Section 4.1.2, specifying the NCII also requires giving the system properties that it must satisfy. Such a list could be made quite long, since there are many desirable properties that a system should have, but here the list is held to just those few properties deemed most important (Table 4.1).

Information assurance appears both as a system property and, above, as a functional capability. It is a system property since it is achieved only when all components of the system are secure and protected. But it is so critical in establishing networks, given the vulnerability the networks could introduce, that it is also explicitly singled out in the supporting resource base.

The other properties listed in Table 4.1 are also critical. The lesson of modern information technology use, as demonstrated for example by the Internet, is that new and useful applications are continually arising. The situation should be no different in support of military operations, so flexibility to accommodate

TABLE 4.1 Important System Properties Required in the Naval Command and Information Infrastructure

System Property	Desired Attributes
Information assurance: assures continued availability of service despite failure of components and attack	The infrastructure should withstand multiple dependent and independent failures, including loss from physical attack. The infrastructure should continue in the face of successful attacks to provide service for the most critical needs. It should be able to isolate the attacker, reallocate resources, repair the damage, and recover. Backup modes of operation should be available.
Flexibility: accommodates new applications	The time required to test or widely deploy a new application should be short and the effort very little. This will encourage creative thinking and the emergence of applications based on user ideas. In particular, the system should include a "sandbox" for testing new applications.
Modular system design: accommodates new technology and software upgrades	Hardware and software will continuously evolve. New applications may require new or expanded software functionality. It is essential that the architecture allow independent upgrades of software modules.
Fast and easy configuration: meets tailored mission needs	It should be possible to generate system software and hardware configurations using system configuration tools. These tools should be capable of automatically generating configuration modules and simulating and testing the composed configuration, and a variety of graphical interfaces should be provided for different users requiring different levels of detail.

new applications is needed. The pace of information technology change is rapid, so it must be possible to incorporate relevant new technologies with minimum cost and time, which calls for modular system design. Finally, the configuration of deployed forces cannot be specified in advance, especially given the variability of military operations in the current and anticipated world environments, so configuration has to be fast and easy to meet tailored mission needs.

These properties require certain design principles and practices and can also be supported by technological innovations. Realizing the system properties should be one of the key considerations in developing the system architecture for the NCII.

4.1.5 Relationship to Other Information Infrastructures

Several information infrastructures are being discussed currently in DOD-wide and Service contexts. This section relates the NCII to them.

The defense information infrastructure (DII) is quite extensive, being regarded as ". . . the sum of all information management assets [supporting warfighters] owned by each of the Office of the Secretary of Defense (OSD) Principal Staff Assistants (PSAs), Joint Chiefs of Staff, Combatant Commanders, the individual Military Services, and Defense Agencies."[4] The NCII and the DII are not mutually inconsistent concepts. However, the DII is much broader in scope and has a nearer-term focus (the planning horizon in its master plan is typically 2 years).

The Department of the Navy's Chief Information Officer (CIO) has led development of the concept of an information technology infrastructure (ITI).[5] The ITI articulates the network connectivity and services needed to support all naval systems that produce, use, or exchange information electronically. In one sense it is broader than the NCII since it also applies to business operations. It is narrower in the sense that it focuses on network connectivity and services, which are in the supporting resource base in the NCII definition. The NCII also includes the functions supporting information gathering and generation and command dissemination (see Figure 4.2). The NCII and ITI are not inconsistent; rather, they have different emphases. In fact, one could imagine the ITI evolving to put more emphasis on upper-layer functions as does the NCII.

The Defense Science Board has articulated the integrated information infrastructure (III) concept.[6] In general terms, it comprises information transport, distributed computing resources, and information services (service agents and application support agents). The NCII and the III are quite similar in spirit, although the NCII goes into more functional detail in the information services (i.e., the upper layer in Figure 4.2).

The concept of the battlespace infosphere (BI) developed by the Air Force Scientific Advisory Board[7] is not a full infrastructure, but rather a combat information management system. Thus, it is narrower in scope than the NCII

---

[4]Defense Information Systems Agency. 1998. *Defense Information Infrastructure Master Plan, Version 7.0*. Department of Defense, Washington, D.C., March 13.

[5]Information Technology Infrastructure Integrated Product Team. 1999. *Information Technology Infrastructure Architecture (ITIA), Version 1.0 Proposed*, 3 volumes (draft). Chief Information Officer, Department of the Navy, Washington, D.C., March 16.

[6]Defense Science Board Summer Task Force. 1998. *Joint Operations Superiority in the 21st Century: Integrating Capabilities Underwriting Joint Vision 2010 and Beyond*. Office of the Under Secretary of Defense for Acquisition and Technology, Washington, D.C., October.

[7]United States Air Force Scientific Advisory Board. 1998. *Report on Information Management to Support the Warrior*, SAB-TR-98-02. Department of the Air Force, Washington, D.C.

but develops the information management concepts in much more detail than does the NCII (or any of the other information infrastructures noted above).

A major point follows from the above discussion. More important than the differences in emphasis among the infrastructures is the fact that several different organizations looking at the problem from different perspectives have all come up with the need for articulating an information infrastructure to support warfighting operations across all echelons.[8] The point for the naval services is to articulate in their planning for network-centric operations a concept such as the NCII. The IT-21 strategy (discussed in Chapter 6) is an excellent start in this regard, although it is not complete. It focuses on connectivity, with less explicit recognition of the upper-layer services depicted in Figure 4.2.

A second important point, made in most of the information infrastructure discussions, is the need for jointness. There are two aspects to this. First, the information infrastructures do not operate in isolation. Rather, information should be shared readily across Service boundaries, as well as across the military-intelligence boundaries. Second, the different information infrastructures should not be developed separately. Given the common need of the Services and the joint community for these infrastructures, there should be cooperative efforts to develop them.

One opportunity for such collaboration has just arisen. The Air Force in its expeditionary force experiments (EFXs) is planning in 2000 to begin exploration and development of the battlespace infosphere and is seeking joint participation.[9] Naval participation, and in particular examination of the NCII concept, would seem to be highly beneficial to both the naval services and the Air Force. By participating, the naval services could benefit from the momentum and significant funding already inherent in the EFX series, and the joint nature of both the NCII and BI could be explored.

A third and final major point to recognize is that all the information infrastructures contain shared and common-use assets. For example, long-haul communications used in the NCII will be based in part on SATCOM assets shared with the other Services and joint community. Software in the NCII—e.g., to support the information gathering and generation functions shown in the upper part of Figure 4.2—will be developed in part for use across the Services and joint community. Some of this software might be unique to naval needs (e.g.,

---

[8]In addition to the information infrastructures noted in the text, mention should also be made of the newly emerging idea of the global information grid (GIG) being articulated by OASD (C3I) and the Joint Staff. Development of the GIG is currently focusing on policy matters, but it is likely that when a technical definition emerges it will in general be consistent with the NCII and the other infrastructures noted here. In fact, it could possibly benefit from some of the concepts being developed for the NCII.

[9]The BI has been renamed the joint BI (JBI), and the EFXs are now called joint EFXs (JEFXs).

information extraction pertaining to sea targets), but a significant portion of it would have more general utility.

Thus, the naval services (and similarly in the case of other joint or Service entities) will not "own" the NCII in the sense of a physical asset, just as no one body "owns" the Internet infrastructure. Rather, those charged with developing the NCII should have a concept of its overall capabilities and then see to what extent these capabilities are unique or, as would more generally be the case, shared or for common use. In this latter case, the NCII's developers would have to interact with the broader community to ensure that developments meet their needs.

It is not possible in this report to indicate which components of the NCII are naval-unique and thus would be developed by the naval services alone and which would be shared or common-use assets developed more broadly. However, the NCII functional architecture (Figure 4.2) does provide a general way for the naval services to proceed. It delineates the necessary functional capabilities, and for each of those capabilities the naval office(s) developing the NCII would address how the capability is realized through naval-unique development or collaborative development for shared or common-use assets.

4.2 TACTICAL NETWORKS[10]

The committee believes it is feasible and desirable for the NCII to include the Navy's tactical networks as well as its operational and strategic networks except in very rare special cases. Such an approach could offer the advantages of a uniform architecture and interfaces, resource management, and information assurance mechanisms and would be a great aid to interoperability. Clearly, a common architecture across the levels would greatly facilitate the rapid and widespread exchange of information that is central to network-centric operations.

A uniform architecture is not, however, the same thing as a seamless network. One concept for network-centric operations allows all data to flow seamlessly through any and all parts of the network. The committee does not agree with this concept. On the contrary, it believes, at least for the current state of technology, that the Navy's tactical networks should be built using a uniform architecture and mechanisms but then deliberately segmented to provide speed-of-service guarantees and some degree of information assurance. Gateways, firewalls, and encryption devices should be interposed between segments.

---

[10]More detail on this subject is contained in Appendix E, "Tactical Information Networks."

4.2.1 Tactical Network Protocols

Network protocols are a key issue in establishing a common architecture. At the operational and strategic level, commercial, Internet-based protocols are being widely used. The question is whether tactical networks must continue to use noncommercial protocols, as is now the case.

In its own architectural diagrams, the Navy and the Assistant Secretary of Defense (ASD) for C3I distinguish those portions of its networked infrastructure based on commercial technology—the Joint Planning Network (JPN)—from the tactical portions of its infrastructure, the Joint Data Network (JDN) and the Joint Composite Tracking Network (JCTN). They provide a rationale for this sharp division: Activities that use the JPN can tolerate delays in information flow, while those using the JDN have tight time constraints and those using the JCTN have very tight time constraints. The argument is then made that commercial technology is incapable of meeting these tight time constraints, and it is concluded that the JDN and JCTN must therefore be military-specific.

This rationale is in some ways sound but it also has a number of weaknesses—details are given in Appendix E. The committee concluded that, on the whole, the disadvantages of noncommercial protocols outweigh their advantages and that tactical information networks should be a uniform part of the NCII architecture.

The committee found two key deficiencies in the Navy's architectural vision for tactical networks:

- It merely renames two existing tactical systems (JDN is really JTIDS, and JCTN is really CEC) and ignores the rest. The architecture omits any reference to sensor system links, e.g., for moving-target imaging (MTIm) and synthetic aperture radar (SAR) data, or to weapons control systems, e.g., ultra-high frequency (UHF) satellite communications (SATCOM) target location updates for Tomahawks.

- It underestimates commercial technology. The committee believes current and emerging trends in commercial off-the-shelf (COTS) networking technology will allow tactical networks to be a uniform component of the NCII architecture without loss of capability. However, some unique tactical problems still will not be solved by COTS techniques and so must be approached by a blend of COTS and military technology. These problems are described below.

4.2.2 Straw-man Architecture

The committee's recommended "straw-man" architecture is a layered architecture with standardized interfaces. The standard network services furnish a broad set of services to the architecture; these include a standardized addressing and naming scheme and data transport using the Internet Protocol. The committee proposes that all new types of data transported between sensors, shooters, weapons, forces, and so forth be IP datagrams. This layering cleanly separates the type of data being transported from the type of radios employed, allowing great flexibility in the types of new systems that can be deployed with a single set of radios. Atop the common service layer ride various domain-specific applications. These applications would of course be quite varied in the tactical systems. As in the existing tactical systems, a straw-man architecture generally segregates the various subnetworks into separate radio channels. This allows the requisite low and bounded latency by ensuring that only specified classes of traffic can transit a given radio channel. And it helps with information assurance by segmenting the overall tactical network into compartments with strictly controlled interactions between them. Figure 4.3 illustrates this architecture, showing the tactical compartments and the points of controlled interaction. The latter also include interaction with operational and strategic parts of the NCII.

Segmentation points demarcate organizational as well as technical boundaries. This is an extremely important point and is perhaps best explained by analogy with the commercial telephony system. The public telephony system is

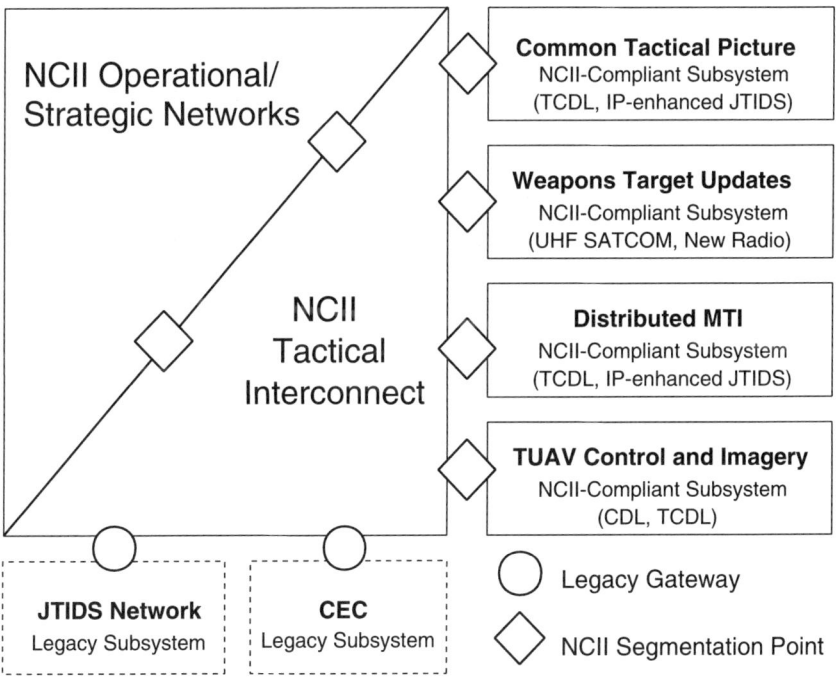

FIGURE 4.3 Straw-man tactical architecture for NCII-compliant and legacy subsystems. Acronyms are defined in Appendix H.

defined by a uniform set of standards with which all equipment must comply, but this does not imply that the local telephone company owns and operates all equipment within the telephone system. Instead, a business will typically own and operate its own internal telephone system (the PBX and office phones), adding or moving phones and cabling as necessary. The business, however, does connect its PBX to the local telephone company's service via standard interfaces. Similarly, the committee believes that tactical subsystems should be owned and managed by the appropriate operational subgroups, not a Navy-wide IT department. These subsystems should, however, be implemented in accordance with NCII standards and connect to the rest of the Navy's networks via NCII standard interfaces.

4.2.3 Challenges

The committee did not find any definitive, Navy-wide list of arguments against using NCII standards and interfaces in tactical networking systems. The challenges of the tactical environment are very real, however. Table 4.2 indicates the most difficult challenges for tactical systems along with the committee's comments on how they could best be approached within the NCII architecture.

TABLE 4.2 Unique Challenges for Distributed Systems in the Tactical Environment

Challenge	Explanation and Approach to Solution
Low delay	A set of distributed systems must perform a complete end-to-end action with a very stringent time budget.
	The major factors in end-to-end system delay are usually overall system design, humans in the loop, and, to a lesser extent, channel access and transmission speeds for the underlying radio channels. Overall system design, which is outside the scope of this study, is probably the most serious issue in practice. The committee's straw-man architecture segments the overall tactical network into subnetworks. This allows a direct mapping of the straw-man architecture onto the capabilities of the underlying radio channel and in turn enables the low-delay bounds that are required for many types of tactical data transport. Thus, use of the Naval Command and Information Infrastructure network services layer will not have much effect in delay in tactical networks.
High assurance	Many lives are at stake in tactical operations, including those of friendly forces and parties not explicitly targeted. In addition, collateral damage must be held to a minimum. Furthermore, the Navy might actually lose a battle should its tactical systems work poorly. Thus, tactical systems must perform with very high

TABLE 4.2 Continued

Challenge	Explanation and Approach to Solution
	reliability. They must be robust in the face both of enemy attempts to disrupt the systems and of collapse or malfunction due to overall system complexity and the chaos of battle. They must also survive enemy infiltration into the active systems and information warfare activities such as planting false information in various databases.
	Information assurance poses a very wide range of difficult problems. The Navy must not accept anything less than the current best-practice in these fields. In particular, it should adopt the Secret Internet Protocol Router Network (SIPRNET) model as an appropriate starting point for how to structure a mission-critical network. Firewalls and packet encryptors will aid in compartmentalizing the Navy's tactical network and ensuring that compromise or denial of one portion has the least possible effect on remaining portions. A number of problems in this area have no known technical solutions. The Navy's approach, therefore, must be to adopt operational methods to minimize the problems. In addition, the Navy should actively participate in, and fund, R&D programs in information assurance. (See Chapter 5 for a detailed discussion of information assurance.)
Low-bandwidth, intermittent connectivity	The various platforms within the battlespace must communicate via radios, which provide only low-bandwidth, often intermittent connectivity.
	These issues, while serious, are generally at the application level and hence with minor exceptions are not affected by adopting the common NCII architecture. The exceptions are transport-related, namely, that Internet Protocol (IP) headers may impose too much overhead for tactical radio channels, and that the Transmission Control Protocol (TCP) will not work well on channels with many dropped packets or highly variable delay. The committee recommends compressing IP headers by standard techniques and not employing TCP on such channels. (See Appendix E for details.)
Ad hoc, self-organizing systems	The group of platforms within a given battlespace is often assembled at short notice, with little or no prior planning for these particular platforms and systems to work together.
	Commercial technology currently has relatively little to offer for solving this problem; in general, most commercial distributed systems require a fair amount of painstaking configuration of host computers, routers, firewalls, application programs, and so on, and few embody the principle that one can throw together a number of distributed entities and simply have them work. (Apple's proprietary network services are a notable exception.) For the time being, at least, this will require manual or at best semiautomated configuration and so will probably continue to be troublesome in practice.

4.2.4 Findings

Based on the discussion of tactical networks above and in Appendix E, the committee arrived at two findings. First the committee found that tactical networks can advantageously conform to the NCII architecture, including the use of IP as a universal bearer. Segmentation, however, will help guarantee quality of service and information integrity.

The advantages of adopting commercial networking protocols include economy in deployment and upgrade and the robustness that results from their use by millions of information systems exchanging many forms of information. However, in this connection the committee also found that the use of standard protocols in tactical information networks creates technological challenges not now faced by other user communities. Meeting these challenges will require defense R&D in wireless networks that addresses such factors as network self-organization in highly variable and degraded environments.

4.3 ARCHITECTURAL GUIDANCE AND DEVELOPMENT PROCESSES

Architecture is defined as ". . . the structure or components, their relationships, and the principles and guidelines governing their design and evolution over time."[11] The NCII concept, as set forth in Section 4.1, is a high-level concept. An architecture providing general guidance for the NCII developers is necessary to implement this concept. It must be developed to ensure that the required functionality is incorporated, the appropriate systemwide properties realized, and the necessary interconnections enabled.

Development of an NCII architecture is an extensive undertaking, well beyond the scope of this report. The focus here is to assess how the current architectural guidance and development processes in the Department of Defense and the Department of the Navy relate to developing an NCII architecture. There are five organizations that develop architectures or architectural guidance relevant to the NCII: the OASD (C3I), the Defense Information Systems Agency (DISA), the Department of the Navy Chief Information Officer (CIO) and the recently established Chief Engineer (CHENG), and the Space and Naval Warfare Systems Command (SPAWAR). The roles of these organizations and their architectural products are discussed in Section 4.3.1, and the key issues in relating the products and processes of these organizations to an NCII architecture are discussed in Section 4.3.2. Section 4.3.1 is rather lengthy because much background material must be provided. Readers who are familiar with the architec-

---

[11]Department of Defense. 1999. *Joint Technical Architecture, Version 3.0*, Appendix F: Glossary. Washington, D.C., November 15. Available online at <http://www-jta.itsi.disa.mil/jta/jtav3-final-19991115/jta30_15nov99.pdf>.

tural products and processes discussed there might want to go directly to the section on issues (Section 4.3.2).

4.3.1 Existing Architectural Products and Processes

4.3.1.1 C4ISR Architecture Framework (OASD (C3I))

OASD (C3I) coordinated the development of the *C4ISR Architecture Framework* by a working group involving representatives of the Joint Staff, Services, and defense agencies. Version 2.0 of the document was released on December 18, 1997.[12] The motivation for development, described in the document, was the statement, "The Defense Science Board and other major studies have concluded that one of the key means for ensuring interoperable and cost effective military systems is to establish comprehensive architectural guidance for all of DOD." The framework is described at some length here because it forms the basis for most architecture development for large-scale information systems in DOD currently.

The overall nature of the framework is illustrated by the following quotes taken from the document:

> The Framework provides the rules, guidance, and product descriptions for developing and presenting architecture descriptions that ensure a common denominator for understanding, comparing, and integrating architectures . . . .

> The C4ISR Architecture Framework is intended to ensure that the architecture descriptions developed by the Commands, Services, and Agencies are interrelatable between and among each organization's operational, systems, and technical architecture views, and are comparable and integratable across Joint and combined organizational boundaries . . . .

> The Framework provides direction on how to *describe* architectures; the Framework does not provide guidance in how to *design or implement* a specific architecture or how to *develop and acquire* systems-of-systems . . . .

The framework then goes on to indicate that three major perspectives, or views, combine logically to describe an architecture—the operational, systems, and technical views. These views are defined as follows:

• The operational architecture view is a description of the tasks and activities, operational elements, and information flows required to accomplish or support a military operation;

---

[12]C4ISR Architecture Working Group. 1997. *C4ISR Architecture Framework, Version 2.0.* Department of Defense, Washington, D.C. Available online at <http://www.c3i.osd.mil/org/cio/i3/AWG_Digital_Library/pdfdocs/fw.pdf>.

- The systems architecture view is a description, including graphics, of systems and interconnections providing for, or supporting, warfighting functions; and
- The technical architecture view is the minimal set of rules governing the arrangement, interaction, and interdependence of system parts or elements, whose purpose is to ensure that a conforming system satisfies a specified set of requirements.

Briefly stated, the interrelationship between these views is that the operational architecture defines the information exchange requirements that the systems architecture must then support, with the systems architecture being developed in accordance with technical criteria specified in the technical architecture.

For each of the views, the framework prescribes a set of products that are to be used in realizing it. These architecture products are the graphical, textual, and tabular items that are developed in the course of building a given architecture description and that describe the characteristics pertinent to its purpose. When completed, this set of products is intended to constitute the architecture description. There are two categories of such products:

- *Essential products.* These products constitute the minimal set of products required to develop architectures that can be commonly understood and integrated within and across DOD organizational boundaries and between DOD and multinational elements. These products must be developed for all architectures.
- *Supporting products.* These products provide data that will be needed depending on the purpose and objectives of a specific architecture effort. Appropriate products from the supporting product set will be developed.

To be more specific, the set of essential products is as follows:

- *All views:* overview and summary information, integrated dictionary (definition of terms);
- *Operational view:* high-level operational concept graphic, operational node connectivity description (activities at each node and information flows between them), and operational information exchange matrix (includes attributes of exchanged information);
- *Systems view:* system interface description; and
- *Technical view:* technical architecture profile (extraction of standards that apply to the architecture).

In addition, there are 19 supporting products. The supporting products may be necessary as intermediate steps leading to the essential products.

The requirement for use of the *C4ISR Architecture Framework* was stated in a memorandum released on February 23, 1998, and signed by the Under Secretary of Defense (Acquisition and Technology), Acting Assistant Secretary of Defense (C3I), and the Director for C4 Systems, the Joint Staff. In particular, the memorandum states as follows:

> We see the C4ISR Architecture Framework as a critical element of the strategic direction in the Department, and accordingly direct that all on-going, and planned C4ISR or related architectures be developed in accordance with Version 2.0. Existing C4ISR architectures will be redescribed in accordance with the Framework during appropriate revision cycles. We also direct all addressees to examine the C4ISR Architecture Framework as a basis for a single architecture framework for all functional areas/domains within [the] Department.

4.3.1.2 Joint Technical Architecture (DISA)

The Department of Defense Joint Technical Architecture (DOD JTA) provides the technical architecture view applicable to all of DOD.[13] Development of the DOD JTA is coordinated by DISA under the direction of OASD (C3I) and involves representatives from across DOD as well as the intelligence community. Version 2.0 of the JTA was released on May 26, 1998. It gives the purposes of the JTA as follows:

- To provide the foundation for interoperability among all tactical, strategic, and combat support systems;
- To mandate interoperability standards and guidelines for system development and acquisition that will facilitate joint and coalition force operations. These standards are to be applied in concert with DOD standards reform;
- To communicate to industry the DOD's intent to consider open systems products and implementations; and
- To acknowledge the direction of industry's standards-based development.

In keeping with the last two items, the standards contained in the JTA are predominantly commercial. As a listing of standards, the document is not, strictly speaking, an architecture.

The standards are broken out into five categories: information processing; information transfer; information modeling, metadata, and information exchange; human-computer interface; and information systems security. Little rationale is given for the organization of the standards into these categories or

---

[13]Department of Defense. 1998. *Joint Technical Architecture, Version 2.0,* Appendix F: Glossary. Washington, D.C., May 26.

within the categories.[14] For each category, the set of mandated standards is given. Also listed are emerging standards, which are expected to be elevated to mandatory status when their implementations mature.

The JTA has now also added annexes organized by defense domains, and within those domains, by subdomains. These annexes give the mandated and emerging standards and associated descriptive material for the subdomains. There are four domains—C4ISR, weapon systems, modeling and simulation, and combat support—and 21 subdomains. At this time, only five of the subdomains have explicit entries. For example, there are six subdomains under C4ISR, but only one (airborne reconnaissance) has entries.

The requirement to use the JTA standards is indicated by the following extract from a memo issued on November 30, 1998, by the Under Secretary of Defense (Acquisition and Technology), the senior civilian official (OASD (C3I)), and the Director for C4 Systems, the Joint Staff:

> Implementation of JTA, that is the use of applicable JTA mandated standards, is required for all emerging, or changes to an existing capability that produces, uses, or exchanges information in any form electronically; crosses a functional or DoD Component boundary; and gives the warfighter or DoD decision maker an operational capability. Use of an applicable JTA mandated standard must consider the cost, schedule, or performance impacts, and if warranted a waiver from use granted as described below. . . . Each DoD Component and cognizant OSD authority is responsible for implementation to include compliance assurance, programming and budgeting of resources, and scheduling. Only the Component Acquisition Executive, or cognizant OSD authority can grant a wavier from the use of an applicable JTA mandated standard. All waivers shall be submitted to the USD(A&T) and ASD(C3I) (the DOD Chief Information Officer (CIO)) for concurrence . . . .

4.3.1.3 Chief Information Officer Architecture Products

Development of architecture and standards products by the Department of the Navy Chief Information Officer is in response to recent legislation, including the Clinger-Cohen Act of 1996 (Public Law 104-106), which requires department CIOs to "develop, maintain, and facilitate the implementation of a sound and integrated enterprise architecture and standards" (Section 5125 (b) (2)).[15] To provide a focus for these efforts, the Secretary of the Navy mandated

---

[14] Also, in providing its total set of standards, the JTA does not distinguish between those standards that are most relevant for interoperability and those that pertain most to constructing open systems. Interoperability pertains mainly to the standards between systems, while open systems considerations also involve the standards within systems.

[15] The Clinger-Cohen Act of 1996, formerly the Information Technology Management Reform Act and the Federal Acquisition Reform Act, requires the CIOs of the federal agencies to establish acquisition and management processes for information technology (National Defense Authorization Act for Fiscal Year 1996. U.S. Statutes at Large 110 (1996); 186).

that Department of the Navy CIO integrated products teams (IPTs) be the only Department of the Navy authorized entities to develop enterprise information management/information technology architectures and standards. These IPTs report to the Department of the Navy CIO Board of Representatives (BOR), which is chaired by the CIO and formed from representatives of Navy and Marine Corps operating forces, major claimants, and program sponsors.

Two products are being developed: the Department of the Navy Information Technology Infrastructure Architecture (ITIA) and the Department of the Navy Information Technology Standards Guidance (ITSG).[16] Volume I of the ITIA, *Network Infrastructure and Services Architecture*, and Volume II, *Enterprise Architecture Framework*, were approved by the BOR and released on March 16, 1999. Volume III, *Governance*, and Volume IV, *Requirements Process*, have not been released yet. The ITSG was released as version 99-1 on April 5, 1999, along with a cover memo from the Department of the Navy CIO.

Volume I of the ITIA is a systems architecture according to the definition in the *C4ISR Architecture Framework*. It provides significant detail and discussion of (1) a connectivity architecture based on modern network concepts and (2) the basic network services (e.g., domain name service (DNS), file transfer protocol (FTP), e-mail, public-key infrastructure (PKI), Web hosting, voice, multimedia). This volume should be useful in accomplishing its stated purpose:

> This document provides guidance for planning, developing, implementing, and operating all activities associated with DON IT [Information Technology] network infrastructure. It is to be used by DON acquisition programs, organizations, working groups, and Integrated Products Teams (IPTs) to facilitate convergence on a single, comprehensive ITI [Information Technology Infrastructure] architecture. This guidance and associated design templates are not intended to be detailed design and implementation plans, but to serve as frames of reference for design and implementation efforts.

Volume II of the ITIA was developed to provide an overall context for Department of the Navy enterprise architecture modeling efforts. It is based on the *C4ISR Architecture Framework*, extending the views contained therein. It also introduces a fourth view, the mission view, to identify strategic mission areas and priorities. At this time, Volume II remains a fairly high-level descriptive document.

The ITSG is a technical architecture according to the definitions of the *C4ISR Architecture Framework*.[17] By its own description, the ITSG is complementary to the JTA and provides additional guidance for applying the JTA. It

---

[16]Chief Information Officer Infrastructure Integrated Product Team. 1999. *Information Technology Standards Guidance, Version 99-1*. Department of the Navy, Washington, D.C., April 5.

[17]Although, correctly so, it avoids referring to itself as an "architecture," since it is formed around a compilation of standards.

notes that if there is a conflict in standards with the JTA, the JTA takes precedence. The ITSG devotes significantly more attention than the JTA to providing an overall structure in which to present the standards and to discussing the standards. In this regard, as well as in discussing aspects of the standards unique to naval use, the ITSG should provide a significant benefit to its users. The requirement for using the ITSG is indicated in the following extract from the DON CIO cover memorandum dated April 7, 1999:

> The ITSG applies to all DON systems that produce, use, or exchange information electronically, and is intended for anyone involved in the management, development, acquisition and operation of new or improved systems. It provides the standards, specifications, best practices and operating profiles required to implement and maintain an integrated, enterprise information infrastructure.
> ... Enterprise-wide use of these DON IT standards will enable coordinated communications across dispersed DON organizations.... All commands in the Navy and Marine Corps are required to consider the standards and guidance in the ITSG to maximize interoperability and enable focused information support across the Department. The ITSG Version 99-1 contains no mandatory requirements, and cannot be used as justification for less than full and open competitive acquisition.

It should be noted that the recently established Navy/Marine Corps intranet program has indicated it will use the ITSG.

4.3.1.4 Chief Engineer Responsibilities

The position of the Department of the Navy Chief Engineer was established and its first incumbent named on April 13, 1999. The responsibilities of that position are indicated by the following extracts from a memorandum issued by the Assistant Secretary of the Navy (ASN) (Research, Development and Acquisition (RDA)) on that date:

> 1. Effective immediately, the Chief Engineer will be the senior technical authority within the acquisition structure for the overall architecture, integration, and interoperability of current and future Combat, Weapons, and C4I Systems used by the Department of the Navy.
>
> 2. The position of Chief Engineer is not intended to dilute any of the traditional responsibility for individual program integrity currently assigned to Program Managers and Program Executive Officers. Rather, the Chief Engineer will be responsible for developing and implementing a process within ASN (RDA) which does not now exist: to assure that component systems are engineered and implemented to operate coherently with other systems as part of a larger force.
>
> 3. The Chief Engineer will be the technical authority for those functions necessary to satisfy this end. These include: (a) leading the functional design for Combat & C4I system functions with respect to the overall warfare architecture; (b) approval of system level interface specifications for all referenced

systems; (c) assessing and approving interface changes that impact interoperability, prior to fleet introduction; (d) assuring that individual programs adhere to the resulting configuration, and (e) recommending investment decision and program priorities to myself and the appropriate service chief concerning fielding systems in balance with their legacy and planned future counterparts.

The Chief Engineer's responsibilities will be exercised through the involvement of and close coordination with the affected PMs, PEOs, SYSCOMs, and where applicable, the Chief Technology Officer.

4. The primary immediate priorities of the Chief Engineer are the successful integration of the Cooperative Engagement Capability (CEC) system with the targeted weapon systems and tactical data links and the development of Navy Theater Ballistic Missile Defense (TBMD) systems.

The Chief Engineer office has existed only for a short period and currently has no staff other than a deputy. For this reason, no detailed architecture products have been produced yet by the office but may well be produced in the future. While activity is currently focused on combat systems interoperability, as noted in item (4) above, and is carried out in close coordination with the Naval Sea Systems Command (NAVSEA), the responsibilities of the Chief Engineer allow consideration of a much broader range of architectural topics. These topics could pertain to tactical operations in addition to theater air and ballistic missile defense, and to the strategic and operational levels of warfare.

4.3.1.5 SPAWAR Navy C4ISR Architectures

SPAWAR has prepared a number of Navy operational, systems, and technical architectures, some of which are drafts. The operational architectures pertain to C4ISR overall and to individual mission areas—air warfare, amphibious warfare, command and control warfare, mine warfare, strike warfare, surface warfare, and undersea warfare. They are lengthy expositions that largely describe the as-is situation, although some indication of potential future modifications is given. After beginning with some statements of the overall operational concept, the architectures go into a detailed listing of such items as the operational nodes that are involved, the tasks of each of those nodes, and the information exchanges among them. As such, the operational architectures tend to contain many detailed tables, with little intermediate-level expression between the high-level concepts and the detailed tables.

There are two Navy systems architectures, the as-is C4ISR systems architecture and the target (to-be) C4ISR systems architecture. The target architecture presents a detailed methodology for developing all the architecture products (see under "C4ISR Architecture Framework," Section 4.3.1.1) and relating them to each other. Each of these products, which are typically detailed items, is then developed. A key aspect of the methodology is its emphasis on system functions, in contrast to the perspective taken in the *C4ISR Architecture Framework*. The systems architecture view given in the framework focuses on the physical implementation of systems, as indicated by its definition of systems architecture as "a description, including graphics, of systems and interconnections" and by the fact that the only essential (required) architecture product is the system interface description. This distinction is explicitly noted in the following excerpt from the SPAWAR *Naval C4ISR Architecture Primer:*[18]

> . . . in transitioning from an operational to a system architecture the Navy approach and the CISA approach are different [CISA is the OASD(C3I) organization that developed the C4ISR Architecture Framework]. Essentially, the Navy's approach includes two phases: a functional analysis and a physical analysis. The CISA approach jumps immediately to the physical analysis . . . . The Navy believes that jumping immediately to the physical analysis is appropriate for "As-Is" architectures, but developing "To-Be" architectures requires a more thorough understanding of systems functions, and therefore the Navy's approach precedes the physical analysis by a functional analysis. As technology and operational requirements change, the Navy approach would make operational, systems, and technical architectures less costly to implement.

In short, articulation of the system functions is a critical intermediate step in moving from the operational view to the physical implementation of the C4ISR system.

The Navy C4ISR Technical Architecture, currently released as Version 2.0, pertains to the inter-platform and intra-platform C4ISR interfaces necessary to support Navy missions and their subordinate functionality and performance parameters. It also pertains to C4ISR interfaces with other Navy systems, such as weapons, sensors, and combat support. This technical architecture contains a subset of the standards in the JTA and additional standards unique to Navy missions and noncompeting with the JTA. It is patterned after the JTA and as such is largely a listing of standards. It divides standards into the same categories as the JTA (see Section 4.3.1.2 for the five categories), except that it also adds the category "intelligence, surveillance, and reconnaissance." In addition, it contains a set of appendixes listing the standards unique to Navy mission areas. A separately published appendix, "Migration Strategy," provides information to aid in the migration from current standards to target standards. The last release of this volume, Version 1.5 dated July 16, 1998, refers to target standards for the year 2000.

---

[18]Deputy Chief Engineer for Architecture and Standards. 1997. *Naval C4ISR Architecture Primer,* final draft, SPAWAR 051-2. Space and Electronic Warfare Systems Command, Washington, D.C., January 17, pp. 4-18.

4.3.2 Application to the NCII

When the various DOD and Navy architectural products and processes described above are related to the development of an NCII architecture, several shortcomings in the current products and processes are evident.

4.3.2.1 Operational and Systems Architectures

The discussion in Section 4.1.2 emphasizes that information needs cannot be regarded as fixed. Users will vary in their needs, will evolve with respect to the type of information they want as they explore and use what is available, and will want to tailor the way information is provided and presented to them. Furthermore, the ways in which new information technology can be applied are often not immediately apparent. One has to work with new technologies to explore their uses, and these uses will suggest additional technologies to be explored, which will lead to further new uses. This continual interplay between the exploration of technology and the evolution of operational concepts will lead to ongoing changes in the way information is used. All this requires a rapid, iterative process of technology exploration and refinement of operational concepts (a theme developed in detail in Chapter 2).

This requirement for flexibility and rapid iteration does not seem well supported by the detailed methodology of the *C4ISR Architecture Framework*. Three points are particularly relevant:

1. The number and detailed extent of the C4ISR architecture products means they take significant time to develop. This is inconsistent with the rapid iteration necessary in introducing technology and refining operational concepts.

2. The rapid iterative process requires the close and frequent interaction of system users and developers. If the operational architecture is developed through to its end and the systems architecture is then begun, which seems to be implied by the *C4ISR Architecture Framework* methodology, then this interaction cannot take place.

3. Articulation of the system functions is a key intermediate step in moving between operational concept and system realization. As noted in the discussion of systems architectures above, system functions do not play a prominent role in the *C4ISR Architecture Framework* methodology.

The *C4ISR Architecture Framework* would seem to have utility in the development of well-understood systems for which the requirements can be laid out in detail in advance of the development of the system. To accommodate the more flexible, iterative process necessary for constructing the NCII, the following significant changes to the framework methodology appear necessary:

- New high-level products must be developed to describe the operational and systems architectures, intermediate in detail between the current operational concepts given on a single chart and the detailed tabular information now produced in operational and systems architectures. Since such products could be modified relatively quickly, they would support a rapid iterative process and allow the close interaction of operational and systems personnel.[19]

- An architectural product must be established to indicate the concept of operations for using information. The current examples of operational architecture all relate to how information supports traditional warfighting tasks, which is obviously required. But there is also a need to consider how the information is treated. For example, is it obtained by push or pull? How does the commander want to promote or limit its distribution? Critical factors such as these need to be recorded for systems to be developed and used properly. For best use, these would be high-level (five-page summary) products. The best course might be to have the operational architecture provide only a general template for this type, with the particular details inserted by individual commanders and their staffs.

- Greater prominence must be assigned to the role of system functions in the system architecture products. As noted, the system functions are central in the relationship between concepts of operation and system realization. Such a system architecture product could be made essential (i.e., required in all system architecture developments). In addition, an intermediate level of detail would serve the flexibility needed in a rapid, iterative development process.

4.3.2.2 Technical Architecture

A technical architecture, as envisioned in the *C4ISR Architecture Framework*, serves important purposes: It promotes the use of commercial products, which can mean lower costs and faster technology refresh cycles; it leads to open-architecture designs, which facilitate interoperability; and it aids modular design practices, which can make technology upgrades easier. Standards typically change on longer time scales than do the individual technologies, so developing architecture products that can be modified quickly is less critical for technical architectures.

However, an important concern in technical architecture products is that the set of required standards be kept as small as possible, for two different reasons: to avoid unduly constraining system developers by mandating standardization where it is unnecessary or premature to do so, and to limit the set of choices for

---

[19]Development of an automated tool to facilitate construction of operational architectures is discussed in Chapter 6, Section 6.2.7.3.

a given standards area so that different developers do not unnecessarily choose different standards, which can adversely affect interoperability. The JTA claims to be a minimal set of standards, and both the Department of the Navy ITSG and the SPAWAR Technical Architecture claim to further refine the standards selection to suit the Navy's needs. However, the limited analysis possible for this report did not allow determining if these three technical architectures were appropriately minimal.

Note that most of the standards in the JTA, or refinements of the JTA for naval purposes, would be of two types: commercial standards (e.g., for network services) or DOD-wide standards (e.g., for specific application domains such as intelligence or logistics). Few, if any, of the standards would be naval unique. Thus, the naval offices responsible for determining the standards would, in general, have to choose from broader sets of standards and not develop their own standards. Relatedly, these offices should interact with the broader communities developing the standards to ensure that naval needs are met.

4.3.2.3 Beyond Current Standards-based Architectures

The systems architectures discussed thus far above would all be based on the interfaces, services, and accompanying standards specified in the technical architectures (JTA, ITSG, and the Navy C4ISR Technical Architecture). That is the current state of practice, and much can be said for it. However, there still are shortcomings. For example, given the pace at which technology is advancing, it is not possible to impose common standards on a wide community (consider, as a possible extreme, a community of U.S. forces and coalition forces). Thus, limitations could occur when elements of this wide community need to interoperate with one another. In addition, even if common standards from the JTA (or a similar source) are used, the definitional consistency of the data that are exchanged must also be considered. Again, for practical reasons, it is not possible to impose common data definitions across wide communities. Furthermore, account must be taken of the fact that excessive imposition of standards will limit innovation.

Thus, one needs to look for advances in technology or in design practice that will lead beyond the current technical architectures to address problems such as those just noted. The JTA and similar documents do have sections on emerging standards and are thus anticipating the future to a limited degree. But it is not the function of these technical architectures to be a vehicle for tracking and promoting revolutionary, but potentially highly beneficial, technologies.

Such technologies are now being pursued, including the following:

- *Semantic interoperability.* Research in this area is aimed at achieving means for common semantic understanding across components developed with different data representations. One example of research in this area was carried

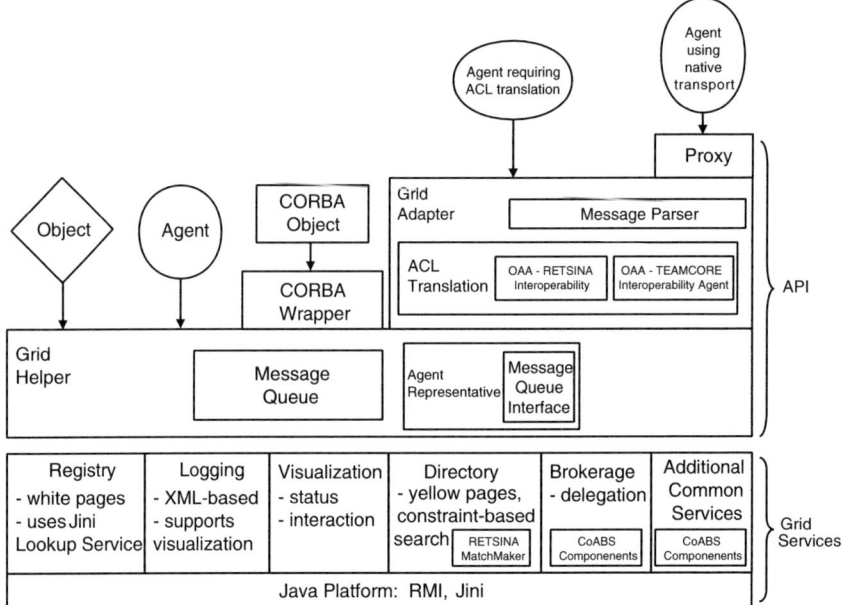

FIGURE 4.4 Schematic of the emerging Control of Agent-based Systems grid architecture, which minimizes integration effort to connect to the grid by adapting the connection mechanisms instead of the client components. SOURCE: Hendler, James, "Control of Agent-based Systems Technical Overview," a briefing presented to the System Architecture Panel on April 15, 1999, Information Systems Office, DARPA, Arlington, Va. Acronyms: ACL, agent communication language; API, application program(ming) interface; CORBA, common object request broker architecture; OAA, over-the-air activation signal; RMI, remote method invocation; XML, Extensible Markup Language.

out in conjunction with the Defense Advanced Research Projects Agency (DARPA) Dynamic Multiuser Information Fusion (DMIF) program and is now being continued under the Air Force Office of Scientific Research (AFOSR).[20]

• *Agents.* The DARPA Control of Agent-Based Systems (CoABS) program is exploring the use of agents as an interoperability mechanism. Interfaces might be more flexibly defined if the agents could negotiate interactions between system components. Currently under development in the program is a metaframework or grid (see Figure 4.4) that will allow agents operating under different agent communities or interagent languages to communicate. In addition, an effort is about to begin to establish a new agent language intended to

---

[20]Krikeles, B., and T. Libby. 1999. *Achieving Information Superiority: Interoperable Components and Battlespace Representation.* ALPHATECH, Inc., Burlington, Mass., March 27.

progress well beyond current Web languages (HTML, XML) that will provide readable (interoperable) semantics, given a developed ontology.[21]

- *Publishing internal properties.* In this instance the components of a system would make known to the broader system certain aspects of their internal composition. This would permit greater flexibility and tailoring in establishing interfaces with those components. One example of this general idea is the terminal access packet discussed in Chapter 6 (Section 6.2.3.3).

While such work is ongoing in the research community, it needs greater recognition and support at senior Navy Department and DOD levels. Officials at those levels should not believe that the DOD and naval technical architectures as they now exist provide a complete solution to the technology specification for architectures. The capabilities envisioned for the NCII cannot be fully realized with current technical architectures. The necessary flexibility and adaptability will require advances in architecture-related technologies such as those noted above.

4.3.2.4 Organizational Responsibilities

According to the discussions above, three naval organizations are actively involved in architecture and thus could be involved in development of an NCII architecture—the Department of the Navy CIO, Department of the Navy CHENG, and SPAWAR.[22] It is thus important to understand what might be the boundaries of responsibility that each organization could have for the NCII architecture. The Department of the Navy CIO would appear to have a significant responsibility given the enterprise-infrastructure architect role assigned it by the Clinger-Cohen legislation. However, the Department of the Navy CHENG would also seem to have significant responsibility according to its charter. And SPAWAR has traditionally been involved with the development of C4ISR architectures. Perhaps the Department of the Navy CIO should be responsible for the general infrastructure aspects, while the Department of the Navy CHENG should oversee the C4ISR specific aspects and interfaces to weapons and other systems, and SPAWAR should develop the detailed C4ISR aspects.

The responsibilities of the Department of the Navy CIO and Department of the Navy CHENG would appear to lie in the systems and technical architecture

---

[21]Note is also made of the Foundation for Intelligent Physical Agents, which is trying to promote open specification of agent systems to maximize their interoperability. Information is available online at <www.fipa.org>.

[22]The combat systems that NAVSEA and NAVAIR develop are outside the scope of the NCII, but since they must interface with the NCII, broader considerations would include NAVSEA and NAVAIR.

areas. Of the three organizations, SPAWAR is the only one that has developed operational architectures, although these are primarily as-is architectures. Organizations from the doctrine and experimentation communities might be appropriate for developing the future operational architectures.

Organizational responsibility for the NCII architecture must be assigned before the architecture can be properly developed and implemented. The matter of organizational responsibility for network-centric operations as a whole, including NCII development and operation, is discussed in detail in Chapter 7. One suggested scheme for assigning responsibilities for the NCII in keeping with the discussion above and consistent with the material in Chapter 7, could be as follows:

- *Resource allocation and requirements sponsor:* OPNAV N6;
- *Operational architecture:* Commander, Operations Information and Space Command, with the support of OPNAV N6;
- *Policy and standards:* Department of the Navy Chief Information Officer;
- *Systems and technical architectures (including enforcement):* Department of the Navy Chief Engineer;
- *Acquisition and procurement:* Program management as designated by the ASN (RDA) (e.g., the PEO-IT); and
- *Operational management:* Commander, Operations Information and Space Command.

Included in this list is a new position introduced and recommended in Chapter 7—the Commander, Operations Information and Space Command. This individual would be a functional type commander analogous to existing type commanders (e.g., for air assets) and would be concerned with the information assets of the fleet just as the other type commanders are concerned with the assets in their areas. SPAWAR, NAVSEA, and NAVAIR would also figure in the assignment of responsibilities by providing support to the areas listed.[23]

4.4 RECOMMENDATIONS

The recommendations presented here are organized according to the three main topics discussed above: establishment of the NCII, tactical networks, and architecture development.

---

[23]Chapter 7 also recommends double-hatting a Navy SYSCOM commander as a deputy to the ASN (RDA) for integration of network-centric operations. This individual would coordinate in the acquisition and procurement decisions.

4.4.1 Establishment of the NCII

The NCII offers a comprehensive, unifying concept. In broad terms, the committee recommends that the naval services adopt and apply the NCII as the overarching concept for providing information support to network-centric operations. In more specific terms, it makes three recommendations:

Recommendation: The Department of the Navy should develop and enforce a uniform NCII architecture across the strategic, operational, and tactical levels of naval forces.

This means that, for all levels, (1) the same set of functions will apply (as defined in Figure 4.2),[24] (2) interfaces and standards associated with these functions will be the same, and (3) consistent definitions will be used for the data exchanged between the functions. Such an architecture would integrate the system capabilities and facilitate the interoperation of forces. As such, it would promote the widespread and flexible exchange of information necessary for network-centric operations.

Recommendation: The Department of the Navy should give an organization the responsibility and adequate resources for (1) establishing a comprehensive view of capabilities and programs necessary to implement the NCII and (2) seeing that these capabilities are realized.

During its information-gathering efforts, the study committee found numerous naval and joint organizations that were able to provide valuable information on various components of information infrastructures. Yet the committee was continually struck by the fact that no one office or organization had a comprehensive view of the capabilities and programs that would constitute an infrastructure such as the NCII. No organization was found, for example, that had an end-to-end systems view of information-handling capabilities, such as that shown in Figure 4.2.[25] Furthermore, many offices provided valuable information on programs relating to individual functional capabilities, but none had an overview of a full set of relevant programs.

Effective realization of the NCII requires that some organization have an overview of its requirements and the programs satisfying those requirements.

---

[24]The tactical domain will, in addition, have its own unique functions that are particular to warfighting mission areas. These are considered in Chapter 3.

[25]There are excellent end-to-end views of communications and networking capabilities, but as Figure 4.2 makes clear, there is also a significant information-handling component that rides on top of the communications and networking.

The set of programs is large and comes from diverse sources (Chapter 6 briefly discusses some of these programs). The organization would not control the programs as does a traditional program manager. Indeed, in modern large-scale information systems, no one controls all the pieces; rather, the managing organization understands what pieces are available and how they fit together. Accordingly, the desired organization would identify the relevant developmental activities (both government and commercial) that could support the infrastructure, track their evolution and progress, and indicate which should be applied in implementing the infrastructure. Furthermore, this organization would identify requirements that are not being met through the activities and establish priorities for meeting these shortfalls. Gaining such an overview is a substantial undertaking, so this organization would require significant resources in terms of both staff and funds.

Recommendation: The NCII should be developed in collaboration with the other Services, the joint community, and National agencies to promote interoperability and build on each other's efforts.

The NCII is presented here as a naval concept, in keeping with the study's purpose, which is to examine naval network-centric operations; nonetheless, much of what is discussed is of interest and use to joint and other Service operations. In fact, the other Services are developing analogous concepts, and many of the supporting programs come from the joint community (e.g., DARPA and DISA), as well as the intelligence community. Thus, to promote interoperability and avoid unnecessary duplication, the naval organization responsible for implementing the NCII should collaborate with the broader community. As noted in Section 4.1.5, one potentially valuable, immediate opportunity would be participation with the Air Force in its expeditionary force experiments.

4.4.2 Tactical Networks

The committee makes two recommendations pertaining to a transition strategy in particular to ensure that tactical information networks conform to the NCII architecture.

Recommendation: With few, if any, exceptions, new tactical information networks should conform strictly to the NCII goal architecture and should use appropriate gateways, firewalls, and encryption devices to ensure high quality of service.

The term "strictly" is used because at any moment there may be within the NCII legacy systems that do not conform to the goal architecture but that have been grandfathered into an interim standard. New systems must not be allowed to perpetuate the characteristics of the nonconforming systems.

Although the committee recognizes that engineering some difficult future communications links may lead to new waveforms and antenna control schemes for transport, it thinks it may be possible and desirable to implement a standard IP bearer in most or all network terminals and to separate that protocol from message content. In some cases the Navy or the DOD has already thought seriously about this possibility. Summarizing the conclusions of Appendix E on this matter, the committee recommends as follows:

Recommendation: Terminals of the Joint Tactical Information Distribution System (JTIDS) and common data link (CDL) families should be modified to use NCII standard protocols. The pros and cons of so modifying the CEC data distribution system should be studied further.

In addition, a further recommendation pertains to necessary research and development.

Recommendation: The DOD, including the Navy, should sponsor a vigorous research and development program aimed at improving the performance of wireless information networks that can self-organize with high assurance despite limited and highly variable connectivity between pairs of nodes and despite likely loss or degradation of some nodes in the network.

4.4.3 Architecture Development

Current architectural guidance and development processes will have to be significantly modified and enhanced to support the development of the NCII. To that end, the committee recommends the following:

Recommendation: The Department of the Navy should work with the ASD (C3I) and the other Services to make the operational and systems architecture products specified in the *C4ISR Architecture Framework* suitable for the flexible and rapidly evolving information support that the NCII must provide.

The discussion in Section 4.3.2.1 indicates some of the additions and changes to current architecture products that should be made.

Recommendation: The Department of the Navy should ensure that the naval technical architectures are minimal necessary sets of required standards.

Recommendation: The Department of the Navy should support efforts to advance beyond standards-based architectures (such as the current JTA).

More advanced architectural concepts are needed to realize the flexible, rapidly configurable information support envisioned with the NCII. Examples of research in this area are noted above in Section 4.3.2.3. The important point is that senior Navy and DOD officials recognize and support the need for such research.

Recommendation: The Department of the Navy, in developing an NCII architecture, should clarify the architectural responsibilities across the various naval offices currently involved in architecture development.

Responsibilities must be clearly delineated. A suggested assignment of responsibilities is given in Section 4.3.2.4.

5

Information Assurance—Securing the Naval Command and Information Infrastructure

The Naval Command and Information Infrastructure (NCII), as a highly networked system, can be vulnerable to attacks against its communications and computing elements. These vulnerabilities would pose numerous risks for network-centric operations (NCO). This chapter discusses those vulnerabilities and possible approaches to minimizing the associated risk. While the risks are significant, the committee believes they are outweighed by the benefits in operational effectiveness to be gained from NCO. However, this does not mean that the risks can be ignored. Vigilance is required on the part of system designers, implementers, managers, and users to anticipate security vulnerabilities and to address them by technical or procedural means. Constant awareness that portions of the system may be compromised will help warfighters react appropriately to situations. Backup plans should be developed for the most likely compromise scenarios, and warfighters should be trained in these procedures.

This chapter briefly sketches the magnitude of the security problem in today's systems; discusses the defense-in-depth strategy of prevention, detection, and tolerance; then, describes and assesses what the Department of the Navy is doing today for information assurance; and finally, identifies needed research and discusses some promising research programs that may produce needed technology.

5.1 INTRODUCTION

The Defense Information Systems Agency (DISA) estimates that there are 250,000 attacks on Department of Defense (DOD) computer systems every year.

Computer attacks against U.S. systems were up 22 percent from 1996 to 1997, according to a survey by the Computer Security Institute and the FBI. The most recent Computer Security Institute/Federal Bureau of Investigation survey, published in March 1999, confirms this trend.[1] The 1999 report notes that denial-of-service attacks were reported by 32 percent of survey respondents, sabotage of data or networks was reported by 19 percent, and virus contamination was reported by 90 percent. Such attacks can be considered as the ordinary background activity that must be dealt with day to day. Some of this activity, when directed against DOD systems, might include information warfare actions to "prepare the battlefield" in the event of a need to interfere with U.S. activity in some future engagement. This is certainly of concern. Of even greater concern, perhaps, is the fact that the United States can expect targeted attacks on DOD systems to increase during hostilities. Both the threat and U.S. vulnerability can be expected to increase, especially as a result of our increased reliance on the technology that network-centric warfare represents. Vulnerability is increasing along with the increasing connectivity among military systems and between military and civilian networks. Thus, vulnerabilities in the networking technology or in any connected system can be exploited by anyone anywhere to penetrate and corrupt DOD systems.

Another source of vulnerability is the increased reliance on commercial products. Commercial security is neither designed nor intended to withstand information warfare attacks, and a large number of exploitable flaws in commonly used products are known to a wide community. Furthermore, the increased homogeneity that results from the nature of today's commercial computer system marketplace leaves DOD open to attacks that can quickly affect a large percentage of its operations. DOD also depends on vulnerable commercial infrastructures such as telephone networks that, although highly reliable, were not designed to withstand information warfare attack. In addition, since the fleet's operational networks and the naval force business networks will of necessity be interconnected, the shore establishment will provide many attractive opportunities for penetration and disruption that can extend to the fleets and even their tactical networks, as well as their essential shore support.

5.2 THREATS TO THE NAVAL COMMAND AND INFORMATION INFRASTRUCTURE

The United States can count upon its adversaries to search for ways to disrupt the NCII. An adversary may be able to perform analysis (such as traffic

---

[1] Rapalus, Patrice. 1999. *Issues and Trends: 1999 CSI/FBI Computer Crime and Security Survey.* Computer Security Institute, San Francisco, Calif. Available online at <http://www.gocsi.com/prelea990301.htm>.

analysis) to identify critical nodes and bottlenecks and may develop attacks on these points. Individual elements attacked to gain access or produce an effect may include links, nodes, people, software, and hardware. Because of the numerous connections, both sanctioned and unsanctioned, with the public Internet that are likely to exist within the NCII, penetration of even a low-level network may permit a skilled information warfare attacker to gain access to far more critical systems.

Because of the drawdown in physical assets and forces, an adversary can choose attacks that have magnifying effects, thus significantly degrading the ability of naval forces to conduct operations. For example, in a battle group, which now often consists of a nuclear-powered aircraft carrier (CVN), Aegis ships, and a nuclear-powered attack submarine (SSN), significantly reducing the communications or computing ability of even a single platform could severely impede operations. Marine Corps plans to project forces directly to objectives without building up ground infrastructure are likewise vulnerable to asymmetrical attacks by adversaries. Thus, adversaries who have no traditional military to engage U.S. forces with any hope of success may nevertheless reasonably expect that information attacks will succeed with little risk. The NCII will be an attractive target because naval forces and the success of their operations will depend on the continued correct functioning of the NCII. Such attacks could be on the NCII alone or could be part of an overall military plan of attack against U.S. forces that also includes traditional physical force.

In the future, naval forces will increasingly be faced with unconventional threats, which could include international criminal enterprises, terrorists, and sometimes also nongovernmental organizations (NGOs). These potential adversaries can rapidly and cheaply obtain IT-based capabilities as a consequence of the globalization of communication, information, and Internet technologies. Expertise in developing and using these technologies is cheap and available worldwide, which is evidenced by the large number of foreigners employed as technical system developers by the U.S. software industry. Even an economically disadvantaged state or nonstate organization can hire criminal elements or disaffected nongovernment members to complement and extend its own ability to attack the NCII. A near-peer power might aid and encourage rogue states or factions or terrorist groups in penetrating or disrupting the NCII.

Enemy ability to penetrate, exploit, and disrupt the NCII could be facilitated by insider support. A malicious insider could, alone or working with outside adversaries, seriously disrupt NCO. Nearly everyone in the naval forces may have access to the NCII, as may interoperable joint peers. As the number of people with access to the NCII grows, it is more likely to include individuals with a desire or motive to cause mischief or engage in sabotage, or who are susceptible to being co-opted by an adversary. Insiders with access to key systems or databases, such as system or security administrators, will be attractive targets for recruiting. One way to minimize this risk somewhat is to reduce the scope of access and control available to any single individual and to require two- (or more) person control of key functions.

5.3 VULNERABILITIES OF THE NAVAL COMMAND AND INFORMATION INFRASTRUCTURE

5.3.1 Use of Commercial Products

The NCII, including its protection functions, will be built largely from commercial software and hardware computing and networking components. These commercial products contain numerous security vulnerabilities, which, as they are discovered, are routinely posted to frequently accessed Web sites (e.g., bugtraq). Attacks are developed against many of these vulnerabilities, and software tools to carry out the attacks are posted to hacker Web sites.[2] Commercial security products are not built to withstand the strength of attack that can be expected for military systems but to provide protection appropriate for business operations. Known vulnerabilities in these security products, as well as attacks exploiting them, are also posted on the Web. Vendors may respond by issuing patches (which may take weeks) or correcting the problems in scheduled new product releases (which can take months), resulting in a period of exposure during which procedural workarounds must be employed to reduce risk as far as possible. Many system operators may not be aware of the vulnerabilities that have been discovered in the products they are using or of the availability of procedural workarounds or patches.

The high rate of release of new products and product upgrades means that at any given time there will be no common software configuration across the NCII. With each new product and product release comes the need to keep up to date on product vulnerabilities and fixes. In addition, policy must be generated about acceptable and safe product configurations, and these configurations must be monitored and enforced across the NCII, because failure to do so would result in unnecessary exposure to vulnerabilities.

Additionally, because so much commercial software, including that from the well-known vendors and manufacturers, is produced overseas or domestically using overseas or green-card labor, it is possible for an adversary to plant or co-opt people in product development positions and have them attempt to include malicious triggerable code in commercial products that will be used in the United States and by the DOD. Such hidden features can easily go undetected by the vendor.

---

[2]See, for example, <http://www.hackershomepage.com/index.html> and <http://www.hackcity.de/programme.shtml>.

5.3.2 Reliance on Unclassified or Low-classification Information for Sensitive Functions

Sensitive NCII functions may rely on unclassified open source information. Such information sources are vulnerable to tampering and insertion of bad information by malicious entities *before* the information enters the NCII.

5.3.3 Outsourcing and Contract Personnel

Outsourcing of certain functions and the resulting introduction of contract personnel who require access to the NCII increase the possibility of introducing individuals who can cause damage or collaborate with hostile outsiders.

5.3.4 Joint and Coalition Member Access

To carry out joint and coalition operations, it may be necessary to give NCII access to joint and coalition personnel. This increases the likelihood of having insiders with motivation to cause damage. This population may also have a much poorer understanding of security, thereby decreasing general security awareness and vigilance among the user and operators of the NCII.

5.3.5 Connectivity to Public Networks

The NCII will have connections to the Internet and the Web to gain access to useful information, such as weather, environmental, news, and personal and recreational information. Attacks on these public databases may hinder the NCII. This connectivity also exposes the NCII to viruses and other information warfare weapons in data and code that enter the NCII. Also, NCII users might download arbitrary code, which could be infected with viruses or worms that could spread and cause damage within the NCII. It is also possible for an adversary to disguise hostile code, such as viruses, in attractive, free, software that NCII users may be tempted to download from the Web, thereby compromising the NCII.

Another risk of connecting to public networks is the increased use of mobile code, for example to implement so-called intelligent agents. With mobile code, users may be importing code into the NCII without being aware of it. While vendors have been adding security capability into the tools and languages commonly used to build mobile code (e.g., Java), such protections are not commonly in use on the Web.

Connectivity to public networks may also allow adversaries to observe the Department of the Navy's activity on the public networks and infer information about the Navy Department's operations and plans. To the extent that the NCII uses public networks to convey encrypted classified information, an adversary will be able to perform traffic analysis and infer useful information, including information that will help it to understand useful targets for denial-of-service attacks. In the DOD Eligible Receiver exercise in 1997, red team attackers penetrated sensitive networks by first attacking less sensitive networks to which they were connected.

5.3.6 Homogeneous Technology

Market forces and consolidation within the computer industry have meant that a few brands of software and hardware are ubiquitous. The NCII will also be largely homogeneous, with common products in use everywhere. Such homogeneity leads to widespread common vulnerabilities that can be exploited by common attacks. Large-scale networks of such systems are particularly vulnerable to virus attacks that can spread rapidly, because every system is vulnerable to the same virus. The devastating consequences of disease that are possible with homogeneous populations have been long recognized by the agricultural industry, which uses the strategy of crop diversity to limit the spread of disease and its consequences. For information networks, the availability of diverse implementations of common protocols and standards could help provide robustness, although such availability is not expected anytime soon.

5.3.7 Vulnerabilities of Tactical Networks

Tactical networks have particular vulnerabilities in addition to those they share with conventional wired networks. Tactical networks are subject to spoofing, jamming, and interception through the air. An adversary can launch a spoofing attack by attempting to introduce false information into a tactical network through false radio transmission. Through-the-air transmissions are vulnerable to jamming by suitably located and directed enemy transmissions. Because tactical transmissions are through the air, they may be subject to interception with greater ease and at a greater distance than those carried over wired networks. An adversary who intercepts U.S. radio signals can attempt to gain an advantage in several ways. It can try to gain intelligence about U.S. forces' status and intentions by reading the data; it can make inferences about present and future activities by noting the source, destination, and volume of radio communications (that is, by performing traffic analysis); and it can geolocate the transmitting platform.

In addition, because tactical networks may be within reach of enemy forces, end instruments are subject to terminal capture. Enemy capture of a network node means that the enemy is inside a naval network. If this seems a remote possibility, it should be remembered that the naval tactical networks will include Marine Corps ground networks and will be closely linked into those for the Army. A tactical node can be overrun as a result of an action as simple as the capture of a single wheeled vehicle. Enemy capture of a functioning network node could take some time to notice and respond to, during which time a great deal of damage could be inflicted, some of which might last far longer than the node itself. For instance, an enemy could spoof the common tactical picture, adding fictitious elements to it, and could also engage in various types of network denial-of-service attacks.

Another problem would arise if the United States were unwilling to share its cryptographic apparatus with coalition partners. Either it would not fully benefit from their data or it would risk the introduction of corrupted data.

5.3.8 Interconnection of Networks of Different Classifications

The NCII will require the interconnection of networks of different classifications, so that information contained in low networks is available to high networks and also so that appropriately sanitized high data can flow to low networks. There is a risk that unless extreme care is taken in the design and implementation of the boundary controllers that connect such networks, high information could leak into low networks. If there is a hostile insider or hostile code on a high network collaborating with an entity on a low network, high information could be sent covertly using steganographic means. There are no means of detecting such an information flow. Man-in-the-loop security release stations are useless against such a covert flow but pose their own risks, since approving information for release is a tedious task and the operator can routinely and unthinkingly approve the release of information that should not be released. In addition, there is a risk that low code and data that enter a high network can be maliciously tampered with in the low network to corrupt high databases or to introduce malicious code into high networks.

5.3.9 Interference with Critical Functions

The indiscriminate interconnection of strategic and tactical information networks with mission-critical networks (e.g., those used for air defense) can have undesirable consequences. First, such interconnection exposes these critical functions to tampering from a large interconnected population. Second, the bandwidth and computing resources for those critical functions may not be available when needed owing to competition from other users and applications. And third, unanticipated interactions between the interconnected networks may result in the failure of critical functions; these interactions can be particularly difficult to diagnose and correct.

5.4 DEFENSE IN DEPTH

Experience has shown that many successful attacks on DOD systems are not detected. In these attacks, an intruder may make surreptitious use of a penetrated system; may silently steal data or gather intelligence; may plant malicious code, perhaps for future use; may alter data, perhaps to lead the user or system to an erroneous decision; or may interfere with or degrade system operation. Such attacks could make systems unusable, degrade performance, lead commanders to make poor decisions due to faulty data, leak valuable secrets, or leave behind code that could provide continuing backdoor access or be activated at the occurrence of a predetermined event to take obstructive action. It is clear that such attacks cannot be prevented or even reliably detected. Thus, in addition to erecting access barriers and deploying detection systems, the Department of the Navy must discover how to design its critical systems, using commercially available components, so that they can be relied on to provide continuous correct operation in situations in which they are successfully attacked.

The notion that it is not possible to discover all vulnerabilities and use this information to guide a protection strategy is contrary to current thinking in DOD, where the emphasis is on discovery of vulnerabilities, so that appropriate protections can be placed to counter them. This popular vulnerability discovery approach puts protections in place only where there are known vulnerabilities. But because there is no way that all vulnerabilities can be discovered, such an approach will leave the system unprotected from its unknown exploitable vulnerabilities, which, if discovered at all, would be found out only during the operational lifetime of the system. This is a dangerous situation, because an adversary may well discover and exploit vulnerabilities that are still unknown to the Department of the Navy. In fact, the situation is asymmetric, because a determined adversary can decide which part of the system it wants to manipulate or exploit, purchase the commercial products that are used in that part of the system, and spend many months deconstructing these products to discover vulnerabilities that can be profitably and surreptitiously exploited. While such an approach is clearly affordable by an adversary, it is not affordable as a defense, since the defender would have to perform a costly analysis for every system component, whereas the adversary can pick and choose its focus of attack.

The Department of the Navy must operate on the assumption that any component of any system may have unknown security vulnerabilities that could be exploited by an adversary. Even the security protections put in place may contain such unknown vulnerabilities. With this in mind, the Department of the Navy will have to add redundant, independent security mechanisms and assume that not all attacks can be prevented, although there must be some means to detect attacks that are successful. Even more so, the Department of the Navy must recognize that its detection technology is far from perfect and that there will be successful attacks that are not detected. Thus, naval systems must be designed, to the extent possible, to be able to continue functioning despite the presence of an attacker. Appropriate strategies such as confusing an attacker who has successfully penetrated naval systems must also be developed.

Defense in depth is a threefold strategy that emphasizes prevention, detec-

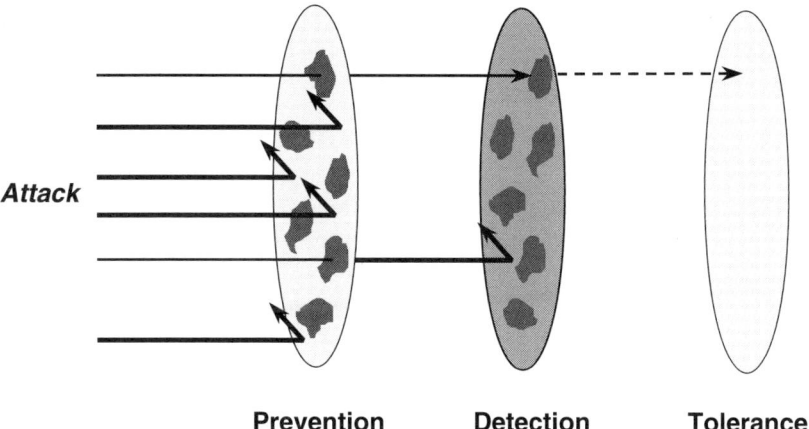

FIGURE 5.1 Defense in depth. SOURCE: Modified from a figure by Sami Saydjari shown during a briefing to the committee, DARPA, Arlington, Va., May 19, 1999.

tion, and tolerance (Figure 5.1). But today such a strategy cannot be completely implemented because of technology shortfalls at each level of the defense-in-depth strategy.

Many people in DOD and the Department of the Navy use the term "defense in depth" to mean what the committee calls a layered defense (discussed in Section 5.4.1.1). Here, a layered defense is considered to be part of a defense-in-depth strategy.

5.4.1 Prevention

To be affordable, naval networks and systems must use commercial products and services to the extent feasible. These products and services were not built to meet the security demands of network-centric operations. They generally have many known security vulnerabilities, and new vulnerabilities are discovered throughout their lifetimes. Because commercial products are so large and complex, it is not generally possible to discover all the vulnerabilities in advance, no matter how much testing is performed. Thus, all components must be treated as vulnerable, and the systems that use these components must be designed bearing in mind that there may be security vulnerabilities in any system component. Even when security functionality is designed into commercial products and services, this security is generally weaker than that required for naval needs.

This points to the need to be able to design systems built from insecure components and services so that the systems are secure against threats to naval forces. While the security research community has long recognized the need for viable approaches to building secure systems from insecure components, very little is known about how to do this, and secure system design remains an ad hoc, poorly understood discipline.

5.4.1.1 Layered Defenses

The committee does not discourage the Department of the Navy from performing vulnerability assessments. The more one knows about a system, the better. The point is not to rely on vulnerability discovery to guide the protection strategy (this is sometimes referred to as a penetrate-and-patch strategy). A better and safer strategy is to assume that every system component contains unknown security vulnerabilities that can be exploited by an adversary. Where to place protections against such unknown vulnerabilities will depend on an analysis of the consequences if any of these unknown vulnerabilities are exploited. In designing protections, it must also be kept in mind that the protections themselves can contain unknown, exploitable vulnerabilities.

Because every system component, including the protection components, must be assumed to contain unknown, exploitable security vulnerabilities, a layered defense strategy must be employed. The idea here is that protections are employed to counter known and unknown security vulnerabilities in the system. Barriers to attack can be employed to counter many types of potential vulnerabilities; however, a single barrier should not be relied on to repel a determined adversary. Because these protection components themselves may contain vulnerabilities (both known and unknown), additional layers of protection must be employed. Such a layered defense reduces the likelihood that an attacker can find an exploitable vulnerability, because for this to happen, not only must the original system contain a vulnerability, but each successive protection layer must also be flawed. Having several layers of defense increases the difficulty for the attacker. Each additional layer increases the odds for the defense and reduces the odds for the attacker. Thus, a layered defense helps to provide additional defenses even if a particular protection mechanism is subverted.

5.4.1.2 Malicious Insiders

One important source of vulnerability is the malicious insider. Working alone or with outside adversaries, a malicious insider could seriously disrupt network-centric operations. Monitoring user activities is another risk-reduction technique. Software that tracks downloading and checks for authorization; required use of biometric as well as password identification for users; and continuous record keeping of user log-on and log-off times with subsequent pattern analysis can all add a measure of security and potentially increase the likelihood of detecting malicious insider activities. All of these steps, especially many cascaded together, will add complexity and delay in many system functions. A balance will have to be struck between reducing risk and supporting functionality. The NCII must be able to accommodate ad hoc adjustment by commanders to account for shifting degrees of urgency and penalty for error, while sustaining some essential safeguards.

5.4.1.3 The Limitations of Commercial Product Assessments

Many in the DOD are concerned about the widespread use of commercial products in DOD systems and the potential vulnerabilities to the missions that rely on the use of these products. Some of these people advocate that the DOD institute a function that would evaluate commercial products to discover their vulnerabilities, so that appropriate defenses or workarounds can be developed. It will probably not be possible to do a good job of this, no matter how much testing is performed, because of the size (lines of code) and complexity of these products.

If an adversary were to plant or co-opt people in product development positions and have them insert malicious, triggerable code into commercial products, such hidden features could go undetected by the vendor's quality assurance and testing. Detecting such codes would be even more difficult for DOD, which has access only to the object code of these products, despite the advanced reverse-engineering capabilities present in certain parts of DOD. The likelihood of such a scenario can be assessed in light of the fact that it is fairly common for popular programs to have undocumented features that survive in final product releases without the knowledge of the product manager (e.g., special key sequences to bring up hidden games or photographs of the developers). The functionality of these products is simply so extensive that unused hidden functionality is highly unlikely to be discovered by any software evaluation techniques.

The committee believes that the Department of the Navy should collect as much information as possible about products it has in widespread use. This information can be collected from an in-house evaluation (of both functional and security aspects) as well as from evaluations performed by other organizations. For example, the National Institute of Standards and Technology has established common criteria for products as well as a common evaluation methodology, and product evaluation information is available for some products.[3]

DARPA's Information Assurance Science and Engineering Tools program[4] is attempting to develop assurance metrics and evaluation tools for systems.

---

[3]National Institute of Standards and Technology. 1999. *Common Criteria Project*. Gaithersburg, Md., November 1. Available online at <http://csrc.nist.gov/cc/>.

[4]Skroch, Michael J. 2000. *IA Science and Engineering Tools*. DARPA, Arlington, Va., January 1. Available online at <http://iso.isotic.org/Programs/progtemp/progtemp.cfm?mode=342>.

However, experience shows that such an endeavor cannot be expected to have a high likelihood of success, for the reasons outlined above, although the committee expects the program to advance the state of knowledge, if not of practice.

DARPA's Information Survivability program[5] developed a tool that can help identify places in source code which, if flawed or tampered with, could result in a security breach or program failure. Because it is generally difficult to obtain source code for commercial products, a similar tool is being developed to operate on object code. These early investigations, using approaches borrowed from the software reliability community, may help vendors build better products and help the Department of the Navy understand the potential vulnerabilities of those products.

5.4.1.4 The Limitations of Red Teaming

"Red teaming" is commonly advocated within DOD as a way of discovering and fixing vulnerabilities. A red team typically takes on the role of an attacker, either with or without the knowledge of the system operators, in order to discover whether available countermeasures for known attack methods are implemented on the system. It is frequently mentioned as the primary way of establishing security in a system. Red teaming is not, however, intended as a means of discovering vulnerability or even as a primary method for fixing discovered security flaws. The value of red teaming is twofold. First, it is arguably the best tool for raising security awareness in an organization. Most red teams discover known security holes for which known fixes, configurations, or patches have not been applied or where compensating security procedures are not in effect or not being enforced. By exploiting these weaknesses, they generally are able to demonstrate that they could have done significant damage to the system's resources and could have had significant mission impact. This is useful in attracting high-level attention to security issues. Second, red teaming is useful for ensuring that correct security configurations are maintained for the system. This type of red teaming can be carried out periodically by the system's security staff. Because red teaming is very useful in these two ways, the committee encourages the Department of the Navy to continue and increase its regular practice of red teaming.

But red teaming is not a good way of discovering vulnerabilities. First of all, most red teams use only known, published attacks that exploit known vulnerabilities. So the knowledge gained is not new, but pertains only to the specific installation. (This also means that a red team exercise is too weak to be

---

[5]For more information, see Information Technology Office, *Proceedings of the DARPA Information Survivability Conference and Exposition (DISCEX 2000)* (Hilton Head Island, South Carolina, January 25-27, 2000), two volumes. DARPA, Arlington, Va., forthcoming.

a useful "final exam" for systems, as is advocated by some in the DOD.) Second, red teaming occurs only after a system has been designed, implemented, deployed, and operated for some time. Even if vulnerabilities were discovered by red teaming, it is too late in the system life cycle to effectively and affordably correct them. Some organizations have put forth the notion of red teaming a system design, which would help to discover security problems early and correct design problems that are not correctable once the system is built. Red teams, or penetration teams, could also be employed to try to discover ways of attacking system or component implementations. This approach would not use known attacks, since there would not yet be any known attacks against these new components, but it would add a security dimension to component and system testing. The committee strongly advocates security red teaming of both designs and implementations before new systems are put into use.

Red teaming should not be thought of as a security panacea or as the primary way of achieving system security. At best, it amounts to reliance on a penetrate-and-patch strategy as the main mode of system defense. Penetrate and patch does not work as a defense, because it can never be expected to discover all system vulnerabilities. Most seriously, it may not uncover the unknown vulnerabilities that could be exploited by an information warfare adversary. Instead, it should be regularly applied as a consciousness-raising tool and for maintaining compliance with correct security configurations in deployed systems.

5.4.2 Detection

Even a layered defense will not eliminate the possibility of successful attack. The Department of the Navy must employ methods to detect attacks that are not successfully repelled. Intrusion detection technologies are commercially available, although these have well-known shortcomings, such as a very high false alarm rate and the ability to detect only a limited class of attacks on a limited set of system components. Specialized attack types on DOD-unique system components will certainly not be detected, and neither will application-specific attacks. Even generic attacks on operating systems and networks cannot be reliably detected. Moreover, today's intrusion detection products generally can detect only attacks that have been seen before, whereas an information warfare adversary can be expected to be developing an arsenal of new attacks that current intrusion detection systems cannot anticipate.

To carry out joint and coalition operations, it may be necessary to give NCII access to joint and coalition personnel, increasing the likelihood of insiders with motivation to cause damage. Intrusion detection should also be directed at detecting malicious insiders. This problem is particularly difficult because insiders may not have to break in or circumvent access controls, so that detection approaches that rely on detecting evidence of such breaches will not be effective against them.

Although DARPA has a sizable research program to develop new intrusion detection technologies, even these still have a long way to go to reduce false alarm rates to acceptable levels, increase detection rates, and detect more types of attack. A joint evaluation on intrusion detection was conducted in 1998; MIT Lincoln Laboratory (MIT/LL; Lexington, Massachusetts) conducted the off-line evaluation, and the Air Force Research Laboratory (Hanscom Air Force Base, Bedford, Massachusetts) did the real-time evaluation.[6] Conducted under the sponsorship of DARPA and the Department of the Air Force, the evaluation showed that the DARPA technology reduced false alarms by orders of magnitude and increased detection rates roughly fivefold. Yet this same evaluation also showed that false alarm rates are still too high, especially when multiplied across a very large system, as would be the case in network-centric operations. The evaluation also showed that detection rates and the number of attack types detected must still be increased significantly. The evaluation showed that most detection techniques are good at detecting certain attack types and not so good at detecting other types. Thus, a reasonable detection strategy is to employ a variety of detection techniques. Figure 5.2 compares the results of the best combination of research prototypes with the results from a key word baseline system that is representative of the many commercially available intrusion detectors that use string matches on key words as the detection method.

The DARPA/Air Force Research Laboratory (AFRL) evaluations were intended to drive improvements in the research prototypes. A second round of evaluations was conducted in 1999, and a third is planned for 2000.[7] The results of the 1998 evaluation were reported in the July 1999 *Communications of the ACM* (Association for Computing Machinery).[8] The 1998 evaluation used Unix audit logs and Transmission Control Protocol (TCP) dump data, which are the sources of data for typical commercial intrusion detection products. The 1999 and 2000 evaluations will increase the number of data sources used in the evaluation. The methodology used employs a live evaluation of the intrusion detection research prototypes on a network at MIT/LL using simulated data similar to data collected at Air Force bases. MIT/LL generated large amounts of realistic background traffic similar to observed and collected traffic at the bases. This

---

[6]Lippmann, Richard, Marc Zissman, David Fried, Sam Gorton, Isaac Graf, David McClung, Dan Weber, Seth Webster, and Dan Wyschogrod, "1998 DARPA Intrusion Detection Evaluation Plans–2/98," a briefing at the DARPA PI meeting, February 3, 1998, Annapolis, Md. Available online at <http://www.ll.mid.edu/IST/ideval/docs/9802-pi/index.html>.

[7]Zissman, Marc A., and Richard P. Lippmann. 1999. *DARPA Intrusion Detection Evaluation*. MIT Lincoln Laboratory, Lexington, Mass., December 22. Available online at <http://www.ll.mit.edu/IST/ideval/>. Also, personal communication from Richard Lippmann, MIT Lincoln Laboratory, February 18, 2000, regarding the 1999 and 2000 evaluations.

[8]Durst, Robert, Terrence Champion, Brian Witten, Eric Miller, and Luigi Spagnuolo. 1999. "Testing and Evaluating Computer Intrusion Detection Systems," *Communications of the ACM*, Vol. 42, No. 7, July, pp. 53-61.

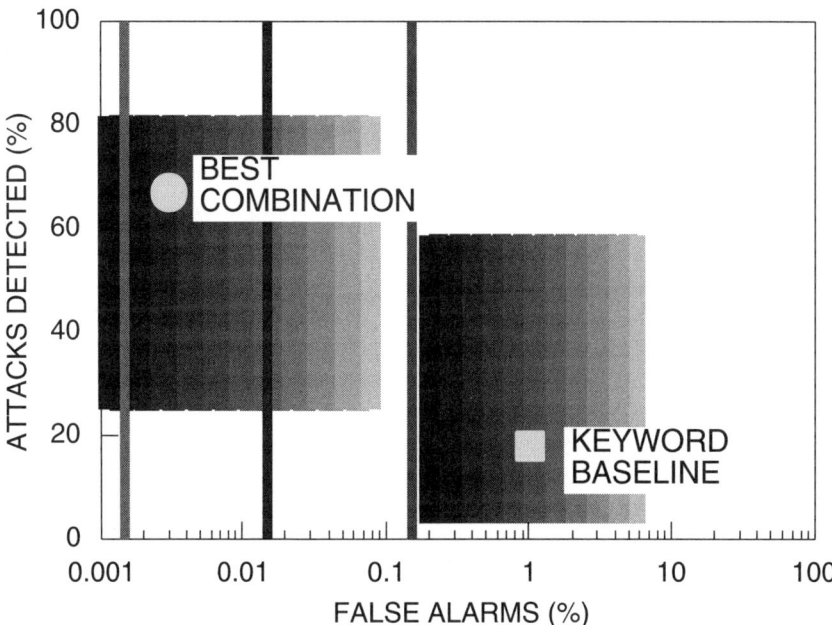

FIGURE 5.2 Comparison of research prototypes and keyword baseline for intrusion detection.
SOURCE: Modified from: (1) Lippmann, Richard, Robert Cunningham, Dave Fried, Isaac Graf, Kris Kendall, Seth Webster, and Marc Zissman. 2000. "Results of the DARPA 1998 Offline Intrusion Detection Evaluation," in *Proceedings of the 2nd International Workshop on Recent Advances in Intrusion Detection* [workshop held September 7-9, 1999], Center for Education and Research in Information Assurance and Security (CERIAS), Purdue University, West Lafayette, Ind., forthcoming. (2) Lippmann, Richard. 1999. "Best Combination System from 1998 Evaluation Compared to Keyword Baseline," *Summary and Plans for the 1999 DARPA Evaluation*, January 28. Available online at <http://www.ll.mit.edu/IST/ideval/summary-html-dir/>.

background traffic was used to measure false alarm rates of the intrusion detection prototypes. AFRL and MIT/LL also created the largest known collection of automated attacks with signatures, for both audit trail and TCP dump data sources. These automated attacks were used to evaluate the ability of the research prototypes to detect known attacks. In addition, the evaluation measured the ability of the research prototypes to detect types of attacks never seen before. To do this, AFRL and MIT/LL developed a set of new attacks. This is the first evaluation that allowed side-by-side comparisons of different detection methods for both false alarm rate and detection rates against both known and previously unknown attack types. The researchers were also provided with extensive training and testing data sets and performed self-evaluations. Some of them have begun to publish their results in conference proceedings and technical journals.

5.4.3 Tolerating Attacks

Because intrusion detection technology cannot be depended on to detect all successful attacks, the third aspect of defense in depth is to design systems so they will be able to tolerate attacks. This means that systems must be able to continue to operate despite a successful attack, by means of attack containment as well as strategies to minimize damage. Today very little is known about how to build a system this way. DARPA has launched a new research program in intrusion tolerance to begin to explore this area.

5.5 ASSESSMENT OF CURRENT INFORMATION ASSURANCE ACTIVITIES

This section gives a general assessment of the state of practice of information assurance. The assessment also applies to the Department of the Navy and briefly discusses specific naval activities. It suggests near-term actions the Navy Department can take to begin to improve its information assurance position.

Like the DOD in general as well as the country at large, within the Department of the Navy available security technologies are not being used widely enough.[9] Even when these are installed, adequate procedures are needed to maintain correct and secure configurations. While this task can be partially automated, a large part is still labor-intensive, requires strict adherence to procedures, and relies for its success on the awareness and commitment of the vast majority of users. A procedural or technical means should be developed and used to verify that the mandated security functionality is correctly applied in all naval systems. Moreover, security is not a part of the readiness assessment of the forces, so that it is not given as high a priority as other considerations, and forces are considered ready even when they cannot defend their information-based operations (on which the success of their missions may well depend).

The Department of the Navy has no coherent plans for information network security in its tactical networks. Security should be part of the foundation of an information network architecture—it is extremely difficult to add after the fact. Years of work have gone into adding security to the Internet, after the fact, but the results are still far from satisfactory. Web sites continue to be hacked and intruders continue to gain access to poorly protected computers. The Depart-

---

[9]The general situation is discussed in, for example, Computer Science and Telecommunications Board, National Research Council, 1999, *Realizing the Potential of C4I: Fundamental Challenges,* National Academy Press, Washington, D.C. (see, e.g., p. 156).

ment of the Navy should take great care that its tactical networks are at least as secure as those provided by commercial vendors.

The difficulty of achieving security is evident when one considers that there are no commercial solutions to many security problems, and that current commercial technology (e.g., intrusion detection, firewalls, and authentication systems) leave much to be desired.

5.5.1 Current Naval Activities

The Department of the Navy has issued guidance for how to secure information systems.[10] It cannot be assumed, however, that all or most naval locations are following this guidance. Moreover, the guidance is only technical. Where no technical solution exists, or the solution is inadequate, additional guidance is needed on procedural security methods. One shortcoming of this guidance is that it may suggest too much reliance on some immature and unproven security technologies. For example, intrusion detection systems are recommended, but their significant shortcomings are not discussed. The guidance also recommends that the so-called "network intrusion detectors" (meaning detectors that look for a set of known intrusions from data sniffed from a communications line) can be used at certain interfaces instead of network firewalls. The document provides guidance for about five years into the future. It is very conservative in its projections and barely considers whether any new technology for security will be available. For instance, the data encryption standard 3DES is recommended for the indefinite future, although it is expected to be replaced by a much better standard in a few years. As another example, the guidance does not consider that there may be new types of intrusion detection available in the next five years. Generally, this guidance document is a good place to get started, and it refers to other documents for more complete guidance in some areas, such as how to configure Unix for security.

The Department of the Navy, and particularly the Naval Research Laboratory (NRL), is engineering solutions for IT-21 that include firewalls; virtual private networks; multilevel security for coalition operations and tactical/nontactical (TnT) interaction; intrusion detection; synchronous optical network (SONET)/asynchronous transfer mode (ATM) security; Class C2 operating systems; secure Web protocols such as Secure Socket Layer (SSL); virus detection; some authentication in routing, switching, and domain name service; and mail guards. Security architectures are being developed to use these protections in a layered defense strategy. Protections and intrusion detection are the first two components of defense in depth.

---

[10]Integrated Products Team. 1999. *Information Technology Standards Guidance*, version 99-1. Office of the Chief Information Officer, Department of the Navy, Washington, D.C., April 5.

The NRL also performs security certification and accreditation, develops embeddable and programmable cryptography for naval cryptographic replacement needs, and performs highly regarded security research.

The Department of the Navy is developing a Navy/Marine Corps intranet (N/MCI) to link naval ashore assets, which is expected to be operational by the end of CY2001. The N/MCI is expected to include commercial virtual private network (VPN) technologies that allow private interactions among identified communities across the common network. Protections for the N/MCI will also include conventional multiple levels of firewalls, intrusion detection, encryption and a public-key infrastructure, incident response teams, red teams, physical and procedural controls, and awareness and training programs. The N/MCI will provide access to naval, joint, coalition, and public Internet sources of information and reliance on a multiplicity of databases and sources, most outside the direct control of the naval forces. Much of the system, including its protection mechanisms, will be implemented with commercial software and hardware products and services. There will be widespread outsourcing of installation, support, and upgrade of enterprise hardware and software.

The Department of the Navy, like the rest of DOD, uses cryptography to separate classification levels in its networks. This allows information of different classifications to share some of the same physical network elements, such as routers. In effect, each classification has its own virtual network. One purpose for the encryption is to protect the classified information from being read by unauthorized parties. However, an additional reason for the encryption (and for classification in general) is to provide a degree of confinement, that is, to minimize the extent of exposure if someone gets access to classified information. The cryptography is good for this purpose only as regards damage from inappropriate information release. Because of the shared network elements, other types of damage are not confined to a given classification. Thus, for example, an adversary who gains access to the unclassified portion of the network may be able to use this access to perform denial-of-service attacks or other manipulations on the higher classified virtual networks by attacking or manipulating the shared elements. These shared physical elements form points of vulnerability, in that even if an attacker could not recover classified data, it could still use such points to manipulate the network.

5.5.2 Risks of Connectivity

Sensitive NCII functions may rely on unclassified, open source information. Such information sources are vulnerable to tampering and insertion of bad information by malicious entities before the information enters the NCII. Countermeasures could include integrity checks on the incoming data that alarm when the data fall out of expected or usual bounds, and comparison of data from a variety of sources to avoid overreliance on unique sources. It may also be possible to maintain a pedigree for the data that would record information about its source, which could be used if the data are discovered to be tainted. These pedigrees could, for example, help to locate other data that might also be tainted and could be helpful in backtracking to discover the source of the tainted data.

Connectivity to external networks, including the open Internet, exposes the NCII to viruses and other information warfare weapons in data and code that enter the NCII. Also, NCII users may download arbitrary code, which may be infected with viruses or worms that can spread and cause damage within the NCII. The NCII should employ the most recent virus detection software on all its systems, including its end user systems, and NCII operators must be vigilant about ensuring that the correct versions are installed and the detection tables are kept up to date. Policies about importing code should also be developed and enforced. It may also be desirable to disallow mobile code (applets) until adequate security solutions are available.

5.5.3 Security for Tactical Networks

Four major vulnerabilities of wireless communication—spoofing, jamming, interception, and terminal capture—are considered here, along with a general discussion of security and lower-layer tactical radios.

5.5.3.1 Spoofing

An adversary who introduces false information into a tactical network may degrade the performance of that network. Such spoofing can be prevented in all cases but one by the use of appropriate cryptographic apparatus at each end of the transmission. The exception occurs when a terminal is captured; reliable methods of personal identification have long been sought, but few, if any, are available for field use.

5.5.3.2 Jamming

There are two main techniques for resisting communications jamming—signal processing and antenna directivity—and a combination of them is often appropriate. By the use of spread-spectrum waveforms, spectrum (but not necessarily data rate) can be traded for jam resistance. If spectrum is limited, data rate is compromised at some point, but trades between data rate and resistance are possible, and, with programmable modular radios, these trades can be made dynamically.

Spread-spectrum options and their antijam performance are well understood. However, most calculations of processing gain—the degree of jam resistance afforded by the waveform—assume that the transmitter and receiver are already synchronized. Designers must take care that the synchronization process itself is not a fruitful jamming target. They must also ensure that their waveforms are not vulnerable to signal-specific techniques such as a jammer following the instantaneous frequency of a slow hopper and concentrating its power there.

Of the existing systems, JTIDS provides robust protection against jamming at moderate data rates and SINCGARS provides some protection against jamming at low data rates. Both systems use nondirective antennas to simplify their use in information networks involving multiple moving units.

Antenna directivity helps defeat jamming in two ways: directive-transmit antennas send more signal power to the receiver than would omnidirectional antennas, and (in most scenarios), directional-receive antennas receive less jamming power than would omnidirectional antennas.

Directional antennas at both ends of a link are appropriate for point-to-point transmission, and directional antennas at diverse locations can be pointed at a single broadcast transmitter. However, implementing a datalink architecture with directional antennas in which every participant can hear every other participant directly would require multibeam antennas on every node. This is the expensive solution adopted for the CEC data distribution system. Prospects for cheaper multibeam antennas are discussed in Chapter 6.

5.5.3.3 Interception

An adversary's attempt to read intercepted radio signals can be foiled by effective encryption. Traffic analysis can be countered by some mixture of having every unit transmit all the time—even when they have nothing to say—and of encrypting packet headers so as to prevent destinations from being deduced. Contrary to first impressions, the transition to network-centric operations will probably make traffic analysis more difficult. Current transmission security devices will likely be augmented with end-to-end encryption, which, like current network encryption system (NES) devices, hide the network addresses. Furthermore automatic relaying of traffic will tend to mix traffic streams and help disguise the actual sources and destinations of traffic flows. That said, this area does require careful analysis and design.

Enemy interception of radio signals for the purpose of performing geolocation, however, can be forestalled only by making the signals hard to detect. There are two trustworthy means of accomplishing this. Directional transmit antennas pointed at the intended receiver and away from areas where adversary interceptors might be located constitute one way. The other is to use a spread-spectrum coding. Less reliable methods include concealing the transmission in the frequency or time shadow of other transmissions, and using infrequent, short-duration signals in the hope that the adversary has not deployed intercept apparatus capable of detecting such signals. It is inevitable that individual platforms will perform more radio transmissions as they become full participants in NCOs. This effect is most pronounced in developing the common tactical picture (CTP), because each platform will need to transmit its geolocation relatively often in order to be properly located within the CTP. There is no obvious answer to this problem beyond making the signals hard to intercept.

Although some spread-spectrum modulations are equally useful for overcoming jamming and preventing interception, they cannot do both simultaneously. To overcome jamming, the signals would have to be sent at a high power and would be interceptable. To prevent interception they would have to be sent at a low power level that will not overcome jamming. Antenna directionality, on the other hand, helps defeat both. Directionality at the transmitter helps overcome jamming and reduces interceptability; omnidirectionality at the receive antenna resists jamming and permits the transmitter to use lower power.

5.5.3.4 Terminal Capture

A particular risk of the NCII in the tactical arena is enemy capture of an NCII end user terminal. Palatable solutions to the problem of terminal capture are not obvious and require further investigation. Expensive biometric authentication technology (voice identification, retinal scan, iris scan, fingerprinting, etc.) is available. Alternatively, frequent reauthentication could be required of users at remote terminals. Neither approach appears attractive in tactical circumstances.

Link encryption plus over-the-air-rekeying is not a sufficient answer to this problem. Full-scale network protection requires a number of additional technologies, including firewalls, intruder detection, layered defenses, and so forth. Commercial technology provides limited help in this area. It can supply firewalls, an infrastructure for public key management, key exchange protocols, authenticated routing protocols, and so forth. But at present no commercial technology can adequately deal with the enemy capture problem. DARPA is tackling a wide range of issues with its Information Assurance program, but even so it is unlikely that tactical networks will be sufficiently well-protected in the near to mid term. This should be of significant concern to the Department of the Navy.

5.5.3.5 Lower-layer Tactical Radios

The Department of the Navy employs a wide variety of radios (Open Systems Interconnection (OSI) layers 1 and 2) in its tactical networks. But it also needs lower layers (radios) of the following broad classes:

- Multiple-access (shared-channel);
- High-capacity, point-to-point; and
- Satellite links.

The radios employed for these classes are likely to be unique to the military. At the very least, the Department of the Navy needs anti-jam (AJ) protection for most of its tactical links. In many cases, it is also desirable that the radios have a low probability of interception (LPI) and a low probability of detection (LPD). None of these features are likely to be available in commercial radios to any significant degree.

Encryption is a murkier issue. Traditionally the only encryption used on tactical radios has been link encryption; in essence, all bits sent over the link are encrypted just before transmission and decrypted just after reception. However, end-to-end encryption devices are beginning to enter the commercial marketplace. These devices encrypt a packet at the source computer, keep it encrypted all the way through the network, across multiple links, until it is safely inside the destination computer or destination firewall. Then and only then is it decrypted. Typically these devices use the IP Security (IPSec) protocol suite or vendor-specific protocols and encrypt the contents of datagrams but not their headers. When used for VPNs they can also encrypt a complete datagram and encapsulate it in another datagram wrapper, which hides the encrypted datagram's header as well as its payload. Since IPSec was designed to handle an open-ended set of encryption mechanisms, it is easy to imagine a type-1 IPSec device. Indeed, such devices may already exist. Tactical networks may benefit from using such an end-to-end encryption strategy. However, this would leave a certain amount of wrapper header in the clear; possibly necessitating an additional layer of link encryption. Needless to say, radio LPI, LPD, and AJ features may also require their own encryption mechanisms. To summarize a complex issue, while it is true that type-1 encryption is needed for the data carried through tactical networks, commercial radios (without embedded cryptos) may suffice for some tactical traffic.

5.5.4 Air Gaps as a Defense

Many believe that for network-centric operations, every network should be connected to every other network. But in fact this may not always be wise. Interconnection brings risk from increased system complexity. It also introduces new weak points for intrusion, interdependencies and, possibly, unintended interactions. An "air gap" (i.e., a physical separation in connectivity) may be necessary in some circumstances. Use of this physical isolation of networks may be preferable in some instances.

Increasing dependence on networked operation can produce greater vulnerability to network intrusion and more serious consequences from such intrusions. Perhaps just as serious is the increased risk of failure propagation because the networks are interconnected. For example, the CEC network is greatly overprovisioned so as to ensure that it is available when needed. It is segregated from other networks so as to eliminate as much as possible any unforeseeable interference from other systems with the reliability of the CEC network and applications. If this network is interconnected with tactical or other networks, there is considerable unknown risk that CEC applications may not operate correctly when needed.

Similar problems are encountered in safety-critical industries, such as the aviation industry, where the solution traditionally has been to isolate mission-critical functions in their own dedicated computers, networks, and power supplies. Cost pressure to share these resources with other onboard systems is challenging the industry to discover approaches to design safety. This problem is also well known in the telecommunications industry, where undesired feature interaction has been studied intensively. Current understanding of how to build complex systems does not allow guarantees that critical functions such as CEC will operate properly in the dynamic computing environment of the NCII. It may not even be possible to estimate the increased risk that such integration will bring, but this should be attempted and validated to the extent possible before any integration is planned.

5.5.5 Virtual Private Networks

Virtual private networks (VPNs) are a popular technology to allow protected networks to be connected to each other by establishing tunnels through the firewalls that use cryptography to protect the communication from tampering or eavesdropping on the intervening network. Commercial VPNs are not standardized, and many use simplistic key-management schemes that are not secure enough to protect classified traffic or sensitive military traffic. Also, they may not support a large enough number of participating organizations (10 to 100 may be required). In the future, the industry may provide enhancements for supporting security management protocols, for negotiating security policies across administrative domains, and for management tools.

DARPA has created the Dynamic Coalitions program to develop technologies that could improve or replace commercial VPNs. This research will secure the underlying group communication technologies and provide services such as authentication and authorization that are needed for secure collaboration in a coalition environment. One aim is for fast set-up of services as new coalitions or collaborating groups are formed. In addition, DARPA's new Survivable Network Coalitions program will develop technologies to enable secure collaboration within dynamically established mission-specific coalitions while minimizing potential threats from increased exposure or compromised partners. The program will develop technologies to support the secure creation of dynamic coalitions, including technologies for policy management, secure group communications, and supporting security infrastructure services. Technologies will be developed to dynamically manage and validate operational policy configurations across multiple theaters, securely manage the dissemination of information within large groups, and augment existing public-key infrastructure technologies to accommodate rapid revocation and cross-certification. These technologies will help coalitions to continue to operate while minimizing potential threats from increased access.

5.5.6 Multilevel Security

The National Security Agency (NSA) Trusted Product Evaluation Program spurred many large vendors of operating systems to build "secure" versions of their systems, most of which were designed to meet the criteria of Class C2 (which does not support multilevel security) of the Trusted Computer System Evaluation Criteria, the so-called Orange Book.[11] Some vendors built products to Class B1 (which includes labels but is not strong enough to separate classified information), and a very few vendors built operating systems for Classes B2 and B3, which can be used for protecting classified information where some users are not cleared for the highest levels of information in the system. These systems have not found widespread use within DOD because of their expense (generally in the tens of thousands of dollars per copy), their reliance on specific hardware protections (and thus lack of portability), their lag in features and performance over mainstream technology, and their inability to support popular application software in use in DOD. These operating systems were not popular commercially either, because their conception of security was too narrowly specific to DOD, and the DOD security policy was wired in, meaning that commercial users would have difficulty using these products to enforce their own security policies. Similarly, the NSA Trusted Database Evaluation Criteria resulted in vendor offerings of products with support for labeling of multilevel data at the granularity of rows in a table.[12] However, these products rely on operating system support for some of their security, and most have not been ported to the few available high-assurance operating systems (because it would not have been profitable to do so).

It is highly unlikely that we will see high-assurance, multilevel secure (MLS) systems in widespread use in the future. Instead, the computing world will be divided into regions or islands of different classifications, where each island or region operates at system high and holds data that is classified at its system high

---

[11]Computer Security Center. 1985. *Trusted Computer System Evaluation Criteria* (DOD 5200.28-STD supersedes CSC-STD-001-83, August 15, 1983). Department of Defense, Fort George G. Meade, Md., December. Available online at <http://www.radium.ncsc.mil/tpep/library/rainbow/5200.28-STD.html>.

[12]National Computer Security Center. 1991. *Trusted Database Management System Interpretation of the Trusted Computer System Evaluation Criteria*, NCSC-TG-021. National Security Agency, Fort George G. Meade, Md., April. Available online at <http://www.radium.ncsc.mil/tpep/library/rainbow/NCSC-TG-021.txt>.

level and lower. However, MLS technologies can be used to interconnect these islands into a system that can exchange information from low islands to high islands. Once low information has entered a high island, it must be treated there as high information, although advisory labels can be used to indicate that the same information is available on low systems elsewhere. Low-assurance MLS technologies (e.g., Orange Book Class B1) can be used to provide such advisory labels, but these technologies are not strong enough to be relied upon to separate data of different classifications with high assurance.

While DOD cannot afford to develop special-purpose multilevel systems for wholesale desktop use, it can afford to develop specialized security technology to interconnect different classification islands and to approach MLS capabilities. A few such technologies have been developed. Starlight is an Australian-developed system that allows a computer monitor and keyboard to be switched from one system high network to another, giving an approximation to desktop MLS processing. The NRL pump is a system that can reliably transport data from low networks into high networks without leaking any high information back into the low network. Secure Information Through Replicated Architecture (SINTRA) is a system developed by NRL that creates the functionality of an MLS database by linking system-high databases in different classification islands in a way that does not leak information. Software Architecture and Logic for Secure Cooperative Applications (SALSA), a similar idea for MLS workflows, was also developed by NRL.

Encryption can also be used for high-assurance separation of information of different classification. It has long been used by DOD in communications networks to provide secure separation of communications of different classifications on a common physical network. Commercial networks, including commercial satellite networks, can be used for carrying multilevel data, since the data are separated by cryptographic means. (However, the use of commercial networks permits vulnerability to denial-of-service attacks and traffic analysis.)

The idea of cryptographic separation can be extended into information repository systems. Yet this does not solve the multilevel security problem, because within a processor, information must be decrypted for processing, so that additional strong controls are needed to prevent leakage of unencrypted high data to low processes or users. Moreover, this approach is not popular commercially for reasons of cost and ease of use.

True multilevel security capabilities, other than networks shared through cryptographic separation, barely exist in the Department of the Navy and in the DOD. However, the NRL pump and the Australian Starlight system are promising for future multilevel processing needs, as are SINTRA and SALSA. NRL should be tasked to develop these and other promising near-term technologies to the point where they are ready for widespread introduction into the Navy Department and for use with a wide variety of technologies.

5.5.7 Traffic Analysis

An adversary may be able to perform traffic analysis to identify critical nodes and bottlenecks and may develop attacks on these points, or it may infer information about naval operations and plans. NRL has developed a technique called onion routing to provide some protection against traffic analysis. It should continue to pursue this and other means of protection against traffic analysis. It should grow these technologies and promote their use within the Department of the Navy.

5.5.8 Other Issues

There is a critical shortage of information assurance expertise both within the Department of the Navy and in the nation at large. The Navy Department should encourage the government to expand the information assurance skill base, perhaps through educational programs sponsored by the National Science Foundation (NSF). Such programs could help produce the naval personnel of the future who will be responsible for maintaining the security of the NCII. There is also a shortage today of good minds working on security research. An investment in research by NSF at universities into the fundamentals of computer security and other areas related to information assurance would begin to produce a new generation of researchers and help to gain a greater degree of academic acceptability for the field. Even so, it is not clear whether progress will be rapid enough to keep up with the threats and vulnerabilities introduced by new technologies.

Information assurance is not receiving appropriate attention at high enough levels within the Department of the Navy. While NRL is developing many promising technical solutions, it is having difficulty getting attention within the Navy Department for the work. The concerns it reports to offices such as N6 get diluted by many other issues. Responsibility for information assurance must be made a key priority and assigned at a high level within the Navy Department. This would ensure that NRL and other naval information assurance research receive adequate resources and that the new technologies will be implemented. It would also create a focal point within the Department of the Navy with sufficient authority to develop and deploy methodologies and tools for assessment and compliance.

Dependability is another requirement of critical naval systems. Redundancy is the standard means for achieving dependability, but it is an expensive approach and provides no protection for correlated failures and little protection for software failures. Like security, dependability is not part of the readiness assessment. Additional near-term measures could include increased continuous monitoring and testing of systems. Most importantly, naval forces should plan ways of operating when there are security and dependability failures, such as outages or service degradation.

Security and dependability should also be an explicit part of the spiral development model discussed in Chapter 2. First, security and dependability should be designed into new technologies so that vulnerabilities can be identified early and corrected in subsequent designs. Red teaming should be an integral component of the spiral development experiments. Second, there should be security and dependability dimensions of experiments that are carried out as part of the spiral process. It is important that the new alternative doctrine, operational concepts, and tactics explored in these experiments take into account the security and dependability characteristics of the technologies. To allow informed choices, it is important that these experiments mimic the hypothesized weaknesses of the technologies, so that strategies can be developed to deal with unavailabilities, degraded capabilities, or misbehaviors due to security or dependability failures.

The necessary protections for the NCII will not be inexpensive to procure, install, maintain, and keep current. In addition to the costs of the technology will be the costs of developing and enforcing essential policies and procedures, especially because there are currently no adequate technical solutions for many aspects of the information assurance problem. Ongoing programs for security awareness and training will also help to keep naval personnel vigilant about noticing and reporting suspicious behavior of their systems. The Department of the Navy can expect to spend at least 5 percent of its budget for information technology (IT), IT maintenance, and IT training on information assurance. Note that industry typically spends about 5 percent of its IT budgets on security, while the U.S. government spends less.

5.6 RESEARCH PRODUCTS SUITABLE FOR NEAR-TERM APPLICATION

In recent years, DARPA and other agencies have funded much promising research in information assurance. This research has produced some promising near-term technologies that the Department of the Navy should evaluate and exploit to the extent possible. Because of its unique expertise in information assurance and its capabilities for technology evaluation and development, NRL should be tasked to mature the most promising and applicable of these and to make them ready for deployment. Examples include encrypted e-mail, secured protocols, better intrusion detection components, digital signatures on software, and security wrappers. Once these technologies have been made ready by NRL, the Navy Department must then help to deploy them throughout the organization. NRL could also be charged with monitoring the state of the art within the Department of the Navy for security and dependability technologies and procedures, this information then being disseminated throughout the Navy Department.

DARPA's Information Survivability program,[13] which finished in 1999, is a good source of technologies that NRL could consider experimenting with. These technologies include early prototypes of hardened network services, early versions of wrapper generation tool kits, prototype local intrusion detectors, and a framework for coordination of intrusion detection and response. The program is divided into four areas. Examples of research results emerging from these four areas that could be considered by NRL are summarized in Table 5.1.

In addition, the DARPA Information Assurance program has taken many of the DARPA Information Survivability technologies in prevention, detection and response, and security management for C4I information systems and has integrated them into a security architecture that, while integrating security and survivability concepts, techniques, and mechanisms, will also provide interfaces for future security upgrades. Access control technologies that have been integrated into the architecture include encryption of message traffic and firewalls. Other solutions include policy-controlled automated security guards and release stations, strong user authentication, and protected execution domains to limit damage from certain types of attack. Intrusion detection and response capabilities have also been integrated into the architecture. In addition, a security management function is being integrated into the architecture. NRL could evaluate and mature portions of this security architecture for possible use within the Department of the Navy.

This collection of research results includes specific prototype technologies that address portions of the Department of the Navy's technical information assurance problems. Most of these technologies (specifically network security technologies, secure computing platforms, and wrapper technologies to embed security functionality into legacy systems) fall into the protection category. Several technologies fall into the detection category. There are few, if any, technologies for the tolerance component of defense in depth. There is the beginning of a security architecture to tie many of these elements together in a standard way. By maturing and deploying selected technologies from this set as well as from other technologies being developed elsewhere, the Department of the Navy can accelerate the application of appropriate technical solutions to aspects of its information assurance problem.

---

[13]See Footnote 5.

TABLE 5.1 Technologies from the DARPA Information Survivability Program to Consider for Near-term Application

Problem	Technology	Description
Secure Networking		
Internet routing infrastructure is subject to denial-of-service attacks.	Secure open shortest path first (OSPF) routing protocol Nimrod routing security	Authentication added to routing
Internet naming and addressing infrastructure is vulnerable to denial-of-service attacks.	Internet Protocol (IP) version 6 (IPv6, also known as IPSec) prototype	Encrypts IP packets
	Secure domain name service (DNS) protocol	Adds authentication and authorization to DNS
Network security services are not easily integrated into applications.	Cryptographic application programming interfaces Simple public-key interface Key management interfaces	Give programmers standard ways of adding security functionality to software
Network security services are not interoperable.	IPSec key agreement protocol Secure mobile IP	Multilayer security negotiation protocol Preserves user identity for nomadic users traveling across different networks
Encrypting group communication and managing membership changes are not scalable to large groups.	Secure fault-tolerant group communication	Scalable algorithms for key management and membership revocation, and language for specifying group communication policies
Secure Computing Platforms		
Traditional operating system security is inappropriate for modern networked environments with complex policies.	Domain-type enforcement "compiler"	Fine-grain access control and an associated policy
Adequate security is available only in special-purpose operating systems with tiny markets.	Fluke operating system	Enables integrating security into commercially viable next-generation operating systems

Table continued on next page

TABLE 5.1 Continued

Problem	Technology	Description
Distributed systems meeting more than one critical property are beyond the state of the art.	Horus distributed computing system	Group communications system supporting secure, real-time fault tolerance
Secure Computing Platforms		
Rigid security and firewalls inhibit tightly coupled cross-domain applications and establishment of temporary security associations.	Adage authorization server	A distributed authorization server and policy language
	Sigma security middleware	Propagation of access control information across enclave boundaries
Detection and Tolerance		
Current intrusion detection techniques are limited to detecting a small set of well-known events.	Emerald intrusion detection system	Detects unknown attack types on networks
Current detection techniques do not scale and do not support automated response.	GRIDS intrusion detection system	Allows attacks to be specified as a graph across a network, thus allowing detection of larger-scale attacks
	Intrusion detection and isolation protocol (IDIP)	Allows coordinated detection and automated response
Current technology does not support incident response, which is a costly, labor-intensive process.	Diagnosis, explanation, and recovery from computer break-ins	Analyzes a system for indicators of an intrusion; explains to the system administrator what it found, what it probably means, and how to recover
Lack of standards impedes adoption of even current technology.	Common intrusion detection framework	Establishes standard interfaces for event generators (sensors) and analysis engines (detectors)
Current intrusion detection detects attacks on isolated machines, not attacks on the network infrastructure.	JiNao intrusion detection system	Detects intruders in network routers, switches, and network management channels; integrated with network management and has reconfiguration capabilities

TABLE 5.1 Continued

Problem	Technology	Description
Multicast group communication protocols are vulnerable to malicious participants.	Secure group and secure ring protocols	Group communication protocols that can prevent a malicious processor from disrupting the correct delivery of messages and from conducting successful denial-of-service attacks
Common programming errors leave most systems vulnerable to buffer overflow attacks, the most common type of attack on the Internet.	StackGuard compiler	Reduces vulnerability to buffer overflow attacks; no source code changes required; executables are binary-compatible with existing operating systems and libraries
Wrappers		
Commercial products introduce unreliability and security risks.	Generic security wrappers	Wrapper technology to augment legacy and commercial off-the-shelf components with security functionality; includes a wrapper specification language and a kernel-resident wrapper system; intercepts system calls to control privileged and nonprivileged programs; demonstrations include control of administrative privileges, access control, and encryption
Integrators cannot assemble components and wrappers into a trustworthy whole.	Composable replaceable security modules	A tool to build security into systems by assembling security functions from a library of reusable, plugable security modules, with standard functionality and interfaces
Vulnerability/resistance of a product or system requires evaluation.	Vulnerability assessment tool	White-box security evaluation tool locates vulnerable points in source code, using realistic attack models and taxonomies of known security flaws

5.7 INFORMATION ASSURANCE RESEARCH

Because there is a large shortfall of security technologies relative to naval needs, the Department of the Navy must be an advocate within the DOD for long-term research in several areas not being addressed by industry:

- Technologies for intrusion assessment to distinguish serious targeted attacks on critical systems from attacks on systems of lesser concern;
- Technologies for intrusion-tolerant systems, to maximize a critical system's ability to keep on providing service despite successful attack and partial system compromise;
- Technologies to limit an attacker's ability to carry out denial-of-service attacks; and
- Technologies to allow existing and legacy systems to be retrofitted with some security and reliability functionality.

Research is also needed in mobile code security, extending the capabilities of virtual private networks, and dependability.

There is, as well, a need to develop DOD-specific solutions for areas that industry is not addressing because there are no common commercial analogues, particularly in tactical networking. One risk faced by NCII in the tactical arena is that of enemy capture of an NCII end user terminal. Palatable solutions to this are not obvious and require further investigation.

5.7.1 Intrusion Assessment

Once local intrusion detectors are widely employed, it will become possible to make use of their collective outputs to try to gain a picture of the overall situation for the NCII. The aim of intrusion assessment is to distinguish serious targeted attacks on critical systems from attacks of lesser concern. This function takes reports from an infrastructure of intrusion detectors as inputs. The idea is to infer probable specific intent to carry out a particular hostile course of action for some specific purpose. This might be accomplished by correlating and analyzing observed or reported events in a way similar to the fusion of intelligence data (Figure 5.3).

When an attack is detected, it is useful to determine its mission impact as well as the potential mission impact of employing a given countermeasure (which may involve closing some connections for a time). To facilitate this, it might be useful for the common operational picture (COP) to include the NCII's computing and networking assets (links, nodes, and critical processes). Then an information assurance analyst could be given access to an appropriate COP to assess alternative courses of actions and plan actions.

For all this to be possible, the quality of the technology for detecting local

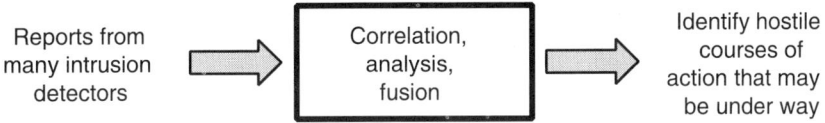

FIGURE 5.3 Fusion of intrusion detection.

intrusions must be greatly improved. The state of the art is illustrated by the following example. In a 2-week period, Air Force Information Warfare Center's (AFIWC's) intrusion detectors at 100 Air Force bases alarmed on 2 million sessions. Manual review reduced these to 12,000 suspicious events. Further manual review reduced them to four actual incidents. Almost every reported event is a false alarm or a trivial incident such as a failed password, and performing intrusion assessment from these alarms over all Air Force bases is a manual and laborious procedure involving hundreds of staff. Intrusion assessment technology should aim for intelligent filtering of events plus automated analysis requiring at most a handful of staff at a DOD reporting center or security anchor desk. Also, the current assessment task is limited to compiling statistics on the frequency of various types of attack, and it is not possible to obtain meaningful, specific, predictive information from this manual analysis.

The number of false alarms in local detectors may be reduced by developing better sensors. The sensors in use today are of two types: operating system audit trails and network sniffers. Both gather far more data than is necessary and omit data that may predict hostile activity. False alarms may be also reduced by adding a capability for peer-to-peer cooperation among local intrusion detectors, so that some limited assessment can be performed among a set of cooperating detectors covering a given region and events of only local concern can be suppressed from being reported to a central point. In support of this, detection components will need the ability to discover each other, negotiate requirements, and collaborate on diagnosis and response.

In addition, the detection technologies must be capable of detecting known attacks with high confidence and unknown attacks with reasonable confidence. The techniques must have very low false alarm rates, especially when they are used in very large systems, and must themselves tolerate missing, incomplete, untimely, or otherwise faulty data.

To analyze a collection of intrusion reports so as to identify specific, imminent, hostile activities of high consequence, it is necessary to infer the intent of hypothesized attackers. It may be possible to use automated planning and plan recognition technology to hypothesize goals for information warfare adversaries, develop plans for accomplishing each goal, and recognize when a plan is being carried out. Analysis techniques to be used at various reporting levels may include report correlation, automatic pattern detection, event classification,

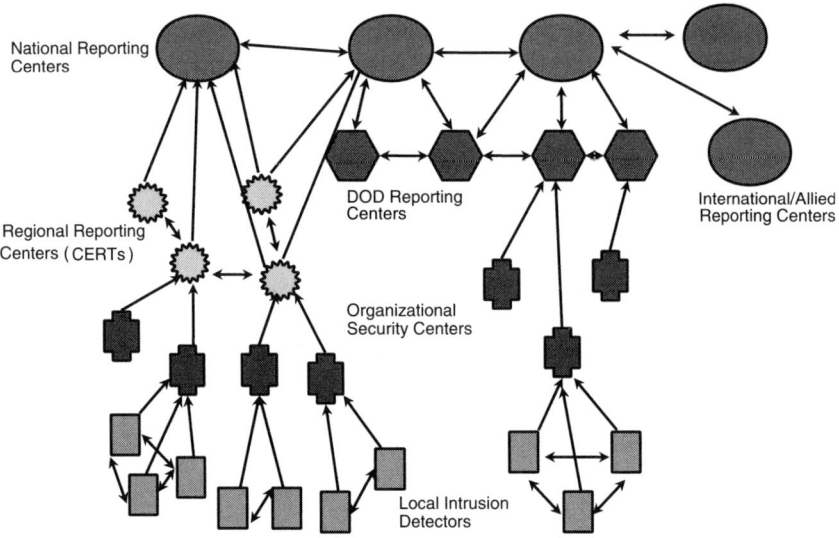

FIGURE 5.4 Network of intrusion detectors and reporting centers.

model-based reasoning to infer intent and predict future status, effect correlation to assess damage, and evidential reasoning to assess certainty.

Figure 5.4 shows a network of intrusion detectors and reporting centers. The reporting centers are organized roughly into layers, with the local detectors reporting into organizational security centers, which in turn report into regional reporting centers, which report into DOD and National reporting centers. At each layer of this network there would be a security detection and response center, which would have a number of functions (see Figure 5.5). The detection function analyzes and filters events reported from lower layers. The analysis is to find items of interest to this layer as well as items for reporting to higher layers. The assessment function attempts to understand coordinated events of interest to this layer and for reporting to higher layers. The tracing function can initiate tracebacks to attempt to discover the source of an event, or may participate in tracebacks initiated by other centers. The event notification function notifies peers or lower layers of hostile events happening elsewhere so they can prepare a defense. The automated response function can initiate local actions or instruct lower layers to take specific recommended actions to thwart a suspected in-progress attack or to implement specific defensive actions. This function can also exchange information with its peers to help decide what actions to take or to recommend to lower layers.

DARPA is beginning to invest in this area with its three new programs: Strategic Intrusion Assessment, Cyber Command and Control, and Autonomic

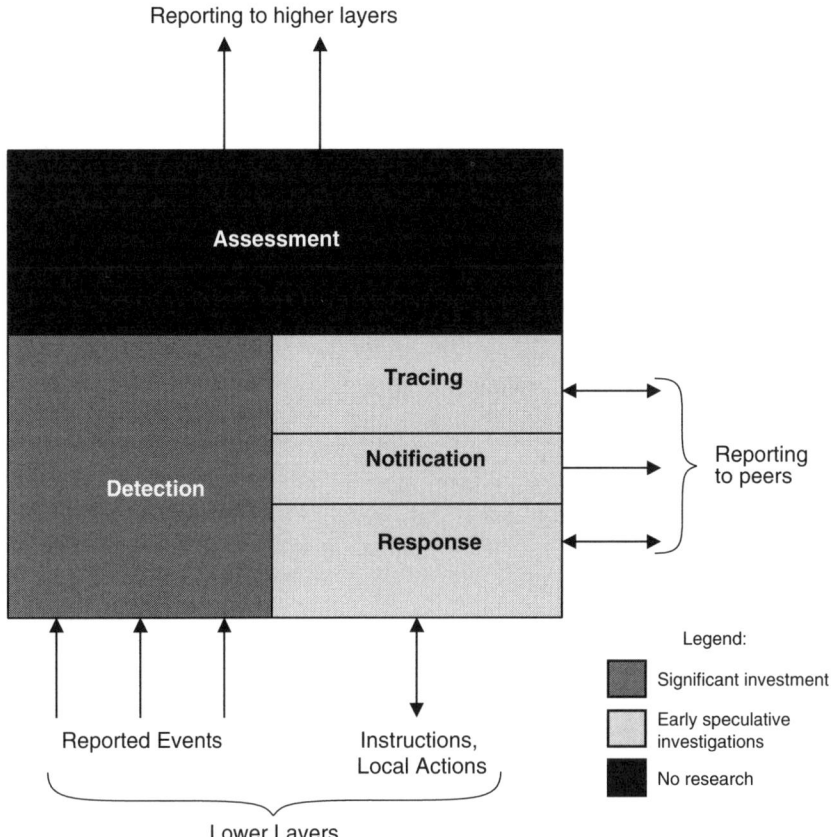

FIGURE 5.5 Security detection and response center.

Information Assurance. The Strategic Intrusion Assessment program aims to develop technologies capable of distinguishing significant patterns of events that cross geographic and administrative domains and that indicate a possible information warfare threat. It will develop a capability for peer-to-peer cooperation among detectors, including the ability for detectors to discover each other, negotiate requirements, and collaborate on diagnosis and response. The program will also develop techniques for correlating the output of intrusion detectors to infer the intent of the attacker. The Cyber Command and Control program will develop decision support tools to allow humans to assess the security status of a command and control system. It will assist human beings in ascertaining the activities and goals of adversaries attacking the system and in determining and carrying out the courses of action to counter them that are most effective and that interfere least with the system's ability to carry out its operational functions. The Autonomic Information Assurance program will develop technologies to allow systems to encode tactics to automatically detect and respond to routine known attacks so that the decision makers can focus on the strategic situation. The program will develop approaches for fast, adaptive defenses against these types of attacks. This should result in the ability to prevent damage from large classes of previously known attacks.

5.7.2 Intrusion Tolerance

Intrusion tolerance aims to ensure the continued correct operation of the surviving portion of a system even when it has been partially compromised. Component technologies include the ability to rapidly recognize corrupted data and programs, intrusion detection to recognize a local attack, techniques to constrain an attacker's resource consumption so as to minimize its opportunity to deny service, resource allocation methods to assign the most important tasks to the remaining resources, and methods to automatically repair damaged processes.

Traditional system designs have "central nervous systems" that, if attacked, can completely disable the system. Corrupted or malicious member entities can lead to incorrect functioning of the system as a whole. A possible architecture for an intrusion-tolerant system is one that is highly decentralized, so that the attacker cannot cripple the entire system by attacking one or a few points. The system could comprise many relatively independent processes that are collaborating to achieve a common goal. In highly decentralized systems, the overall behavior is the result of many small decisions made autonomously by member entities. One example of such a system is the Federal Aviation Administration free-flight air traffic control system currently being developed, in which individual aircraft will be able to negotiate with one another for their desired airspace rather than being controlled centrally from the ground. The Internet is another example of a system in which individual entities act autonomously yet collectively provide a global function. Such systems are inherently survivable in that a subset of the entities may be corrupted or lost and the overall system can continue to carry out its global function. (Note that if decentralized control is deemed to be too costly for routine system operation, the system can be designed to switch from centralized to decentralized operation when a threat is detected, and various degrees of decentralization may be designed in.) Protocols among the individual entities can allow them to cooperate to detect and isolate corrupted or malicious member entities. Survivability can be strengthened by the use of artificial diversity techniques, so that no single attack type can disable a large fraction of the processing entities. The system should be capable of self-monitoring, so that it can ensure that the most critical tasks get access to the uncorrupted resources. Self-monitoring can also recognize and trigger appropriate action when the collection of local responses results in some undesired system-level behavior. This collection of strategies can make systems inherently resistant to attack, much as the immune system makes humans resistant to disease. These defenses limit the spread or impact of an attack.

Intrusion tolerance requires an intrusion detection and response capability. This in turn requires targeted and possibly redeployable sensors to perform security monitoring; an infrastructure of intrusion detectors that performs analysis on data collected by the sensors; and response elements capable of reacting to alerts issued by the detectors to isolate the attacker, assess the damage, and recover.

Economic forces are driving out computational diversity. Market forces and consolidation within the computer industry have resulted in a few major brands of software and hardware being ubiquitous. The NCII will also be largely homogeneous, with common products in use everywhere. Such homogeneity leads to widespread common vulnerabilities that can be exploited by common attacks. DARPA is beginning to investigate means of artificially introducing diversity into homogeneous systems. Introducing diversity into highly decentralized systems can limit the subpopulation susceptible to any given attack and can cause attacks to have only local effects or to die out before they spread widely. Diversity is also a hedge against unanticipated means of attack; at least some system elements will survive and provide a basis for reconstitution. Means under investigation for artificially introducing diversity include self-specializing software that could reconfigure itself for a new specialization in response to attack; compilers that could vary the location of buffers, data structures, and code sequences, making attacks against them nonrepeatable; and multiple diverse implementations of the same functions.

Executing processes may fail or behave incorrectly owing to a corrupted program state. Individual processes in a survivable system should be able to detect and repair a corrupt program state so as to tolerate attack and maximize their running lifetime. Processes could check on their own integrity, send out alarms, and restore themselves to a safe state. For example, they could repair dangling pointers, change variable values to maintain invariants, restart failed connections, purge filled buffers, resynchronize with other processes, close unresponsive wait states, restore to a good checkpoint, and reload code or data from disk.

DARPA's new Intrusion Tolerant Systems program is investigating innovative designs for decentralized systems that are inherently resistant to attack. The research will continue and build on research into artificial diversity techniques that were initially investigated in DARPA's Information Survivability program. It will also develop methods to enable member entities to detect and isolate corrupt entities. Survivable resource allocation methods are being developed to assign surviving resources to the most critical functions so as to allow damaged systems to continue functioning despite the loss of significant resources and even of critical centralized control functions. Monitoring approaches are being developed to give systems the ability to monitor global behavior and to take local action to prevent undesired emergent effects. Integrity techniques are being developed to allow continued correct operation of the surviving portion of the system even when an attack has compromised data and code. These will include techniques to rapidly distinguish intact from corrupted information after penetration, to protect mobile code from corruption, and to maintain the integrity of systems in the presence of attack.

5.7.3 Preventing Denial of Service

Attacks that consume system resources or make them unavailable to the legitimate users and processes of the system are known as denial-of-service attacks. These attacks waste system resources and can exhaust them. They can cripple network elements and disrupt network operation. For example, Internet routing protocols commonly in use today have vulnerabilities that allow malicious entities to spoof router table updates. This could result in the propagation of bad routing information throughout the network, with consequences ranging from poor network performance, through the inability to deliver messages to certain destinations, to the collapse of portions of the network. Similar attacks on the Internet domain name service could make the network unable to deliver messages. Attackers also could take advantage of high-cost protocol checks (such as authentication) to consume resources.

Denial-of-service attacks can be thwarted by constraining an attacker's consumption of resources. Techniques are needed that would constrain denial-of-service attackers to a small percentage of system resources and that would slow such attacks sufficiently so that they can be detected.

DARPA has announced the new Fault Tolerant Network (FTN) program[14] to develop focused technologies that support continued operation in the presence of successful attacks. The program will address in particular the vulnerabilities and issues expected to arise in the highly networked environments envisioned for future warfighting operations.

Many network services represent potential points of failure. The FTN program will apply fault tolerance ideas to network services to reduce the amount of damage sustained during an attack and to enable continuous correct service delivery even when the services are under successful attack. This should help ensure continued availability and provide a technical basis for graceful degradation of service under successful attack, as well as help maximize the residual capacity available to legitimate users.

---

[14]This program within the Information Assurance and Survivability program suite focuses on three areas to be studied and evaluated: (1) fault tolerant survivability, (2) preventing denial-of-service attack, and (3) active network response.

5.7.4 Hardening Legacy Systems

Naval systems contain a high percentage of legacy and off-the-shelf components that do not provide the properties needed for information survivability. Methodologies are needed to allow the insertion of security functionality into legacy systems to address the vulnerabilities of the large in-place legacy base, which will persist for years to come. Also, because commercial products will not have the security and survivability features required for critical naval applications, a means is needed to allow designers to selectively harden such products.

The concept of security wrappers appears promising for retrofitting some security functionality into legacy systems. Wrappers encapsulate an existing component by intercepting all input to and output from the component (see Figure 5.6). This is done so as to require little or, ideally, no change to the wrapped component or to the components that interact with it. Indeed, the wrapped component, as well as those components with which it interacts, will be unaware of the wrapper. The wrapper interposes additional security and survivability functionality. For example, a wrapper may be added to create a security log, which may be analyzed by a separate intrusion detector. Or, wrappers on sending and receiving components can perform encryption/decryption or message integrity checks. As another example, a wrapper can be used to perform access control by filtering access requests to a component. To add fault tolerance, a wrapper could transparently replicate a component.

A number of security and survivability functions, such as encryption, access control, security monitoring, message integrity, management and control, and replication, could be provided by wrappers. These basic functions could be implemented in reusable building blocks that could be automatically provided by a wrapper generator according to a wrapper specification. The building blocks might come in several varieties for the same function; for example, there might be several implementations of authentication (passwords, secure ID, Kerberos, Fortezza, and so on), each with different specified strengths and costs. A wrapper specification could provide requirements for strength of mechanisms and constraints on their costs. Such modularity would also lend itself to future upgrading to stronger mechanisms without having to completely reconstruct the wrappers.

A significant research challenge is to provide some assurance that the wrappers cannot be bypassed. Another challenge is to develop a methodology that would allow an integrator to predict the security and survivability characteristics of components and wrappers, as well as tools to assist an integrator in assembling components and wrappers in a trustworthy manner.

The DARPA Information Survivability program has investigated the use of wrappers for security and survivability and has funded a collection of security wrapper projects that are developing tool kits that allow a developer to automatically generate security wrappers from a set of wrapper specifications.

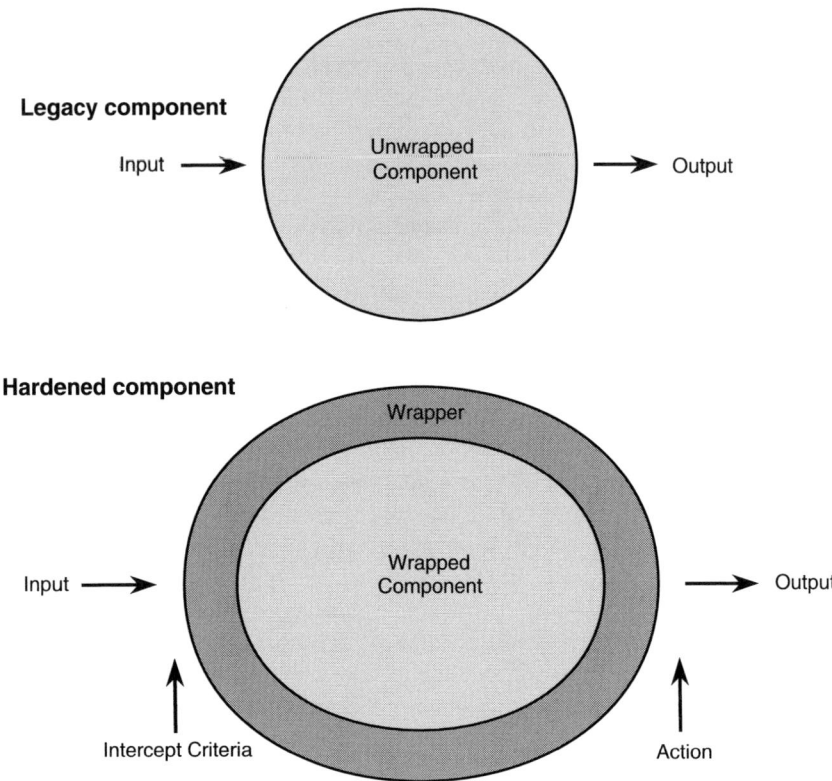

FIGURE 5.6 Concept of security wrappers.
SOURCE: Developed from information in Oostendorp, Karen A., Christopher D. Vance, Kelly C. Djahandari, Benjamin L. Uecker, and Lee Badger. 1997. *Preliminary Wrapper Definition Language Specification,* TIS Report No. 0684, Trusted Information Systems, Inc., Glenwood, Md.

5.7.5 Mobile Code Security

In the future, mobile code is expected to be used to deploy new services quickly and cheaply. While there are obvious security challenges inherent in doing this, commercial systems such as Java are rapidly developing many of the necessary security underpinnings for such uses. Continued attention is required to ensure that such security mechanisms are adequate.

Mobile code could also be used to more easily deploy new security functionality, as well as to upgrade existing functionality. This will make it easier to evolve and maintain systems. Active networking technology being developed by DARPA will extend similar capabilities to the network, resulting in networks that are extensible. For example, network functionality could be added to log events so that intruders can later be tracked, or to choose routes based on security considerations. To enable future naval uses of such technology, continued research is needed in mobile code security.

5.7.6 Dependability

While hardware dependability is a fairly well-understood science, this is not the case for software dependability. Software can have numerous bugs, and systems with large software components frequently fail, lock up, suffer unexplainable performance degradations, and can loose data. Dependability methods for hardware do not work for software, for many reasons. Hardware faults are often the result of parts wearing out or are due to random effects such as stray radiation. Since these types of faults tend to be independent, in that failure in one part does not affect the likelihood of failure in another part, failure models can be developed for commonly used hardware components to predict such characteristics as mean time to failure. The independence of the faults makes the use of redundant hardware components an effective fault tolerance strategy.

Software does not share these characteristics of hardware. It does not wear out. The causes for failure come from the humans who designed and wrote it. Failures cannot be assumed to be independent, especially because many copies of the same software are used in a system, but also because it is likely that different software developers will make the same kinds of errors. This makes it is very difficult to use redundancy as a means of software fault tolerance. Detection of and recovery from software faults is a research area that is not receiving much funding in the United States. Much of the research is motivated by the need to build ultracritical systems whose failure has very costly consequences, such as nuclear power generation, flight control, or air traffic control. Solutions that are acceptable for such systems are generally too costly for less critical systems, such as command and control systems. This means that designers of most of the large-scale computing systems in use today must use ad hoc means of obtaining reliability in those systems.

5.8 RECOMMENDATIONS

While a technical solution for the information assurance problem does not seem possible in the foreseeable future, the benefits to be gained from network-centric operations nonetheless make such operations imperative. A program of vigilance, testing, and continuing information assurance research will therefore be required. The committee recommends as follows:

Recommendation: The Department of the Navy should endeavor to use all available technologies to secure the NCII.

NRL could be tasked to help the Department of the Navy maintain an awareness of available technologies and also to follow in-progress research and select promising new technologies to develop and disseminate throughout the Department of the Navy. Earlier in this chapter in section 5.6, the committee identifies possible near-term technologies for consideration.

Recommendation: The Department of the Navy should take steps to ensure that systems are continuously maintained in a secure state.

Even with suitable technologies installed, continuous effort is required to maintain systems in a secure state. This can be accomplished by a combination of technical and procedural means, including regular internal use of red teams to test system configurations. System managers must keep up to date with the latest patches, fixes, and recommended configurations. A central source of this information within the Department of the Navy would help with this effort.

Recommendation: The Department of the Navy should take steps to increase and maintain security at all levels of personnel.

Red teaming exercises are valuable as a security awareness tool, especially to bring security to the attention of management. In addition, all naval personnel should be made aware of and trained in information assurance.

Recommendation: In designing the NCII, the Department of the Navy should use a defense-in-depth strategy to address unknown vulnerabilities.

In designing the NCII, it is not safe to take a risk management approach that assumes that all system vulnerabilities can be discovered and thereby will provide a basis for protection strategy. Instead, all components must be treated as vulnerable, including any protection mechanisms. Layered protection and a defense-in-depth strategy are required. Procedural and physical security measures should be developed to further reduce the risk where the available technology is not adequate.

Recommendation: The NCII should be designed to address security and assurance for tactical links.

Although cryptography makes spoofing difficult, attention must be paid in the tactical portion of the NCII to reducing the hazards arising from an adversary's capture of equipment, to providing a way to incorporate coalition partners in the NCII, and to defeating traffic analysis. The roles of spread spectrum and directional antennas in defeating jamming and interception are well understood, but existing spread-spectrum devices and multibeam directional antennas are expensive. In the near term, programmable modular radios should be programmed to adapt their waveforms and data rate to instantaneous jamming conditions. In the long term, affordable, high-gain, multibeam antennas should be sought.

Recommendation: The CNO and the CMC should take steps to ensure that fleet and Marine training encompasses situations with impaired information and NCII functionality, and that fallback positions and capabilities are prepared to meet such eventualities.

Network-centric operations will depend for their success on the correct functioning of the NCII. But NCII functionality may deteriorate or malfunction or be unavailable due to attack or failure. Because it is likely there will be attack on the NCII, it is imperative that naval forces train for situations with impaired NCII function. By this the committee means not only that the system staff should train to quickly restore service, it also means that operational forces should train to deal with system failure situations. The Department of the Navy has a tradition of developing operational workarounds for loss or degradation of radio frequency communications in tactical operations. The same should be done for the NCII. Without this, naval forces will be unprepared to deal with these situations, which are likely to arise.

Recommendation: When mission-critical networks and functions are considered for interconnection, the Department of the Navy should weigh the risks against the benefits.

A prominent feature of the NCII is its ability to connect disparate networks and functions. The NCII designers should recognize that while this brings great benefit, it also brings risks. When mission-critical networks and functions are considered for connection, it is essential to weigh the risks against the benefits. For example, certain functions may be considered so critical that any risk to their timely and correct functioning during a crisis is intolerable. In such cases, the decision may be to not connect, and to use an air-gap defense.

Recommendation: The Secretary of the Navy, the CNO, and the CMC should assign responsibility for information assurance at a high enough level within the Navy and the Marine Corps, and with sufficient emphasis, to ensure that adequate and integrated attention is paid to all aspects of this problem in the design and operation of the NCII.

Information assurance is not receiving appropriate attention at high levels within the Department of the Navy. Currently no single individual within the Department of the Navy has this responsibility and authority.

Recommendation: The Department of the Navy should push for research to address its critical NCII information assurance needs.

Because there is a large shortfall of current security technologies relative to naval needs, the Department of the Navy must be an advocate within the DOD for long-term research in several areas not being addressed by industry, including intrusion assessment, intrusion-tolerant systems, prevention of denial of service, approaches to retrofitting legacy systems with some security and reliability functionality, mobile code security, extending the capabilities of virtual private networks, and dependability. There is also a need to develop DOD-specific solutions for areas that industry is not addressing because there are no common commercial analogues, particularly in tactical networking.

6

Realizing Naval Command and Information Infrastructure Capabilities

This chapter provides a high-level assessment of the ability of the Department of the Navy to realize the functional capabilities that the Naval Command and Information Infrastructure (NCII) must provide, as defined in Chapter 4. The chapter begins with a discussion of the baseline naval systems contributing to these functional capabilities (where baseline is taken to mean what is planned over the next few years). IT-21 is the Navy's major strategy for realizing NCII-like capabilities, and so the baseline description presents an overview of IT-21 followed by more detail on certain aspects of it (e.g., communications, Global Command and Control System-Maritime). The chapter considers each of the functional capabilities,[1] discussing where they are likely to be in the near term (next several years) and in the longer term. Based on that assessment, the committee's findings and recommended approach to achieving the NCII functional capabilities are presented in the concluding section.

6.1 BASELINE NAVAL SYSTEMS

6.1.1 Introduction

The communications and information needs of the Navy and Marine Corps follow from the unique characteristics of and tasks assigned to warships and Marine units. The maritime environment and the requirement to operate with

---

[1] Information assurance, however, is treated separately in Chapter 5.

Army, Air Force, and allied forces further shape the configuration and capacity demands of the naval services.

Communications to and from ships are constrained by the limited space available for antennas and equipment and by the fact that such hardware is built in. As a consequence, ship communications suites are not readily reconfigurable to meet changing needs and, in general, a ship's communications capability is largely fixed from the moment of deployment for wartime operations or routine peacetime presence missions. Additionally, antenna placement is a crucial factor because shielding by the superstructure, the motion of the ship induced by high seas, and even routine course changes can adversely affect communications connectivity.

Amphibious ships pose a special case. These vessels of course have the communications and information needs characteristic of any warship. Additionally, the requirements of embarked Marine units, which are wholly dependent on host ships for planning and executing landing operations, must be reflected in the design of amphibious ship communications suites and information systems.

For both routine peacetime deployments and combat, Marine Corps units are organized in Marine air-ground task forces (MAGTFs) in a form and in numbers that depend on the anticipated situation and mission. Once ashore, a MAGTF may be the sole ground element present or it may operate in concert with U.S. Army or coalition forces. During its movement from ship to shore and once established there, the MAGTF employs its own organic communications and information resources to link to Navy ships at sea and to neighboring land forces, if present. As time passes and dependence on immediate fire and logistics support from the sea diminishes, the MAGTF communications architecture takes on a form not unlike that of the Army, with ties to adjacent land force elements and higher-level commanders in theater.

Just as a MAGTF organization is tailored for a particular mission, the communications and information systems to be employed are specially shaped as well. Subject to lift constraints on the weight and cubage that can be transported during an operation, the Marines can and do supplement standard allowances of communication equipment to meet the requirements of the tactical situation the MAGTF expects to encounter. In general, therefore, Marine units are not subject to the kind of built-in communications limits of Navy warships. However, the special needs generated by new tactical concepts, such as Operational Maneuver From the Sea (OMFTS) make reliable connectivity a very real challenge and clearly call for enhanced capabilities.

Because of tightly coupled lift, communications, fire support, and logistics dependencies, it is hard to imagine the Navy and Marine Corps operating in a forward area in isolation from one another, although they may well operate independently of Army and Air Force units under some circumstances. Increasingly, however, naval forces must fit into a greater joint forces construct, and this, in turn, requires enhanced communications to assure connectivity with other forces and the joint command structure and to profit from information collected by National and other Service intelligence, surveillance, and reconnaissance systems.

What follows is a review of the status of the Navy's IT-21 initiative, the new Navy/Marine Corps intranet, the communications and network posture of the Navy and Marine Corps, and the Global Command and Control System-Maritime, which is the key command and control tool for the naval forces. The committee focused on operations afloat and, for Marines, operations ashore in theater, in assessing the current status and adequacy of communications and information systems. While the committee recognized the essential support role played by Navy and Marine Corps commands in the United States, it decided to concentrate on the more challenging environment characterized by an absence of fiber-optic land lines and severe constraints on space for computers, servers, antennas, and communications equipment.

6.1.2 IT-21: The Navy's Principal IT and C4I Thrust

The Navy's reliance on and investment in communications, particularly satellite communications (SATCOM), has increased dramatically in recent years owing in great measure to the need to exploit the benefits offered by long-range precision weapons and by information from a variety of ISR systems, all in support of new concepts such as network-centric warfare and OMFTS, as well as the need to operate effectively with joint forces. The positive impact of these investments was first felt in Navy shore command centers when improved intelligence and situational awareness as well as enhanced connectivity to national and theater commanders were introduced. Additionally, special efforts were made to enhance command, control, communications, and computing (C4) capabilities and the availability of ISR information aboard aircraft carriers and fleet flagships, the major afloat command nodes for naval forces.

To realize the benefits offered by synergies between all ship types in a battle group or amphibious task force, it soon became evident that the communications and command and control (C2) capabilities of vessels other than nuclear-powered aircraft carriers (CVNs) and amphibious assault ships, general purpose and multipurpose (LHA/LHDs), would also have to be upgraded. And here the Navy was faced with the challenge of making hardware and software changes to a variety of ship types, each of which had an overhaul and maintenance schedule different from the schedules of other ships in the battle group with which it was to deploy. Only by solving this problem and deploying ships with matching capabilities could the battle group commander be assured of having ships with communications and information systems that permitted full exploitation of the potential of the naval weapon systems embarked.

As a consequence, IT-21 was born in 1998 as a fleet-driven initiative to coordinate and accelerate the installation and testing of modern information technology (IT) and command, control, and communications (C3) systems already in the acquisition pipeline as well as the training of personnel to operate them. The goal is to ensure that capabilities are in place at deployment time to effect a bridge between ships afloat, space assets, and command centers ashore. Because of funding constraints, the initial focus was on ships at sea, and investments in shore infrastructure were limited to those necessary to support forces afloat and Marine operations ashore.

The principal elements of IT-21 are as follows:

- Full SATCOM capability for all surface combatants;
- Major capacity enhancements to amphibious ship communications;
- Improved shipboard command and control capabilities such as GCCS-M and improved planning and decision tools;
- Enhanced support communications, processing, and storage;
- Robust shipboard local area networks;
- Modern personal computer workstations and commercial-based operating system;
- Matching capacity upgrades at shore communications hubs; and
- Measures to improve information assurance and security.

It is important to note that this initiative is not a program in the sense of an acquisition but, rather, a strategy to install improved C4 hardware and software, most of it already being procured, in an orderly and controlled fashion. The goal here is to install enhancements such that all ships in a carrier battle group and amphibious task group will have compatible C3 capabilities upon deployment.

The mechanism employed in IT-21 is essentially a spiral installation process whereby the configuration is fixed well in advance of the deployment date so that sufficient time is available to install, test, and train personnel to operate the new hardware and software elements. Recommendations derived from operating new systems during deployment are then integrated with assessments of the value and availability of new technologies and components from the acquisition pipeline. From this process emerges a configuration that is specified for installation in time for the next deployment of the battle group and, depending on its complexity, possibly earlier in groups deploying sooner.

If a ship type is said to be IT-21-capable, this does not mean it has the same C4 capabilities as, say, an aircraft carrier. Rather a smaller IT-21-capable ship will have modern IT inserted and enough SATCOM capacity for it to apply its weapon system capabilities in support of the overall battle group mission.

A prototype IT-21 suite was deployed in 1998 in the *Abraham Lincoln* battle group and was the basis for refining the concept and developing a standard installation. The first of these deployed in summer 1999, and the spiral upgrade and installation process will continue through 2002, at which time full IT-21 capability will have been realized for all ships. The spiral process could very well continue after 2002. However, plans for the future are not determined at this point and probably will be keyed to the evolving Navy/Marine Corps intranet (N/MCI) initiative discussed below.

6.1.3 Navy/Marine Corps Intranet

At the time IT-21 was initiated, funding constraints precluded inclusion of the business side of the Navy and its shore support infrastructure in a comprehensive IT upgrade program. Nevertheless, the need to upgrade and integrate the several shore networks that have developed over the years was recognized. These networks include those built around regions or base areas, the Marine Corps enterprise network, and the Naval Air, Naval Sea, and Naval Supply Systems Commands (NAVAIR, NAVSEA, and NAVSUP) networks. All of these, and others, were developed independently; they do not interoperate well (or at all), they lack adequate security provisions, and in aggregate they are expensive to operate and maintain. The Navy/Marine Corps intranet (N/MCI) concept was thus developed to address the goal of having a federation of networks and computers that work as a single integrated system.

The N/MCI concept has been approved by the Secretary of the Navy (SECNAV), the Chief of Naval Operations (CNO), and the Commandant of the Marine Corps (CMC), resources for implementation are being identified, and design and procurement actions are under way. As stated in a recent briefing to industry, the N/MCI is the Department of the Navy enterprise-wide network capability that will provide end-to-end, secure, assured access to the full range of voice, video, and data services by year end 2001, as depicted in Figure 6.1. A coherent department-wide network is the goal, resulting in increased efficiencies and enhanced business and warfighting processes.

The task of implementing the N/MCI is to be given to industry. Bidders for the integration and system operation contract have been informed that they are not bound by any preconceived architecture and are not required to use existing information system or technology infrastructure. The DOD architectural framework will be followed and the resultant network or system is to be defense information infrastructure/common operating environment (DII COE) compliant. The precise relationship of this initiative to IT-21 is not yet clear, but was to be elucidated before the contract bid package was issued in late 1999.

6.1.4 Communications and Networks

Communications and networking services have evolved over the history of the Navy and Marine Corps as a critical component in the accomplishment of any and all assigned missions. This capability extends from the days of visual means (signal flags and lights) for communicating between and among various command elements, high-frequency circuits using Morse code, frequency shift key-

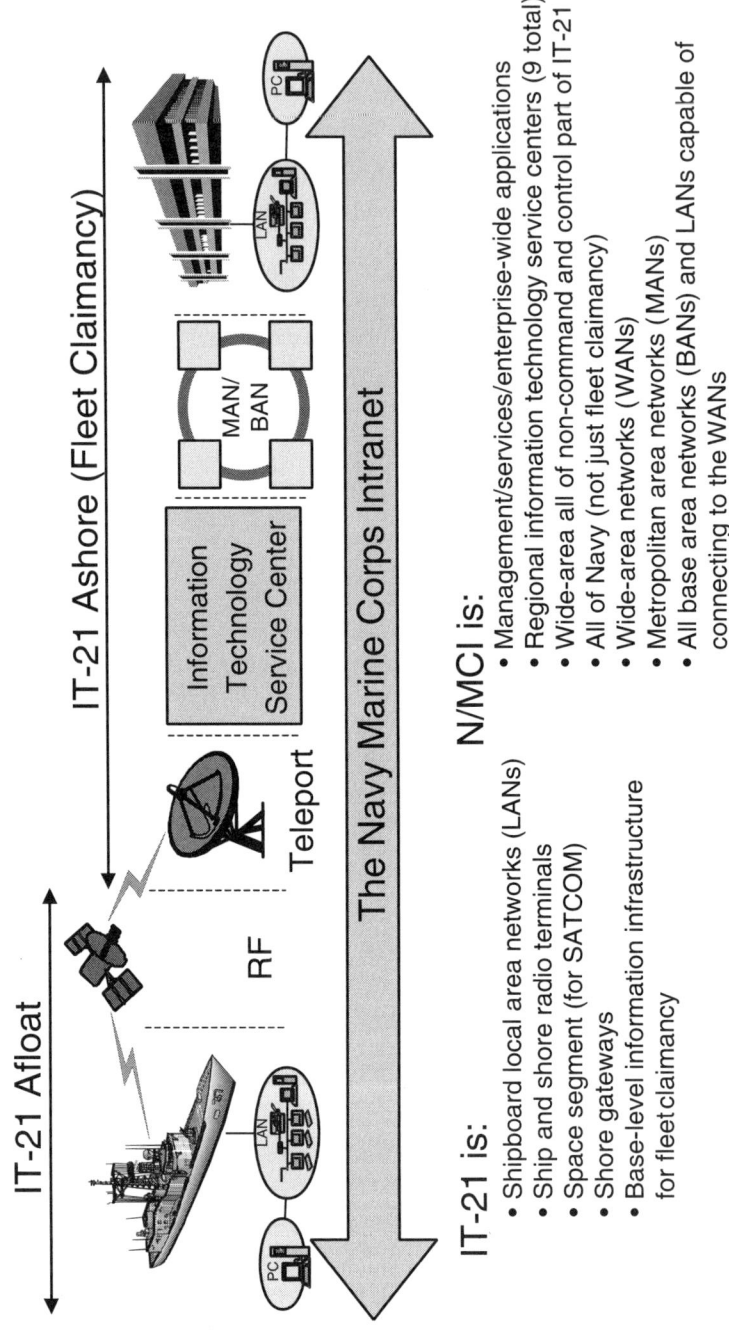

FIGURE 6.1 Naval and Marine Corps intranet conceptual view. ITSC, information technology service center.
SOURCE: Fleet and Allied Requirements Division (N60), Office of the Chief of Naval Operations. 1999. "End-to-End Capability," in *Information Technology for the 21st Century* [IT-21 Generic Brief]. The Pentagon, Washington, D.C., June 18. Available online at <cno-n6.hq.navy.mil/n60/documents.html>.

ing (FSK) and other modulations for long-haul communications, to more modern, high-capacity multimedia (voice, data, video) terrestrial-, satellite-, and airborne-relay and line-of-sight tactical connectivities. It is safe to say that virtually all usable regions of the physical frequency spectrum (acoustic, electromagnetic, and optical) have been and are continuing to be employed by naval forces for basic communications in all types of operational and physical propagation environments. These capabilities have been operated in combinations of network configurations, including point-to-point, broadcast, and multicast, using a wide range of protocols for access and use of the network.

Because of the need for mobility, much of the naval communications infrastructure is provided by radio-frequency circuits and networks that operate over a wide portion of the electromagnetic spectrum, from extremely low frequencies (ELFs) at tens of hertz (Hz), to extremely high frequencies (EHFs) at tens of gigahertz (GHz). Radio frequency propagation characteristics, information bandwidth, and operational posture are the key parameters for selecting the frequency band of operation for a particular application. For example, communications to submarines use the lower frequency bands (ELF, VLF, and LF) to allow the signal to penetrate seawater or reach floating wire or towed buoy antennas at long distances (thousands of miles) when the platform is submerged. The information bandwidth at these frequencies permits only low data rates, however, generally from a few bits per minute to roughly 50 bits per second (bps).

Operation in the high-frequency band allows increased data rates (up to several kilobits per second) at beyond line-of-sight distances using both ionospheric and ground wave propagation modes. One must move to the ultrahigh-frequency (UHF), superhigh-frequency (SHF), and EHF bands to realize high information throughput (tens to thousands of kbps). In doing so, however, the operator must be willing to deal with line-of-sight distances (requiring relays for long distance connectivity), point and tracking systems because of the narrow antenna beam widths, and various deleterious effects from atmospheric attenuation due to water vapor and scintillation.

The myriad communications paths linking ship to shore, ship to ship, and one Marine unit to another carry a variety of information (Figure 6.2). This includes urgent command orders, critical intelligence, tracking data on friendly and enemy forces, information drawn from data repositories remote from the requesting ship, routine peacetime "business" communications, urgent requests for spare parts, and also quality-of-life items (e.g., personal phone calls and e-mail).

It is, however, the increasing call for and availability of synthetic aperture radar (SAR) and electro-optical/infrared (EO/IR) imagery collected by airborne and space sensors that is a principal driver in determining shipboard communications capacity and equipment needs. On the other hand, moving-target indicator (MTI) and signal intelligence (SIGINT) data from platforms such as the Joint Surveillance and Target Attack Radar System (JSTARS), the U-2, and Rivet

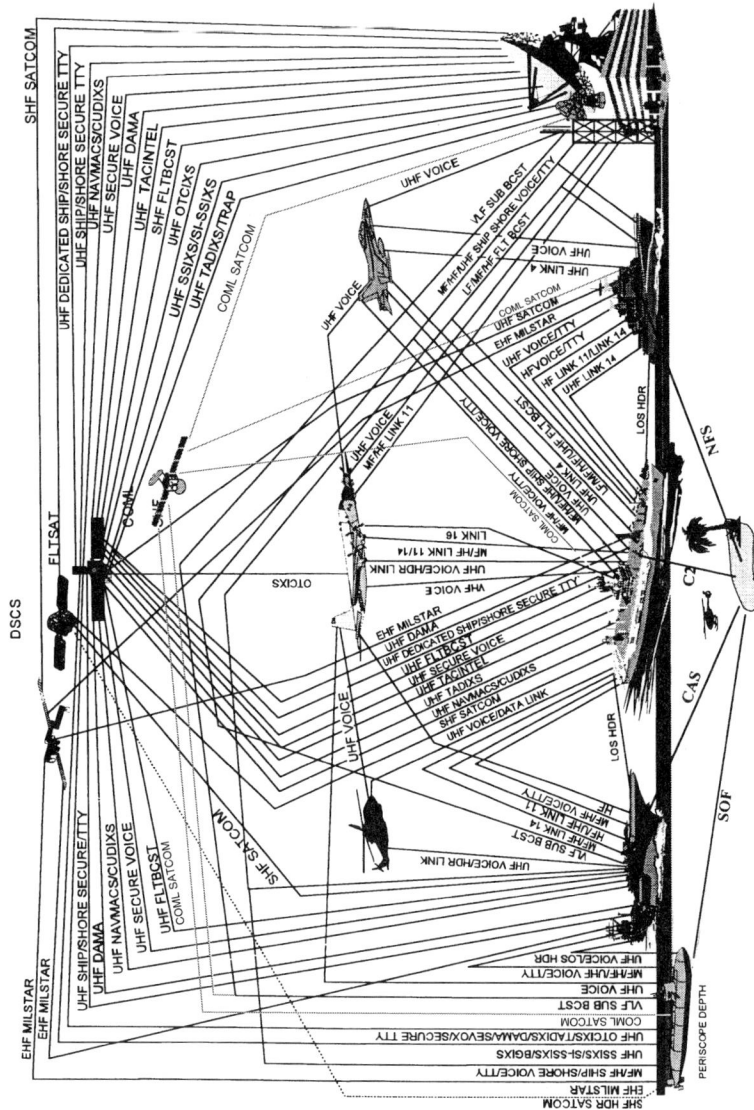

FIGURE 6.2 Communications paths and connectivity today. Courtesy of the Director of the Space, Information Warfare, Command and Control Directorate (N6), Office of the Chief of Naval Operations, Washington, D.C., March, 1999.

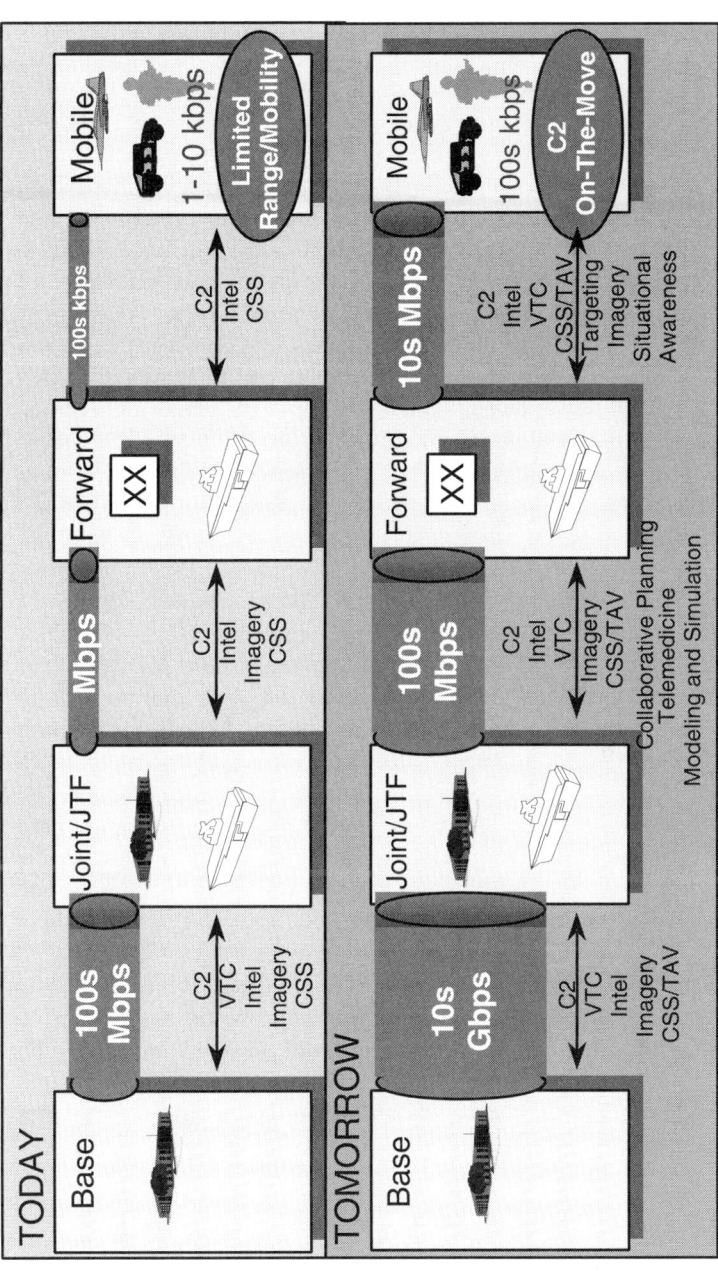

FIGURE 6.3 Communications bandwidth trends. CSS, combat support systems; TAV, total asset visibility; VTC, video teleconference.
SOURCE: Defense Science Board. 2000. Modified from Figure 3.2, "Findings: DoD Requirements; DSB Assessment," in *Report of the Defense Science Board on Tactical Battlefield Communications*, Office of the Under Secretary of Defense for Acquisition and Technology, Washington, D.C., February, p. 53.

Joint require comparatively little bandwidth even though they need special data links and surface terminals. Figure 6.3 shows this trend in bandwidth, brought on, in great measure, by the need for imagery.

Figure 6.4 depicts many of the networks and communication paths to, from, and between ships at sea, their transport capacities, and the services that travel over them. Of particular note is the significant overall increase in SATCOM capability compared to the limited UHF SATCOM bandwidth available during Desert Storm, as well as the increasing use of commercial satellites. Not shown are line-of-sight UHF/VHF circuits, tactical data links, and special links with airborne imagery platforms. Also not indicated is HF radio, which continues to play an important communications role, today carrying some 10 percent of all traffic, including supplying essential connectivity to allied forces.

The communications configurations of individual ship categories are shown in Table 6.1. Capabilities being installed incident to the IT-21 initiative vary between ship types and are dependent on mission needs. Not shown are certain mine warfare vessels and MSC-operated logistics support ships; however, these are being equipped appropriately as well. Of particular note are the special data links that, if installed and matched with appropriate terminals and exploitation segments, can provide real time, direct imagery feeds, along with SIGINT and MTI data, to ships so equipped.

Turning to the Marine Corps, a MAGTF commander is able to communicate with ships and between his units using UHF/VHF line-of-sight and HF radios during the early phases of a classic amphibious operation. But because of the fluid nature of such operations, establishing and maintaining communications between units has always been challenging. Now, two new factors have added to the difficulty: (1) the need for imagery feeds and products that demand much higher frequencies and greater bandwidth and (2) implementation of the OMFTS concept, which calls for over-the-horizon operations by dispersed units, at least during the early phases of a campaign, as shown in Figure 6.5. As a consequence, forward small units will require SATCOM and airborne communications relay resources, items that have not heretofore been included in a standard equipment list.

Table 6.2 shows the networks applicable to selected MAGTF units moving to or operating ashore, along with the equipment considered standard for a given level of command. As noted above, however, communications suites for MAGTF units can be, and usually are, tailored for the particular tactical situation. Capabilities can be added, subject to the availability of transport to lift the equipment into the objective area.

6.1.5 Global Command and Control System-Maritime

Global Command and Control System-Maritime (GCCS-M) is the principal Navy command and control tool for commanders and ship commanding officers.

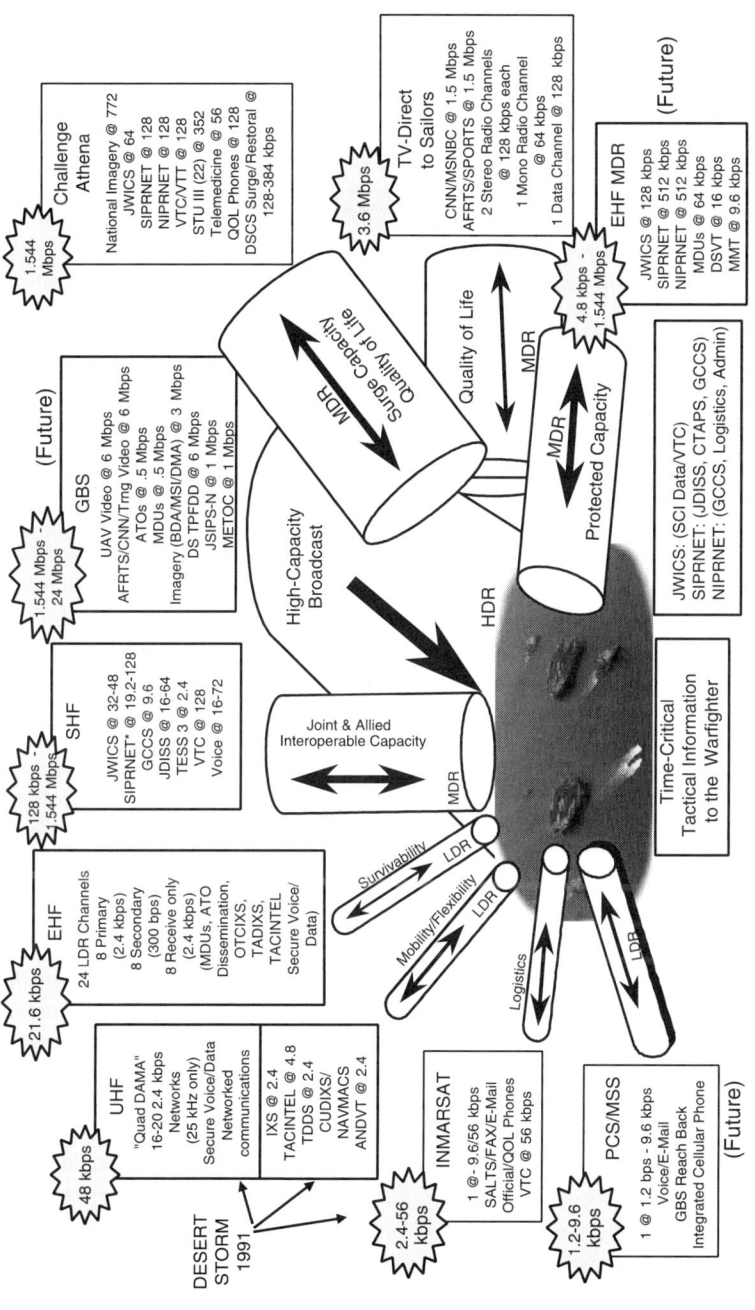

FIGURE 6.4 Satellite communications services and systems. Acronyms are defined in Appendix H.
SOURCE: CNO N6 PRO1 Baseline Assessment Memorandum (Draft), December 17, 1999.

TABLE 6.1 Communications Configuration by Ship Type

Ship Type/Class	Satellite Communications												Non-SATCOM Radio Frequency Communications											Data Links									
	EHF (MDR & LDR)	GBS (Receive)	SUB HDR Antenna	SHF (DSCS X Band)	ITP (C/X Band)	CWSP (CA III) (C Band)	INMARSAT B HSD, B, M	SUB UHF MDR Asymmetric	IBS (Receive)	UHF DAMA (5/25 kHz)	DWTS	UHF (LOS/HQ/NTDS)	VHF Bridge to Bridge	VHF Aeronautical	MCIXS	SINCGARS	HF E-mail (STANAG 5066)	HF Modems	HF Transmit	LF/MF/HF Receivers	LF/VLF (Receive)	ELF (Receive)	LINK 4A	LINK 11	LINK 11 Satellite	LINK 16	LINK 16 Satellite	LINK 22	CDL-Navy	TCDL	CEC	EPLRS
Command Ships																																
AGF	D	D		2		X	N			12	1	15	2	2	BS	12	1	8	19	31			1	1	—	—	—	1	X	X		
LCC	D	D		2		X	N			12	1	28	1	2	BS	16	1	12	32	49			1	1	—	—	—	1	X	X		
Carrier Battle Group Ships																																
CV/CVN	D	D		2		X	N			12	1	37	2	2	BS	2	1	7	17	30			1	1	—	1	—	1	X	X	X	
CG	S	S		1			D+			8		18	1	3	TS	2		6	11	16				1	—	1	—	1		X	X	
DDG	S	S			X		S(I)			8		18	1	3	TS	2		5	9	14				1	—	1	—	1		X		
DD	S	S					S(I)			4		9	1	2	TS	2		5	8	12				1	—	1	—			X		
FFG	S	S					S			4		8		3	TS	2		4	5	9				1	—	—	—			X		
AOE	S	S		1			S			4		8	1	2	TS	2		4	6	8				1	—	—	—					
Amphibious Ready Group Ships																																
LHA	D	D		2		X	N			12	1	19	1	2	BS	16		9	23	35			1	1	—	1	—	1	X	X	X	X
LHD	D	D		2		X	N			12	1	23	1	2	BS	16		9	24	35			1	1	—	1	—	1	X	X	X	X
LPD-17	S	S		2		X	D+			8		15	1	2	TS	16		7	20	31				1	—	1	—	1		X		X
LPD-4	S	S		2			D+			8		15	1	2	TS	12		7	19	31				1	—	—	—			X		X
LSD	S	S		2			S			4		1		1	TS	10		5	9	16				1	—	—	—			X		X
Submarines																																
SSN	S	S	X	1					X	4		2	1				1	2	2	2	7	1		1		—				X		
SSBN	S	S	X	1				X	X	4		2	1				1	2	2	2	12	1		1						X		

SOURCE: Compilation of data courtesy of the Office of the Chief of Naval Operations, Space, Information Warfare, Command and Control Directorate (N6), Washington, D.C., 1999.

Legend:
X, System/capability planned
D, Dual terminal required
S, Single terminal required
(#), Number of transmit/receive channels
L, Internet protocol package
NIS, National Input Segment

D+, Dual International Marine/Maritime Satellite (INMARSAT) until SHF installed, then single
N, INMARSAT B, non-HSD
M, INMARSAT M Terminal
S(I), SNFL Flagship Dual INMARSAT HSD Required
@, Requirement for SNFL Flag Ship
SATCOM, Satellite communications

BS, Maritime Cellular Information Exchange Service (MCIXS) Base Station
TS, MCIXS TELELOR Station
NTM, National technical means
TCDL, Tactical common data link
CDL-Navy, common data link-Navy

TABLE 6.1 Continued

Ship Type/Class	Switching			Applications					Connectivity					Direct Imagery/SIGINT Reception*									
	TRITAC	ADNS	ADNS SCI	NAVMACS/SMS/DMS	SCI TAC VTC	TAC VTC	GCCS-M	NTCSS	Coalition WAN	NIPRNET	SIPRNET	JWICS	PIER	NTM (indirect via NIS/CA)	U-2	JSTARS	River Joint	F-14/F/A-18	P-3/EP-3	Global Hawk	Predator	Pioneer	Future UAVs (via TCS)
Command Ships																							
AGF	x	x	x	x	x	x	x	x	x	x	x	x	x	x	x	x	x	x	x	x	x		x
LCC	x	x	x	x	x	x	x	x	x	x	x	x	x	x	x	x	x	x	x	x	x		x
Carrier Battle Group Ships																							
CV/CVN	x	x	x	x	x	x	x	x	x	x	x	x	x	x	x	x	x	x	x	x	x		x
CG		x	x	x		x	x	x		x	x	x	x				x	x	x		x		x
DDG		x	x	x		x	x	x	x	x	x	x	x				x	x	x		x		x
DD		x	x	x		x	x	x	x	x	x	x	x				x	x	x		x		x
FFG		x		x		x	x	x	x	x	x	x	x				x	x	x		x		x
AOE		x		x						x	x	x	x										
Amphibious Ready Group Ships																							
LHA	x	x	x	x	x	x	x	x		x	x	x	x	x	x	x	x	x	x	x	x		x
LHD	x	x	x	x	x	x	x	x		x	x	x	x	x	x	x	x	x	x	x	x		x
LPD-17	x	x	x	x	x	x	x	x	x	x	x	x	x				x	x	x		x	x	x
LPD-4	x	x		x		x	x	x		x	x	x	x				x	x	x		x		x
LSD	x	x		x						x	x	x	x										
Submarines																							
SSN		x		x			x	x		x	x	x	x				x	x	x	x	x		x
SSBN		x		x			x	x		x	x	x	x				x	x	x	x	x		x

*Direct reception contingent on installation of data links and exploitation systems that match sensor platforms (e.g., CDL-Navy and Joint Service Imagery Processing System-Navy/Tactical Input Segment (JSIPS-N/TIS) for U-2 and Global Hawk). TCDL will be the principal enabler for all ship types.

SOURCE: Compilation of data courtesy of the Office of Naval Operations, Space, Information Warfare, Command and Control Directorate (N6), Washington, D.C., 1999.

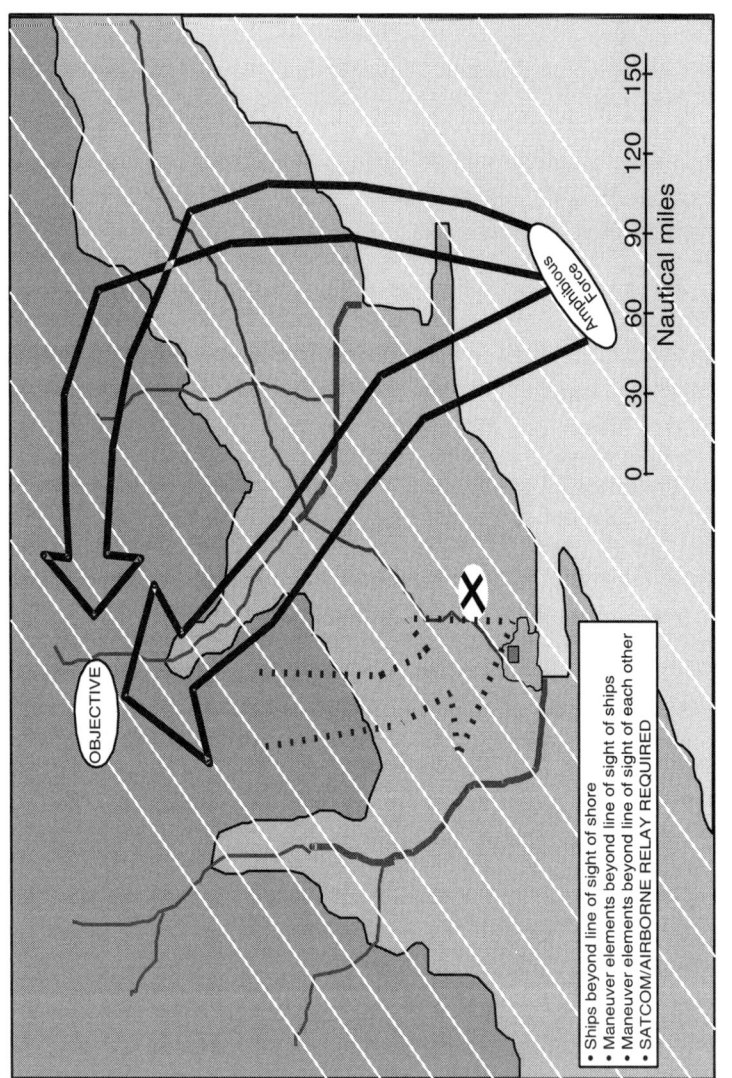

FIGURE 6.5 High-level OMFTS operational concept. SOURCE: Marine Corps Combat Development Command, Quantico, Va., 1999.

TABLE 6.2 Marine Corps Communications Network and Equipment Capabilities, Current (FY98) and Planned (FY06)

	Networks	MEF (FWD) CE FY98	MEF (FWD) CE FY06	MEU CE FY98	MEU CE FY06	Regimental CP FY98	Regimental CP FY06	Battalion CP FY98	Battalion CP FY06	Company CP FY98	Company CP FY06	Platoon CP FY98	Platoon CP FY06	Air Element FY98	Air Element FY06	FSSG (FWD) CP FY98	FSSG (FWD) CP FY06
1	SIPRNET	G	G	G*	G	R	G	R	G	R	G			G	G	R	G
2	NIPRNET	G	G	G*	G	R	G		G					G	G	R	G
3	JWICS	G	G	G*	G												
4	Secure Switched Voice (TRITAC Voice)	G	G	G*‡	G	G	G	G	G		G			G	G	G	G
5	HF (Long-haul Combat Net Radio)	G	G	G	G	G	G	G	G	G	G			G	G	G	G
6	VHF Freq-Hopping (LOS Combat Net Radio)	G	G	G	G	R	G	G	G	G	G	G	G	R	G	R	G
7	UHF TacSat (DAMA) (Satellite Combat Net Radio)	G	G	G	G	G	G	G	G	G	G	G	G	G	G	G	G
8	UHF LOS	G	G	G*	G												
9	SHF SATCOM (Stepsite/Teleport)	G	G														
10	SHF Broadcast	R	G	R	G	R	G							R	G	R	G
11	EHF LDR/MDR SATCOM (Intra/Inter-MAGTF)	R												R	G		
12	JTIDS/MIDS LVT													R	G		
13	VTC	G	G	G*	G												

	System/Equipment	Networks(s) Supported	MEF FY98	MEF FY06	MEU FY98	MEU FY06	Reg FY98	Reg FY06	Bat FY98	Bat FY06	Co FY98	Co FY06	Plat FY98	Plat FY06	Air FY98	Air FY06	FSSG FY98	FSSG FY06
	STAR-T	1, 2, 3, 9, 13	R	R	R	G												
	SMART-T	1, 2, 3, 9, 11, 13	R					G										
	TROJAN SPIRIT	3																
	AN/TRC-170(V)5	1, 2, 4	G												G	G		
	MRC-142	1, 2, 4	G				G		G						G	G	G	G
	DWTS	1, 8	G				G		G		G				G	G	G	G
	EPLRS	1, 8			R		R		R		G							
	AN/PSC-5	7, 8	Y	G	Y	G	Y	G	G	G		G			Y	G	Y	G
	SINCGARS	6	G	G	G	G	G	G	G	G	G	G	G	G	G	G	G	G
	JTRS	5, 6, 7, 8	R	G	R	G	R	G	R	G	R	G	R	G	R	G	R	G
	GBS Receive Terminal	10	R	G	R	G	R	G	R	G					R	G	R	G
	JSTARS CGS	8	R	G		G												
	JTIDS/MIDS LVT	12													Y	G		

NOTE: R, Capability not fielded (pre-initial operational capability); G, capability fielded; Y, between initial operational capability and full operational capability; *capability exists if equipped with the AN/TSC-93B and ancillary equipment; ‡capability exists if equipped with the SB-3865 switch and ancillary equipment.
SOURCE: Compilation of data courtesy of Marine Corps Combat Development Command (MCCDC), Quantico, Va., 1999.

According to the system mission statement, GCCS-M is intended to provide commanders with a single, integrated command, control, communications, and intelligence (C3I) system that receives, processes, displays, and maintains current geolocational information on friendly, hostile, and neutral land, sea, and air forces as well as intelligence and environmental information. In addition to receiving data from sensors and maintaining a common operational picture, all system variants are required to communicate with other GCCS-M locations and transmit, receive, review, and record message traffic. GCCS-M systems interface with a variety of communications and computer systems. As shown in Table 6.1, GCCS-M is currently operational on most Navy ships and, where germane, includes MAGTF command, control, communications, computing, and intelligence (C4I) software applications. It is installed at major fleet headquarters ashore and at tactical support centers (TSCs) to support antisubmarine and surface warfare missions. GCCS-M is also available in several mobile configurations.

GCCS-M is a system in transition—from a number of Navy-unique command and control systems to a single system fully compliant with the DOD's common operating environment (COE) and able to interface seamlessly with regional CINCs and Army and Air Force units. The transition to today's GCCS-M began after the Gulf War, when three principal families of Navy C2 systems were brought together under the Joint Maritime Command Information System (JMCIS) umbrella. At that time, many elements of existing C4I systems were aging and becoming expensive to operate and maintain. And because most of these systems were based on proprietary hardware, operating systems, and standards, the exchange of data among them was difficult and expensive, generally requiring unique communication interfaces to achieve interoperability within the Navy. So, rather than further the development of new stovepipe systems, the evolution to JMCIS was initiated. Finally, upon successful completion of a comprehensive operational evaluation, JMCIS '98 was renamed GCCS-M in recognition of the Navy's intent to bring its C2 system into DII-COE compliance, with a goal of full joint interoperability.

The JMCIS '98 program, now GCCS-M, set forth several key tenets that were to guide future system development. Three of them are worthy of mention:

- *Migration from a Navy-unique JMCIS COE to the DII COE.* GCCS-M today (version 3.1.x) is a level 5 system, the minimum required DII-COE compliance level. All Navy ships, Marine Corps units, and Navy and Marine Corps headquarters are fully interoperable. They also have seamless message traffic connectivity with other military services and joint headquarters and can exchange the track data needed for the common operational picture, provided a common COE software version is installed. Some ships have certain segments installed that are now DII-COE level 7 (fully interoperable), and the Navy has plans in place to migrate GCCS-M to full compliance (level 8) over time, subject to funding constraints.

- *Migration from Navy-specification UNIX-based hardware to commercial PC workstations, servers, and Windows NT operating systems.* This transition is under way and, as a component of IT-21, is planned to be completed by the end of 2002.
- *Combining tactical and nontactical networks, thereby permitting fleet personnel to perform both tasks on a single workstation.* This, too, is planned to be achieved by the end of 2002.

There are three basic GCCS-M system variants: afloat, ashore, and tactical/mobile, each with a heritage linked to the three Navy C2 system categories that existed before JMCIS. The afloat system variant is installed in some 250 ships and submarines and at certain shore sites. As noted previously, its purpose is to provide commanders afloat with a timely, authoritative, fused, and common tactical picture, along with integrated intelligence services and databases. It disseminates intelligence and surveillance data in support of mission planning, execution, and assessment. While core capabilities are identical for all afloat systems, such items as databases, support applications, and mission applications are tailored by individual class of ship.

At major headquarters ashore, the GCCS-M ashore variant is a C4I system that receives, processes, displays, maintains, and assesses the unit characteristics, employment scheduling, materiel condition, combat readiness, warfighting capabilities, positional information, and disposition of U.S. and coalition forces. It provides current geolocational information on hostile and neutral land, sea, and air forces integrated with intelligence and environmental information and near-real-time weapons targeting data to submarines.

Tactical/mobile variants are fielded at shore sites to provide commanders with the ability to plan, direct, and control the tactical operations of forces. These systems tend to be tailored for special purposes. One subvariant, for example, is a complete mobile command center for use by a naval component commander in joint operations. It provides connectivity with the joint task force commander, other component commanders, and afloat naval forces.

GCCS-M comes with a number of mission and support applications. Some common mission applications, not all of which would be installed in every class of ship, follow:

- Additional communications and messages
  — Tactical Information Broadcast System
- Integrated imagery and intelligence (I3)
  — GCCS-M tracks associated with modernized intelligence databases
  — Naval intelligence reference databases
  — Graphical plotting national intelligence reporting
  — Imagery displayed within the common operational picture
- Command and control links and nodes

— Spectrum management
— Tactical electronic order of battle
- Tactical warnings
  — Detect and display threat information
- Mine and antisubmarine warfare (ASW)
  — Sensor Performance Prediction Expeditionary Decision System
  — Integrated Carrier ASW Prediction System
- Meteorology
- Water space management
- Theater ballistic missile defense (TBMD)
  — Correlation of Theater Event System tracks
- Air tasking order (ATO) support
  — Parse and store incoming ATO messages.

6.2 FUNCTIONAL CAPABILITIES ASSESSMENT

This section elaborates on the description of the functional capabilities required in the NCII (see Figure 4.2 in Chapter 4) and discusses how and the extent to which these capabilities are expected to be realized in the near-term and more distant future. Addressed first are functions of the NCII's supporting resource base—communications and networking, plus system resource management. (Information assurance is discussed in depth in Chapter 5.) The treatment of communications and networking is divided into two parts, general considerations and some important particulars of wireless transport.

6.2.1 Communications and Networking—General

Three kinds of requirements for communications and networking are fundamental for the considerations of this section:

- *Connectivity and configurability:* the ability to establish communications among the required parties in a timely manner;
- *Capacity:* the availability of bandwidth for the voice, data, and video information that must be transferred in operational missions; and
- *Interoperability:* the ability to exchange information with other parties, including with other naval forces, with other Services and joint elements, and with coalition partners.

Security of the information transferred and assurance of its delivery, which are also critical, are discussed in Chapter 5. Affordability is also an important factor but is beyond the scope of this study, other than to note that increased reliance on commercial technology can help to reduce costs.

6.2.1.1 Near-term Assessment

6.2.1.1.1 Connectivity and Configurability

Current and planned long-haul communications consist mainly of DOD and DOD-leased commercial satellite and terrestrial (wire and fiber) communications systems. These systems provide connectivity via dedicated trunk circuits. For both satellite and terrestrial services, most existing network control facilities do not allow for a timely (or adaptive) precedence-based preemption of service (i.e., bandwidth on demand). They do, however, support the requirement for transparent connectivity through the use of interfaces at the gateways, wherein multiplexing and demultiplexing of trunks and protocol conversion is achieved. Tactical networks continue to be typically unique systems, usually dedicated to specific applications and not amenable to precedence-based preemption of service. They are generally not transparent to the user, nor are they easily reconfigurable.

Thus, current long-haul and tactical naval communications networks can generally be characterized as dedicated circuits and rigid network structures. Planned systems will increase the inherent flexibility of some networks. In particular, the Military Strategic, Tactical, and Relay (MILSTAR) satellite communications system will decentralize the control of subnetworks, so that they can be rapidly reconfigured by tactical control terminals. The UHF follow-on system will maintain centralized control from gateways (e.g., the Naval Command and Telecommunications Area Master Station–Atlantic), but is implementing demand assigned multiple access (DAMA), which will allow the rapid reassignment of satellite channels to tactical users. This will allow the assignment of dedicated circuits for relatively short duration (hours rather than weeks or months).

6.2.1.1.2 Capacity

Leased commercial terrestrial communications systems are generally capable of providing long-haul circuit-switched and dedicated links between user nodes at throughput rates ranging from tens to hundreds of megabytes per second. DOD and DOD-leased satellite communications for long-haul services support trunk rates between tens of kilobytes per second (UHF follow-on and MILSTAR) and tens of megabytes per second (Defense Satellite Communications System (DSCS) and commercial satellites). Tactical data links typically have data rates well below 1 Mbps, although some direct tactical feeds from sensors have data rates up to hundreds of megabytes per second (e.g., the common data link (CDL)).

Application of near-term technologies and advanced waveforms will provide significant increases in data rates for the communication systems (such as taking 16 kbps rates up to nearly 100 kbps for the UHF follow-on system, and 2.4 kbps rates up to 1 Mbps for MILSTAR). In addition, the Global Broadcast System (GBS) will add spot beam capabilities in the range 1.5 to 24 Mbps. The scope of this study did not allow for any detailed analysis of communications capacity requirements. However, an attempt was made to compare the projected enhancements against stated naval requirements. As reflected in the following, the committee experienced some difficulty in understanding the requirements.

Naval forces assess their communications bandwidth requirements annually as part of the program objective memorandum (POM) process, with fleet commanders playing a key role. The POM results are summarized in the Navy's bandwidth baseline assessment memorandum.[2] The requirements submitted by CINCPACFLT[3] figure prominently in that memorandum. Other statements of requirements presented in the POM bandwidth memorandum derive from naval SATCOM user requirements[4] and *Emerging Requirements Database (ERDB) for Satellite Communications, Version 6.*[5] These requirements studies considered the satellite communications requirements of naval forces deployed in a single major regional conflict (MRC), as well as in two simultaneous MRCs in two different theaters. The requirements reflected full-time communications use under wartime conditions (although one peacetime case was also given) and refer to the maritime component of a joint task in each theater. The naval task forces included a carrier battle force consisting of five carrier battle groups (CVBGs) and an amphibious task force consisting of five amphibious ready groups (ARGs). In addition, the analysis considered one Marine expeditionary force (MEF) and one to two independent operations groups (IOGs) depending on the specific theater.

Individual Ship Requirements. The requirements from the above-noted documents for each type of ship are summarized in Table 6.3 and are seen to vary widely. For example, the CINCPACFLT study recommends equipping every ship with a minimum core SATCOM capability of 128 kbps, thereby providing what was indicated to be sufficient for essential services such as the common tactical picture, record traffic, and command voice links.[6] Large ships were indicated as having greater needs, up to 1.28 Mbps, for activities such as collabo-

---

[2]Director, Space Information Warfare Command and Control, N6. 1999. "Program Objectives Memorandum (POM) [20]00 Bandwidth Baseline Assessment Memorandum," Office of the Chief of Naval Operations, Washington, D.C.

[3]Commander in Chief, U.S. Pacific Fleet, "FY 98 Theater C4I Bandwidth Assessment," a briefing presented to the Resources, Requirements Review Board (R3B), Office of the Chief of Naval Operations, January, 1998; Munns, RADM(S) Charles L., USN, Deputy Chief of Staff for C4I, "Knowledge Centric Future," a briefing presented to the Committee on Network-Centric Naval Forces, January 26, 1999.

[4]Naval Space Command, "Naval SATCOM User Requirements," a briefing presented by LT Michael Finnegan, USN, on Naval SATCOM Industrial Day, Dahlgren, Va., October 22, 1997.

[5]Joint Staff, J6. 1999. *Emerging Requirements Database (ERDB) for Satellite Communications, Version 6,* Washington, D.C.

[6]Note also that the minimum IT-21 core capability for all types of ships is 128 kbps.

TABLE 6.3 Communications Requirements (Mbps) by Ship Type

Type of Ship[a]	CINCPACFLT 2000	NAVSPACOM 2005		Joint Staff 2010
		Wartime	Peacetime	
CVN	1.28	13.4	10.6	30.6
LHA/LHD	1.28	13.4	10.6	29.0
CG	0.512	2.84	4.95	23.0
DDG	0.128	2.84	12.40	10.7
SSN	0.128	1.51	3.77	5.16

[a]CG, guided-missile cruiser; CVN, nuclear-powered aircraft carrier; DDG, guided-missile destroyer; LHA/LHD, amphibious assault ship (general purpose)/amphibious assault ship (multipurpose); SSN, nuclear-powered attack submarine.
SOURCE: Data courtesy of Marine Corps Combat Development Command, Quantico, Va., 1999.

rative planning and receiving imagery. On the other hand, the Naval Space Command (NAVSPACOM) and Joint Staff requirements figures given in Table 6.3 are much greater than the CINCPACFLT figures, typically by a factor of at least 10 and sometimes much more. The difference is perhaps explained by the later time frame of the NAVSPACOM and Joint Staff figures, but there was no discussion to that effect in the POM bandwidth report that presented all these figures (see footnote 3). Overall, the committee believes that the CINCPACFLT requirements could be a significant underestimate, especially if the increasing demands of imagery are considered.[7]

Aggregate Requirements. The aggregate communications requirements from the above-referenced documents, by composite force and total theater, are summarized in Table 6.4. They show, in the FY05 case (wartime), that for the most stressing theater of operations, each CVBG requires a data throughput of roughly 32 Mbps, or 162 Mbps for the carrier battle force (five CVBGs). Similarly, the amphibious task force total throughput requirement is 76 Mbps, including 15 Mbps for each ARG. For IOGs the total requirement is 25 Mbps, while for the fleet broadcast and tactical networks it is 31 Mbps. Adding all these figures up produces the fleet theater requirement of 293 Mbps for total channel capacity.

The briefing by a CINCPACFLT representative (see footnote 4) indicated that in FY03, the planned SATCOM capacity would be approximately 185 Mbps

---

[7]For example, if a 1 square nautical mile area is resolved into 1 ft increments in each of its two dimensions, it would take 3.3 min to transmit the image of this area (uncompressed) at 1.5 Mbps.

TABLE 6.4 Aggregate Communications Requirements (Mbps)

	POM 00 Bandwidth Baseline Assessment Memorandum	NAVSPACOM 2005	Joint Chiefs of Staff Emerging Requirements Database 2010
Carrier battle group	—	32.3	71.0
Carrier task force	—	161.7	—
Amphibious ready group	—	15.2	47.8
Amphibious task force	—	76.0	—
Independent operations groups	—	24.9	—
Fleet broadcast and tactical networks	13.8	30.5	—
Total theater requirement	45.9	293.0	—
Total ships	133 (ships)	92 (ships)	—

SOURCE: CNO N6 PR01 Baseline Assessment Memorandum (Draft), December 17, 1999.

per theater.[8] This is stated to be a 500 percent increase in global capacity and a 2,500 percent increase in total capacity including spot beams, relative to FY98.[9] This 185 Mbps figure is roughly consistent with the FY05 fleet theater requirement of 293 Mbps listed in Table 6.4 (assuming that most of the 185 Mbps theater capacity is allocated to the fleet). Thus, in contrast to the situation for individual ships, the CINCPACFLT figures for aggregate capacity are roughly comparable to the NAVSPACOM figures.

The committee was not able to resolve this discrepancy—that is, the comparable aggregate figures but the very different individual ship figures. The communications capabilities of the individual ships could be a factor. That is, a ship cannot make full use of available SATCOM capacity unless it has adequate antennas and terminal equipment, and antenna space is well known to be at a premium on ships. Still, while this is an important factor, it does not seem to adequately explain the discrepancy.

---

[8]The explicit figures are as follows: Southwest Asia (SWA): 41.1 global + 145 spot = 186.1; western Pacific (WESTPAC): 40.2 global + 145 spot = 185.2; continental United States (CONUS): 44.2 global + 49 spot = 93.2; Mediterranean (MED): 40.1 global + 145 spot = 185.1, where all figures are in megabytes per second.

[9]The timeliness of data delivered via spot beams (GBS) warrants further examination. For example, recent tests showed that data transfer from the Global Hawk unmanned aerial vehicle (UAV) to forward elements took 3 h, 22 min, via GBS and 2 min, 45 s via a direct down link (High-Altitude, Endurance (HAE) UAV ACTD Quick Look, ER 4.1, 19/20 Oct 1999). Mention has also been made that routine dissemination over GBS required scheduling several hours in advance.

Tactical Line-of-Sight Communications. The above discussion pertains to SATCOM. It is also necessary to consider tactical line-of-sight communications, which are divided into two major types—the tactical digital information links (TADILs) among platforms and the direct data links to platforms from theater sensors (e.g., the common data link—CDL). The best current capability among the TADILs, and all that is planned for near-term enhancements, is given by TADIL J (Link-16). Its capacity range is 28.8 to 115.2 kbps. This is a low data rate compared to what has become the norm for commercial use (although there are stressing demands in the tactical environment not met in commercial use). If network-centric operations are going to involve significantly increased information transfer over the tactical data links, increased capacity could well be required. The committee was not aware of any naval assessment of future requirements across all mission areas for TADILs. Available and emerging technology would seem to be available to support increased capacities (see Appendix E).

Direct real-time data links from in-theater sensors to ships might be particularly relevant in scenarios involving strike warfare or warfare in littoral regions. Analyses such as those associated with Tables 6.3 and 6.4 apparently did not consider such requirements. These data feeds are sizable, ranging, for example, from the Pioneer and Predator UAV video at 4.5 Mbps (analog equivalent rate) through Global Hawk's 24 Mbps SAR interleaved with 24 Mbps EO/IR, up to 137 to 274 Mbps for U-2 imagery (see Appendix E). Data link receivers and terminals (processors) exist for such sensor feeds (e.g., CDL and JSIPS-N) and in some cases are unique to the sensor. This equipment is being deployed to the fleet (see, for example, Table 6.1). The question is whether these receivers and terminals will be deployed to a large enough set of platforms and in a timely enough manner to meet anticipated tactical needs.

Summary. The discussion above indicates that the sources differ greatly in their estimates of requirements for a given type of ship for SATCOM capacity, and that the lower of these estimates (as expressed in the CINCPACFLT requirements) could significantly underestimate future needs. The matter is further complicated by the fact that it might be possible to satisfy increased individual ship requirements (relative to the CINCPACFLT requirements) by virtue of the significant increase planned in total theater SATCOM capacity. The discussion above also pointed to the importance of assessing future tactical line-of-sight communications requirements across all mission areas. Thus, the committee strongly suggests that the Navy take a systematic and comprehensive look at future communication requirements and the projected ability to fulfill them. The goal of this assessment should include reconciling the SATCOM estimates and providing a broad look across all tactical missions.

6.2.1.1.3 Interoperability

Interoperability among the current and planned long-haul communications systems still relies heavily upon multiplexing and demultiplexing and protocol conversion at communications gateways. This situation will be improved somewhat by a reduction in the number of Service-unique tactical communications systems. An important step forward is expected with the development of the Joint Tactical Radio System (JTRS) family of radios. The future Navy/Marine Corps intranet development effort and the ongoing IT-21 effort will not only provide an expanded range of communications services to extended user population, but will also enhance interoperabilty among naval forces. Also, the evolving deployment and use of MILSTAR terminals will produce significant improvement in joint interoperability since all MILSTAR terminals share a set of common modes (along with Service-unique modes). Likewise, the planned use of both 5 kHz and 25 kHz UHF follow-on channels by all Services (contrary to the current circumstance, in which most Services use only one type of channel) will greatly enhance joint interoperability. Further increase in interoperability will also come from the use of JTIDS terminals by all the military services. However, current and planned naval communications capabilities have very limited potential for interoperability with foreign government and commercial communications systems. Several recent military operations witnessed the use of U.S. Navy communications systems by allied forces in order to achieve the necessary exchange of information.

6.2.1.2 Future Capabilities

Because of the Joint Technical Architecture (JTA) requirement for adherence to commercial standards and products, the future of commercial communications systems will strongly influence the future of DOD communications systems. Both evolutionary and revolutionary changes will occur in commercial systems in the future, and a reasonably clear vision of that future is forming. Commercial terrestrial systems will continue to evolve toward the establishment of virtual private networks and multimedia services relying on asynchronous transfer mode (ATM) switching and multiplexing technologies and optical-fiber-based link technologies. The revolutionary aspect of the future of commercial systems rests on the establishment of constellations of low Earth-orbiting satellites implementing fast packet switching via onboard ATM switches and laser cross-links. Already, two such constellations are being planned, Teledesic and Celestri. Both plan to employ hundreds of satellites providing multimedia services with data rates between tens of kilobytes per second up to hundreds of megabytes per second. The result will be to extend ATM-switched virtual private networks (VPNs) and their associated multimedia services to globally distributed fixed and mobile users.

The extremes of the vision for DOD communications in the midst of this burst of commercial communications innovations are relatively clear and depend on the extent to which the different communications infrastructures can access and employ the new commercial services. If little use of commercial services can be made, long-haul communications systems will evolve with the development of the MILSTAR follow-on program, advanced EHF, which will probably emulate commercial advances by using onboard switches (most likely ATM switches). As the need for gateways is diminished, the use of cross-links as well as onboard ATM switches will significantly blur the distinction between the sustaining base and long-haul links, as provided by advanced EHF. However, since the DSCS follow-on program, Advanced Wideband, will probably not include onboard switching because of the enormous investment already made in DSCS ground terminals, DSCS gateway terminals and associated terrestrial switched networks will still be required. In addition, because DOD communications satellites will continue to be in geosynchronous orbit, terminal antenna sizes will continue to limit the application of these systems to mobile units.

On the other hand, should it prove possible for the DOD to make extensive use of the future commercial communications infrastructure, the distinction between sustaining base, long-haul, and tactical communications systems will fade, and a majority of naval forces on land, at sea, or in the air will become members of a universal naval multimedia VPN, which will be a subnetwork on the commercial infrastructure. In either case, a critical systems engineering task facing the Department of the Navy is to determine the best mix of commercial communication services required to supplement the military communications infrastructure so that established requirements are met.

As a practical matter, as pointed out in an earlier Naval Studies Board study,[10] bandwidth requirements will inevitably push the naval forces toward the use of commercial satellite and other communication links (as has happened already in the Balkan operations, for example). The naval forces will do best by adopting the commercial systems without change and adapting naval uses and operational approaches to them. Provision will have to be made for security, priority, and preservation of access and service continuity in emergencies.

6.2.2 Communications and Networking—Wireless Transport

This section is titled "wireless transport" rather than the more commonly understood "radio" for two reasons. First, there may be a role for acoustics in underwater wireless transmission. More important, however, is the commingling of various layers of the protocol stack into what are usually called radios or

---

[10]Naval Studies Board, National Research Council. 1997. *Technology for the United States and Marine Corps, 2000-2035: Becoming a 21st-Century Force, Volume 3, Information Warfare.* National Academy Press, Washington, D.C.

datalink terminals. This section focuses on wireless transport of bits and discusses three topics: waveform interoperability, antennas, and terminal equipment for dismounted forces.

6.2.2.1 Waveform Interoperability

The transport mechanism need not (and should not) have to know the meaning of the bits. Once the bits are moved, interoperability depends on the applications and network control processes and on adherence to common standards. But a radio transmitter cannot move the bits unless it emits an electronic waveform that the intended receiver can demodulate to produce the bit stream that will travel up the protocol stack. Waveform characteristics include carrier frequency, signal bandwidth, signaling rate, and modulation method.

6.2.2.1.1 Present Situation

Military radios use a wide variety of waveforms. The reasons for this variation include the following:

- The allocation of RF spectrum to different communications services;
- Differences in required signaling rate;
- Differences in required resistance to interference;
- Differences in required resistance to interception;
- Different needs for directional and omnidirectional antennas; and
- Improvements in modulation schemes.

Consequently, the likelihood that two randomly selected radios will have the same carrier frequency, bandwidth, modulation scheme, and signaling rate is low. Instead, clusters of similar users using common radio equipment to ensure waveform compatibility are seen. This arrangement is satisfactory within the cluster, but if that cluster has to interact with another cluster that uses different waveforms, direct communications between members of different clusters is impossible.

Outside of the datalinks engineered as closed systems, there is little uniformity of modulation type. Simple, binary frequency shift keying was once widely used, but differences in carrier frequency, bandwidth, and signaling rate have long been with us. Furthermore, new modulation schemes are being introduced to increase the efficiency of the transmission (the number of correct bits per joule) or to increase resistance to interference or interception.

The usual approach to obtain waveform interoperability is to provide a gateway, that is, a pair of radios that understand both waveforms and transfer the bits between the two domains. Sometimes the wireless gateway is combined with a protocol gateway, such as in the Navy's equipment that interfaces Link 11 and

Link 16. One difficulty with gateways is that they increase the total number of radios in the system. Another is that the gateway must be part of the forward footprint of the force. A third is that by introducing additional wireless hops, the gateway raises the probability of accidental or deliberate adversary interference with communication.

One possibility would be to make every node a gateway by equipping it with a large set of radios compatible with those on the units with which it may have to interoperate. However, many of the radios are of old design and are unnecessarily heavy and large. The number of radios needed by a unit would be determined by the total number of the types of equipment used by the set of entities with which the unit must communicate. Furthermore, sometimes these legacy systems are out of production. Even if they are in production, there may be little competition, which could mean a low-production-volume environment and high prices.

6.2.2.1.2 Modular Radios

Waveform interoperability will not be achieved by legislating and enforcing a universal waveform; there are too many legacy systems to be ignored, and there are legitimate reasons to use different waveforms over different paths. Rather, a solution is needed that respects the legacy waveforms but uses modern technology to produce a wide variety of waveforms from a single equipment set that can be produced competitively in large numbers. The most promising way to achieve this is through modular radios. If the major components of a radio—RF amplifiers, detectors, modulators, demodulators, cryptographic apparatus, power supplies, and so on—are modules with carefully defined interfaces, respecting a new waveform involves changing only modules, not radio sets. Producing more radios (compared to legacy systems) offers the possibility of significant cost savings. The Navy's Joint Maritime Communications System (JMCOMS) program has been pursuing a "slice radio" to achieve flexibility and cost savings.

The scheme just described involves personalizing a radio to a waveform. If the personalization is done by plugging modules into a backplane (common communications distribution framework) before an operation, then the number of modular radios that will be needed will be at least the number of waveforms that might be encountered. But if this personalization can be done dynamically, the required number of modular radios may be as low as the number of waveforms that must be understood simultaneously, which may be a much lower number. This suggests that the personalization should be done by loading software rather than by unplugging and plugging modules. The term "programmable modular radio" (PMR) describes a modular radio with this capability.

The PMRs should have the following characteristics:

- *Modular.* Hardware and software must be constructed to allow functions to be added or replaced without redesigning the entire radio. Repackaging to suit platform-specific needs should involve minimal change;

• *Open.* Proprietary software or devices must be forbidden, unless generic substitutions are available. All information necessary to add waveforms or other functionality should be expressed in accepted federal or industry-wide standards, preferably available online to approved users;

• *Cost-effective.* Any component or module must be acquired under free and open competition. Ownership and configuration management of software must rest with the government;

• *Assured.* Message content, traffic flow, and routing information must be protected against interception, intrusion, and attack even if equipment or personnel are captured; and

• *Ready to support tactical operations.* Modular radios must be programmable and adaptive to allow bit-by-bit assignment of the next bit to the next available link in accordance with the instantaneous operational need, and they must respond to distributed, survivable, information network management.

Most of these goals are achievable today. Consumers routinely download software to improve the performance of their programmable modems used over commercial telephone lines.

6.2.2.1.3 Programs Developing Programmable Modular Radios

New single-purpose radios are to some extent both modular and programmable, although they are seldom programmable by the user. They are usually microprocessor controlled and have a "front panel" that is little more than a graphical user interface that superficially resembles the familiar array of radio knobs and switches. Internally, the hardware is modular, in the form of either boards or chips, arranged by familiar functions according to the designer's taste. Very wideband RF amplifiers are hard to build, so there would have to be a family of amplifier modules covering different bands that could be incorporated into the radio.

Any well-qualified engineer can come up with workable design for a PMR that uses some number of predetermined waveforms, and it is not surprising that a number of such initiatives have arisen in military and commercial laboratories. But there has not yet been a consensus on a common architecture that would allow any vendor to supply hardware or software components that could be plugged into an existing PMR or be assembled to create a new combination of functions that would interoperate with other components of the architecture.

The committee considered three PMR programs, although it is aware that there are more. The first was the Joint Tactical Radio System (JTRS) program, initiated after DOD recognized that there was no mechanism to maintain synergism among the diverse Service programs. The first phase, now completed, was to define a program framework. The second phase, expected to be under contract soon and completed in FY01, will refine and finalize the core architecture. At that point, the Services will have a JTRS standard they can use in acquiring radios. The goal of the JTRS program office is to produce a documented and supported architecture, not a physical radio. A physical JTRS radio is one that is compliant with the architecture. Prototype radios that will support heritage waveforms will also be developed in the second phase, primarily to verify the architecture but secondarily as a source of potentially deployable JTRS radios. The third phase will be the acquisition or development of additional waveform descriptions.

The second PMR program considered was the Digital Modular Radio (DMR) program, a project of the Space and Naval Warfare Systems Command (SPAWAR), PMW-176. Although some limited production prior to enunciation of the JTRS standard is likely, the DMR will become JTRS-compliant before full production. The committee does not know the extent to which DMR modules are programmable.

The third PMR considered was developed by the Naval Research Laboratory (NRL) Joint Combat Information Terminal (JCIT) program for use in the helicopter-borne Army Airborne Command and Control System and is now in limited production. It is probably the most advanced of the Service programs in terms of implementation. Figure 6.6 is a block diagram of a JCIT supplied to the committee by the project office.

JCIT, as it now exists, can interoperate with a wide variety of waveforms, and its embedded information security (INFOSEC) can interoperate with a wide variety of cryptographic systems. However, it cannot currently interoperate with JTIDS terminals.

The claims made for the JCIT hardware architecture are as follows:

- JCIT transceivers are software reconfigurable;
- New waveforms are incorporated by adding software;
- Radio suites can be tailored to meet a specific mission's requirements;
- JCIT INFOSEC embedded on Standard Electronic Module, Format E (SEM-E) cards emulates all three families of INFOSEC;
- JCIT processors process, correlate, and fuse incoming data; and
- New modules can be developed to meet a specific user's requirements and mixed with existing modules to form an entirely new "box," maximizing technology reuse.

These attributes are well aligned with the committee's stated goals,[11] and JCIT

---

[11]Because JCIT is a terminal, not a radio, the JCIT processor module may violate the confinement of the radio to transport. In addition to translating various transport standards, it apparently participates in the AAC2S application.

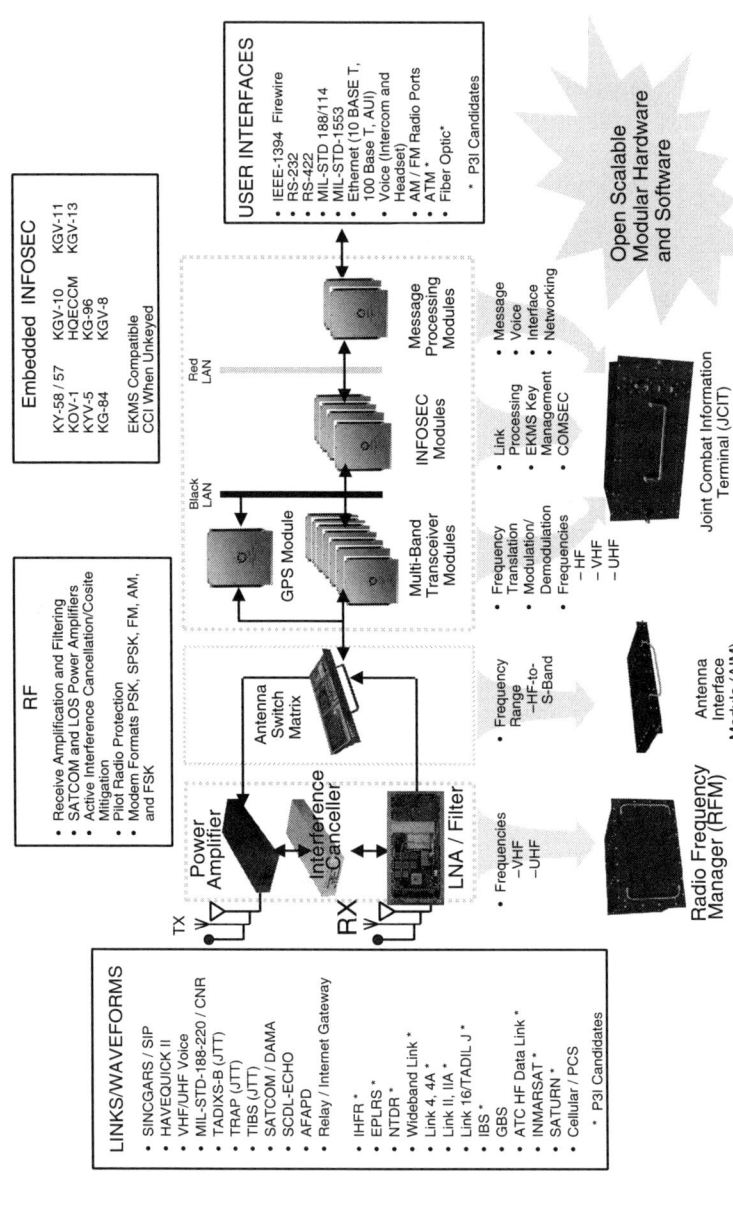

FIGURE 6.6 Components, technical functions, and linkages of the Naval Research Laboratory joint combat information terminal. SOURCE: A2C2 Program Office, Space Systems Development Department. 1999. "JCIT Provides Integrated C4I Architecture (updated)." Army Airborne Command and Control System (A2C2S), Naval Research Laboratory, Washington, D.C.

may be a point of departure in achieving them. Many JCIT waveform descriptions are being adopted by JTRS, and the JCIT radios will become fully JTRS-compliant after the standard is announced.

6.2.2.1.4 Achieving Waveform Interoperability

Many legacy radios will continue to be used in the years ahead. The Army is completing a large purchase of the Single Channel Ground and Airborne Radio System (SINCGARS), and the Navy is planning a large purchase of the Management Information and Distribution System (MIDS) JTIDS terminals. All planned modular radios are intended to interoperate with SINCGARS as well as older Navy datalink terminals, and the JTIDS waveform is included in the JTRS goal set. There are both technical problems in achieving the required high dynamic range and hopping rates and, at least in the past, political problems in seeming to compete with the MIDS terminal acquisition program.

However, the committee believes that PMRs are achievable for most communications waveforms. The technical problems in handling the JTIDS waveform in a modular radio can be overcome with a relatively modest investment; the political problem can be defused by pointing out that modular radios are just another way of implementing a JTIDS radio. The committee suggests that preference should be given to modular radio programs whose individual modules can switch dynamically among multiple waveforms. All modular radio programs should include modules capable of processing the JTIDS waveform.

PMRs can be used in three ways: to reimplement the waveforms produced by obsolescent legacy equipment, to implement highly flexible gateways, and to create direct transmission paths between dissimilarly equipped forces. The first use will occur to the extent that makes economic sense; the others will evolve with experience.

Thus, to take full advantage of the potential value of PMRs, an experimental program is needed to explore how this new capability can best be used. This suggests that, using JCIT terminals, the Marines should experiment with simultaneous interoperation with the Navy, Army ground units, and Army airborne units.

Finally, a strategy is needed to ensure future compatibility and to prevent developers from introducing new waveforms that transfer costs to the information infrastructure. This strategy can be addressed by requiring acquisition agencies contemplating the introduction or further purchase of radios whose waveforms are not emulated by existing PMRs, absent rarely granted waivers, to develop the PMR software that permits the emulation of these waveforms. The waivers would be needed for very high performance radios used in narrow niches as well as for commercial off-the-shelf (COTS) handsets. The deployment of single-purpose nonprogrammable radios would not be prohibited, but the code would have to be delivered that would permit existing PMRs to interoperate with the new radios. If the JTRS program is successful, a single code would be applicable to all PMRs.

6.2.2.2 Antennas

6.2.2.2.1 Shipboard and Aircraft Antennas

Microwave communication antennas need to be large for two reasons. First, for a given transmitter power and service area, sustainable information rates are a first order linear function of antenna area and are independent of carrier frequency and transmitter-receiver distance.[12] If the service area is to be large, the antenna aperture will have to be large. Also, large-aperture antennas are directional, and directionality helps resist jamming and passive exploitation.

Antennas aboard Navy ships have to be stabilized, and mechanical stabilization of large antennas is expensive. Furthermore, there are many radio circuits connecting a ship to the rest of the network, and not enough real estate is available to accommodate multiple, large, mechanically steerable antennas. (The alternative of bringing all sensor flows to sanctuary ashore and providing a single robust link from ship to shore is unattractive for high-data-rate, in-theater sensors.)

The Navy has for some time recognized that it needs multibeam electronically steerable antennas of moderate-to-large aperture that cover at least its microwave communications carrier frequencies, that is, frequencies used for high-speed data satellite relay and for sensor links. Because such antennas will use phased-array technology, it is possible that in aircraft use they could be distributed over the body of the aircraft, thereby permitting larger-aperture antennas than would be possible with reflectors.

Among the difficulties to be overcome are bandwidth, dynamic range, and cost. Dynamic range requirements can be reduced by physically separating transmit and receive arrays. Bandwidth and cost remain serious problems. The committee reviewed the Office of Naval Research's (ONR's) ambitious Advanced Multifunction Radio Frequency System (AMRFS) and noted that even the test articles had unfilled apertures because the program could not afford to fill them.

There are two approaches to reducing cost. One is to utilize whatever commercial designs prevail, accepting whatever bandwidth and aperture these designs provide. Should Teledesic succeed, its mobile subscriber antennas would

---

[12]The maximum sustainable rate is a linear function of received power. The power received is equal to the power transmitted multiplied by the fraction of the service area intercepted by the antenna. For further elaboration, see Wald, B., 1997, *Trade-offs Among UAV and Satellite Communications Relays*, CNA Research Memorandum 97-84, Center for Naval Analyses, Alexandria, Va., October.

be electronically self-steered and support T-1 transmission rates, but only at the Teledesic frequencies.

Antennas now exist for passenger aircraft to receive satellite-relayed TV, although the price of these antennas is consonant with airline rather than consumer use. In the consumer market, Hitachi is developing a system that will permit the receipt of satellite TV in moving cars; presumably the receive antenna will be priced to satisfy the consumer. However, satellite TV antennas are likely to have single beams, so the cited examples solve only the stabilization problem.

The other approach is to search for new technology that will make electronically steerable multibeam antennas of reasonable bandwidth affordable. The committee does not know which technologies will ultimately be successful, but it has been suggested that low-loss microelectromechanical switches make it possible to reduce greatly the number of RF amplifiers needed to feed an array; moreover, because they are switching delay lines, not phase shifters, they operate over relatively wide frequency ranges.

The committee concluded that the Department of the Navy is correct in continuing to give priority to the search for multifrequency, self-stabilizing, multibeam, electronically steerable shipboard and aircraft antennas. However, developmental antenna systems may not be affordable unless requirements are tailored or a breakthrough technology appears.

In the light of this conclusion, the committee suggests that the Department of the Navy, while continuing to push available technology in programs like the AMRFS, should also seek to validate potential breakthrough technologies and should attempt to adapt its transport architectures to the use of future low-cost electronically steered antennas developed for commercial applications.

6.2.2.2.2 Submarine Antennas

Submarine microwave antennas are smaller than their shipboard and even their airborne counterparts. Deployed submarine dishes have a 5-inch diameter; plans for a 16-inch foldable plate are uncertain. In contrast, shipboard antennas will have an aperture of a meter or greater. It is likely, therefore, that submarines, even when willing to broach the surface with an antenna and even when willing to radiate, will be at a communications disadvantage. They will not be able to receive some broadcasts. Other services will be possible only if a precise (small-service-area) beam is pointed at the submarine.

Historically, antenna sizes were constrained by the desire to mount them atop periscopes. Larger antennas would require a very different mount, and their provision in new submarines would be very expensive, while backfitting to existing submarines would probably be prohibitively expensive, if not impossible. However, the number of new submarines being constructed is very small; most of the force will consist of improved SSN-688s for a long time to come.

In the past, the limited data rate of submarine communications was not considered a liability by the "silent service." On many missions, the submarine preferred not to communicate at all, either because of fear of detection or because it needed to stay at depth to perform its mission. Today, many submariners are interested in close interaction with other naval forces; more than one submarine commander has been heard to say, "I should be as well connected to the officer in tactical command (OTC) as any surface ship commander and I should be able to participate in the OTC's videoconference." The committee does not know how important these expressed needs are, but it does know that they would be very difficult to fulfill.

The need to communicate while at depth may be less pressing in littoral warfare than in open-ocean ASW. Submarines engaged in intelligence missions will be near the surface anyway. However, options exist for communications with a deeply submerged submarine.

For decades, submarines have been towing buoyant cable antennas, and nuclear-powered ballistic missile submarines (SSBNs) have been towing receive buoys. Two-way communication has been demonstrated but not deployed. Data rates are low. Towing places some restrictions on a submarine's speed and depth.

DARPA is pursuing the exploitation of modern signal processing to equalize the dispersive acoustic channel and increase the bandwidth that can be transmitted acoustically over modest distances. The idea is that the antenna would not be towed by the submarine but would be part of an autonomous vehicle acoustically linked to the submarine.

Overall, the committee concluded that the submarine will always be at a disadvantage in terms of maximum communications rate unless its antenna aperture can be made comparable to that of a surface ship, but that would be a very expensive undertaking. Two-way communication to a submerged submarine would be possible through the use of towed buoys or an acoustically linked autonomous vehicle.

One response to this situation would be to perform system engineering to quantify the effect of an improved communications rate, for both periscope depth and deeply submerged submarines, on the effectiveness of the entire network in relation to the cost involved. Based on those results, investments should be made as appropriate in improved submarine antennas.

6.2.2.3 Terminals for Dismounted Forces

The Operational Maneuver From the Sea and the Ship to Objective Maneuver (STOM) concepts contemplate deployment from the sea of ground forces that will not have large fixed bases ashore and that will penetrate deep inland while depending on support from the sea. The challenges in expeditionary warfare are discussed in more detail in Chapter 3 and in Appendix C; here only the wireless terminals that will be used by dismounted forces will be considered.

The terminals will have to provide highly reliable, over-the-horizon communications between dismounted forces and supporting ships to receive situational awareness and to send requests for fire and other support. They should also provide reliable position location information (PLI) to avert friendly fire and to facilitate force synchronization. Neither current architecture nor current physical terminals support these functions well.

6.2.2.3.1 Physical Properties

Terminals used by dismounted forces must function as radios and as personal digital assistants and must be affordable in large quantities. They must be light, rugged, and operable in bright sunlight or total darkness and must not reveal the user's position. They should have long battery life. Preferably, they should allow hands-free operation or at least have a more convenient input device than a keyboard. Their transmissions should be cryptographically secure and they should support PLI. Their antennas must not constrain the position or activities of their wearers. They must operate well in both open terrain and urban areas.

6.2.2.3.2 Architecture

Commercial wireless LANs supplied by the extended littoral battlespace (ELB) program were used in the 1999 Kernel Blitz exercise, where they demonstrated the promise and limits of the technology. For reasons discussed in Chapter 3, small, local networks with relatively constant connectivity performed reasonably well; extended networks with variable connectivity performed poorly.

It is likely that this exercise exceeded the ability of single-level, peer-to-peer implementations, and that a hub-and-spoke architecture would have been more effective. In the latter, the hub, perhaps mounted on a light vehicle and equipped with PMRs or possibly a variant of the VRC-99, would be robustly connected to remote information networks but would use single-level wireless networking to interoperate with nearby handsets. Single-purpose terminals from both military and commercial sources could be used at the end of the spokes, with the choice determined by range, rate, information warfare threat, and so on; the radio at the hub would be able to deal with all the waveforms.

This would not be a subversion of the uniform NCII—each handset would be addressable through the system-wide NCII scheme—but the hub could perform routing and buffering functions and act as a proxy for the handsets, to simplify the handsets and lower their weight and power requirements.

6.2.2.3.3 Implementation

The committee found no Department of the Navy program dedicated to developing architecture and apparatus to permit dismounted troops to interoperate well with other component systems, although multiple technology and PLI pro-

TABLE 6.5 Link Resistance to Adversary's Actions

Service	Rate ($\log_{10}$(bps))	Method[a]	Resistance to Jamming	Resistance to Exploitation
Terrestrial				
Link 4, etc.	4	None	Poor	Poor
SINCGARS	4	SS	Fair	Poor
JTIDS	5	SS	Good	Fair
Satellite				
UHF	4	None	Poor	Poor
Challenge Athena	6	DIR	Fair	Fair
DSCS	5	DIR	Fair	Fair
MILSTAR	4	DIR, SS	Good	Good
GBS	7	DIR	Fair	N/A

[a]SS, spread spectrum; DIR, antenna directivity.

grams exist. This suggests that agreement should be obtained between MCCDC and the Army's Training and Doctrine Command (TRADOC) on the characteristics of terminals for dismounted troops, and on an architecture that will permit interoperability in communications and PLI. Further actions would be to experiment with hub-and-spoke implementations of this architecture and to procure appropriate terminal equipment jointly.

6.2.2.4 Resistance to an Adversary's Actions

Connectivity must be maintained in the face of an adversary's attempts to disrupt or exploit the wireless signals. The principles involved are discussed in Sections 5.5.3.1 and 5.5.3.2. Table 6.5 assesses some current systems in this regard. It should be noted that directional transmit antennas do not support broadcasts and that spread spectrum radios occupy much more spectrum than their data rate would suggest, probably requiring unavailable spectrum to convert many high-data-rate unprotected services to spread spectrum.

6.2.3 System Resource Management

6.2.3.1 Introduction

The NCII is intrinsically a shared network, including many resources that will be shared by users and applications. These shared resources include not only the transmission links, multiplexers, and routers that move packets from one location to another but also servers that perform such functions as translating domain names into Internet addresses, authenticating users, and providing directory services. As with most modern networks, the NCII should be designed for intermittent bursts of traffic, so that network resource capabilities (e.g., link capacity) are shared in time and space on an ad hoc basis and not typically reserved in advance. In this way, users can have an effect on one another, particularly if the aggregate demand from the users in a geographical region of the network is anywhere near the peak capability of the local network links, routers, or servers. Mechanisms must be provided to manage overloads and to grant and implement priorities.

Thus, the allocation of limited, shared resources (sometimes referred to as traffic management) is one of the key resource management functions required in the NCII. Other equally important functions include the following:

- Monitoring the network (all shared resources) to detect problems (e.g., equipment failures) and to take necessary actions to repair or work around those problems (e.g., network reconfiguration);
- Authorizing users and applications to access shared network resources;
- Defining and updating closed user groups; and
- Maintaining directories needed by a large number of applications (e.g., information about the current network locations and the technical characteristics of specific end systems, required to establish communication).

The NCII network must react explicitly and instantaneously to rapidly changing demands for transfer of time-critical information. In some important cases, network resource management policies may have to be adjusted or negotiated in near-real time without the intervention of human users and managers. Clearly, not all users will be equal in their ability to obtain critical resources. In addition, the network must be self-configuring and self-adapting in the face of the dynamics anticipated in naval missions, to include adapting to natural or man-made disruption. This requires, for example, an ability to recognize multiple, nearly simultaneous discontinuities of service, locate the sources of the interruption, and invoke work-arounds that minimize the disruption.

Quality of service (QOS) is a network property receiving increased attention. QOS is a property of a network enabling it to ensure one or more attributes for a particular end-to-end connection. Those attributes may relate to the transport itself, such as delay, throughput, or accuracy (packet loss rate), or to more mission-relevant properties such as security, priority, and availability. Included in QOS could be the ability to request (reserve) infrastructure capacity, either indirectly through the needs of specific traffic types or directly through the assignment of priority or capacity. In short, QOS is a means by which users can specify needed NCII performance, including availability or capacity.

6.2.3.2 Near-term Assessment

Resource management systems that can perform some of the functions noted above exist today in telecommunications networks. These commercial systems must respond to changing demand and disruptions, although the rapidity of change and severity of disruption for which these systems are designed are much less stressing than might be encountered in a military situation. These systems are expensive, complex, and custom-designed, costing on the order of $10 million to $100 million, although their cost is spread over many users and applications since they are usually sold (or licensed) on a systemwide basis. Typically these network resource management systems are found in traditional circuit switched networks, but as the Internet grows and matures, such systems are beginning to appear in the networks of various Internet service providers.

The Internet today has self-healing properties, in the sense that routers (switching nodes) will slowly adapt their routing tables as paths to neighboring routers come and go and as neighboring routers change the information they advertise about which destinations they can reach and in how many hops. However, the Internet and intranets, as they are implemented today, generally cannot assign priority to a specific traffic type or a set of unusually important users.[13]

Commercial requirements to assign priorities to certain types of time-sensitive or critical applications are leading to the introduction of various mechanisms that can be used to assign priorities, although these mechanisms are not yet widely deployed. Over the next few years new versions of the underlying Internet protocols may provide the means to offer new network resource management capabilities of military importance. For example, version 6 of the Internet Protocol (IPv6) has recently been introduced. Networks that implement IPv6 will provide new QOS-related capabilities such as the specification of flows of multiple, related packets traveling between the same source-destination pairs.

Resource Reservation Protocol (RSVP), a new capability within the Internet Protocol suite, is now being tested. It should enable end systems to specify QOS requirements to an IP network. If implemented, RSVP will enable a network to tailor a user's connection in specific ways, such as latency (delay and delay variability), guaranteed/reserved capacity, and packet loss rate. When the use of this new IP software migrates throughout the packet switched networks of the world, including any NCII-like networks, it should be possible to satisfy military QOS specification needs relatively easily.

How long it will take the commercial QOS offerings to emerge is uncertain. Before such graded service is offered commercially, a philosophical change in the use of the Internet must occur. Costs to users now vary only with the peak and/or average transmission rate to the host Internet service provider and are not content-dependent. To change to a more dynamic QOS, billing systems able to recognize, record, and bill the dynamic and QOS-specified use of capacity or other traffic-dependent network services will have to be in place.

---

[13]Firewalls and virtual private networks can be utilized to manage which users and applications may access protected domains.

The committee concluded that there is now technology to provide some degree of system resource management for an NCII. The Navy and Marine Corps should take advantage of this technology in developing their networks. One example where that is the case is the Automated Digital Network System (ADNS), which allocates IP traffic from ships to the available SATCOM channels. Emerging Internet technology (e.g., IPv6, RSVP) should provide significant new QOS capabilities suitable to military needs. The naval services should take advantage of this technology to provide system resource management for the NCII. In fact, to ensure that these emerging technologies will best offer the means for introducing military-unique capabilities, the Department of the Navy should be tracking and influencing the course of their development.

Even if the emerging technical capabilities become available, the Navy and Marine Corps will still have to confront difficult questions in applying them to their needs. How access privilege gets assigned correctly to a limited set of users when people's lives are at stake is one such difficult example.

6.2.3.3 Future Capabilities

6.2.3.3.1 DARPA Programs

The need for QOS guarantees for future DOD networks has been recognized at DARPA for some time. The Quorum program, which has been under way for several years, addresses many of the needs stated above. Figure 6.7 illustrates the goals of Quorum, expressed primarily in terms of assured dynamic response, arguably one of the most critical tactical needs.

Another DARPA program directly related to resource management and QOS is called the Agile Information Control Environment (AICE). The goal of this recently started program is to seek near-optimal algorithms for network resource management that will increase dramatically the operational value of the information received. Its construct is hierarchical, as shown in Figure 6.8, where the requests for QOS are imprinted via a meta-level virtual network that lies above the physical elements of what would be an NCII-like system.

These two DARPA programs are seeking to provide important services for system resource management in a military common-user infrastructure such as the NCII. Quorum, which has been ongoing for some time, may be furnishing capabilities that could be realized in the NCII in the near term. Because of the general resource management capabilities promised by the Quorum and AICE programs, the Navy should follow them closely and become actively involved. The Navy's new class of destroyers has already been identified as one of the primary Quorum applications. More generally, the naval services should identify which Quorum and AICE capabilities are particularly important to their needs and beyond what is expected from commercial developments (see preceding

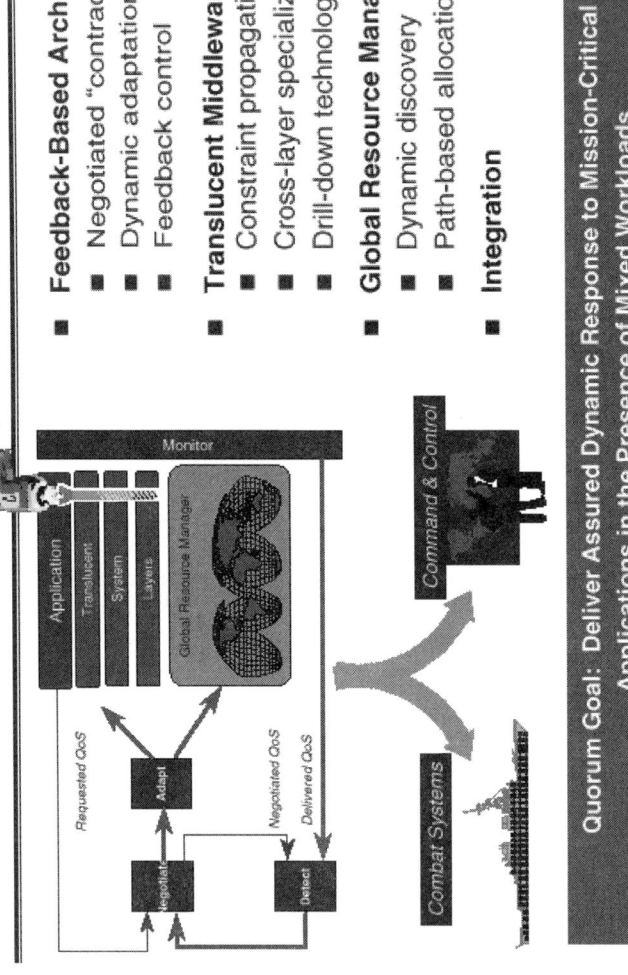

FIGURE 6.7 DARPA Quorum program. SOURCE: Koob, Gary; figure modified from Quorum Figure (as of October 14, 1999), Information Technology Office, DARPA, Arlington, Va. Available online at <http://www.darpa.mil/ito/research/quorum/index.html>.

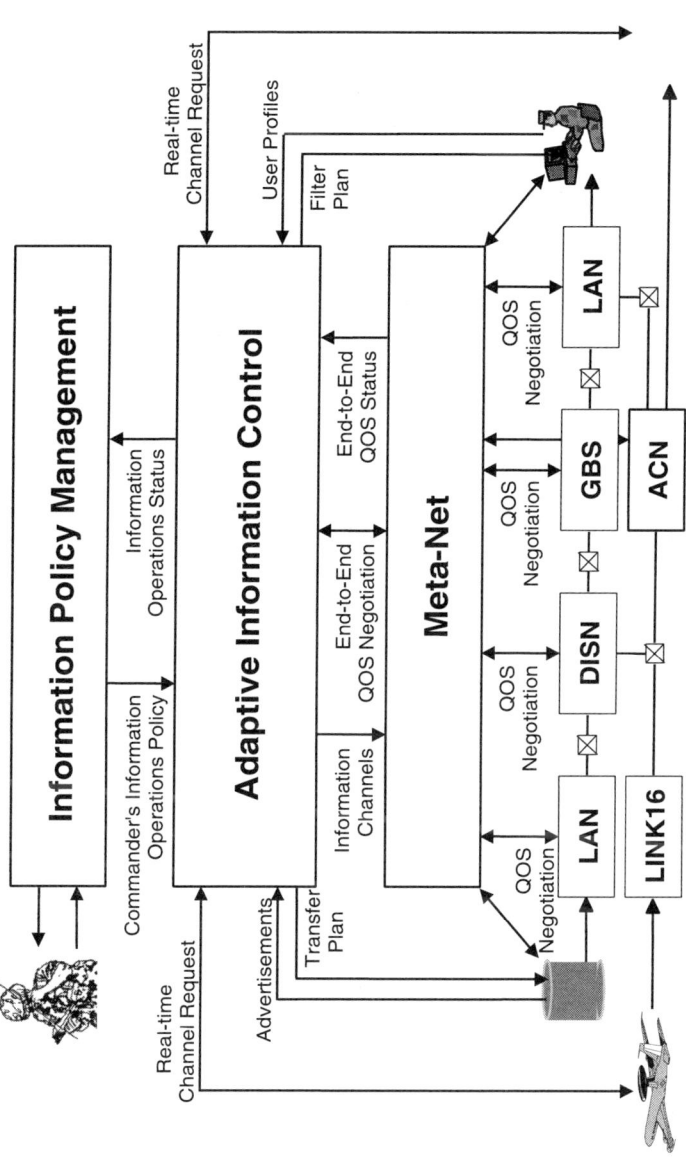

FIGURE 6.8 The hierarchical structure of DARPA's Agile Information Control Environment program. Acronyms are defined in Appendix H.
SOURCE: Beaton, Robert. Information Policy Management (IPM) viewgraph, from "Agile Information Control Environment," Defense Advanced Research Projects Agency, Arlington, Va., no date.

subsection), and then develop a plan to incorporate these capabilities into the NCII as soon as feasible.

6.2.3.3.2 Knowing Terminal Attributes

The ability of terminals (end systems) to explicitly advertise their attributes to authorized network entities without disclosing those same attributes to unauthorized entities is an interesting illustration of future capabilities. As such, it represents an example not known to be within current research programs but that the naval services should be considering and possibly pursuing for the NCII. Although it will have implications beyond resource management, the idea proposed here represents a major new adjunct to present infrastructure systems that will facilitate the military use of packet systems. This capability in a military network can offer the following end-to-end advantages:

- An awareness of what devices are connected to the network at all times;
- A knowledge of the capabilities of a terminal regarding its storage capacity, displays, input/output, location;
- A knowledge of the status, health, or readiness of the terminal or other end device (e.g., battery life remaining);
- An ability to know at the source just what kind and amount of information the destination device can accept, thus limiting the introduction of extraneous, unusable information into the network;
- An assurance that the terminal is an authenticated attached device; and
- With some assumptions, an authentication of the user as well.

As can be seen, many of these advantages are related to increased network assurance, at least by knowing what is connected to the network. Conceivably, a continuous monitoring of the network, its attached devices, and set of users would be possible. Given the ability to quickly validate both a terminal and a user identification (ID), additional network security may be possible.

Today's Internet protocols (IPv4) are very limited in terms of the information they provide to both users and providers about what is attached to the network. In today's IP connections, the attached end-system device or terminal has to be explicitly aware of the IP address of the port to which it is attached. Beyond that, the network is almost totally unaware of that address or its capabilities. This ignorance, then, is shared not only across the transport layers of the network (and below) but often at the middleware and application layers as well. Given this lack of awareness, sources intending to convey information to a given terminal cannot learn, from the present network, the attributes or readiness of a destination node.

However, the basic concept of advertising attributes is not new. Something like it was used in the packet radio network at the genesis of the Internet, in 1977.

It could be called a "terminal awareness packet" (TAP). TAP would allow the network to become aware not only of the presence of a unique attached end system but also of the capability and readiness of that device. To an extent, the trend to unique terminal devices has already begun. The European Groupe Special Mobile (European cellular system) is a good example, with its unique cell phone ID plus user ID (sometimes including a PIN number). This concept tries to identify the user and to assure that the person is an authorized user. Another trend that has surfaced lately is the unique ID embedded in each Pentium III processor, a number that can be sensed remotely.

Given a uniquely numbered end device and an agreed-on set of device characteristics describing its capability and readiness, a communication of these parameters to the network is possible, either at connect time, boot time, or on demand using loop-back probes at run time. This information is the content of the above-mentioned TAP. While this kind of information may not be of interest for commercial and consumer use, it seems appropriate for military operations to know the location, status, and capability of a node or individual. In general, this information facilitates the achievement of interoperability and the efficient use of resources. Quite naturally embodied in this capability is a continuous depiction not just of network connectivity but of user connectivity for all or any subset of the network.

Using the existing Simple Network Management Protocol (SNMP), end systems and servers can be configured with management information bases that describe their current capabilities and configuration, for example, and can make those attributes available to other network entities. IPv6 will have some features that could be adapted to military information needs. There, the so-called "flow ID" is tied to QOS and has about 20 bits to describe a wide range of traffic types including, say, their importance. Mobile hosts in a network that implements the emerging "mobile IP" will originate packets that can alert the network to their current network location. This dynamic name-address binding could be a natural component of the proposed TAP.

Thus, following through the above arguments, one can see that it might be feasible to implement a TAP capability. Because these new aspects of networking offer important features, the military needs to track and influence them. Additionally, they must also be examined closely for any vulnerabilities they add.

6.2.4 Collection Management

The collection management process determines the data collection plan for intelligence, surveillance, and reconnaissance (ISR) assets, based on needs specified by the operational commanders. The sensor assets involved can be global (e.g., space-based), regional (e.g., surveillance aircraft), or local (e.g., UAVs).[14]

---

[14] Sensors are discussed in detail in Chapter 3.

The data collection plan will specify such factors as sensor flight path, flight times, and orientation. Overall, the collection management process is made quite complex by the need to balance tasking assignments among competing requests and to optimize the use of available sensor assets. Such optimization should occur across all available assets and not just in the tasking of individual sensors. Ideally, the data collection plan should be rapidly alterable to respond to information gained by the sensors, changes in the operational situation, and the possible malfunctioning of some sensors or platforms.

6.2.4.1 Near-term Assessment

Significant effort is now being devoted to collection management. Needs are managed by collection management tools unique to a given sensor and supported by many manual processes, although they are computer-assisted. Tactical, theater, national, and even commercial collectors are characteristically tasked with tools that manage a queue of needs for their own individual collection asset. There is little opportunity for a commander to influence the collection of platforms of all types and numbers through a collection manager who has a single (or even a few) tools to do so. Rather, the manager must submit a commander's needs using many processes and many different systems. This means there is only a limited ability to deal with assets as an integrated set and realize the synergies of cross-cueing and improved responsiveness to the commander's request.

Today, the approval and prioritization of needs for collection typically occur through a hierarchical process and a chain of command. Feedback is usually provided in days rather than minutes or hours. Consequently, redundant nominations occur, particularly across multiple platforms, and the collection process cannot be responsive to rapidly changing situations or newly gained information. In summary, current collection management capabilities have significant limitations and fall well short of the timely and flexible information management capabilities envisioned for the NCII.

The Joint Collection Management Tasking (JCMT) system and the Requirements Management System (RMS), both currently operational, provide good examples of what is available today. Modest enhancements in capabilities will be realized across the Future Year Defense Program (FYDP). They will provide some increased integration of assets through a shared requirements database and offer access to commercial imagery platforms. They will also enable order entry and tracking of requests for products and services, with more frequent reports on status.

The tasking, processing, exploitation, and dissemination (TPED) baseline/modernization plan being developed by the National Imagery and Mapping Agency (NIMA) is an example of how collection management systems are being evolved. Enhancements, programmed over the FYDP and beyond, will include access to emerging imagery architectures and to high-resolution commercial imagery platforms through a shared requirements database. Migration to an order entry and tracking capability should facilitate the submission of collection and production needs by the users and provide them faster feedback of results. Special enhancements are being developed for the tactical user. New work-flow management capabilities will provide faster distribution of collected data to expedite the production of intelligence and geospatial information. Expanding interfaces to other collectors, including airborne sensors, to promote cross-cueing will enhance responsiveness to user needs.

6.2.4.2 Future Capabilities

In the future, data collection capabilities will grow. The nature and number of collectors will make new data available to the NCII, such as from hyperspectral and ultraspectral imagery platforms, in greater volumes. The collectors available to strategic, operational, and tactical commanders—commercial, National, airborne, manned and unmanned, still and motion, and macro and micro—will grow in number and capabilities. There will also be an increasing demand to acquire mobile targets within shorter time lines. Future users will therefore require a new generation of collection management tools that treat ISR platforms as assets that are fully integrated and provide feedback within tactical time lines. This will require an investment of military research and development resources. New methods and new tools must be developed to provide intelligent cross-cueing, such as between signals and imagery collectors, between different imagery collectors, or between MTI and SAR, so as to acquire difficult, mobile targets. The methods and tools are needed to enable tactical echelons to get priority needs tasked and produce feedback sufficiently responsive to enable dynamic retargeting and retasking. They should be able to accomplish this retasking automatically.

The TPED baseline/modernization plan mentioned above should help steer current collection management systems toward such capabilities. An example that takes a dramatically different approach to next-generation tools is the DARPA Advanced ISR Management program. The intent of the program is to develop capabilities that would allow ISR confederations to operate so as to improve the capacity for synergistic collections. This would enhance time-critical targeting and battlefield awareness. The project is initially focused at the joint task force level. It addresses three major issues: the need for the development of information to factor in a commander's intent, addressing those needs using large-scale optimization techniques, and synchronizing the multiple assets available in order to satisfy users' needs in a near-real-time environment. To do so, the tools must include learning and inferencing capabilities to interpret complex plans and situations and predict and assess progress, as well as to optimize strategies over a massive decision solution space. Prototype products are scheduled for analyses in demonstrations through special projects from 1999 to 2001. The successful products could transition to various requirements and collection management systems and future architecture developments, or through the Community Integrated Collection Management program and the airborne/overhead integrated task forces.

6.2.5 Information Exploitation

6.2.5.1 Introduction

6.2.5.1.1 Definitions

Information exploitation as used here covers all aspects of how sensor and other data are processed to gain intelligence and geospatial information. The phases of information exploitation are depicted in Figure 6.9, which is intended to show that any or all of the individual steps can contribute to a consistent, structured set of information used to represent the current situation. However,

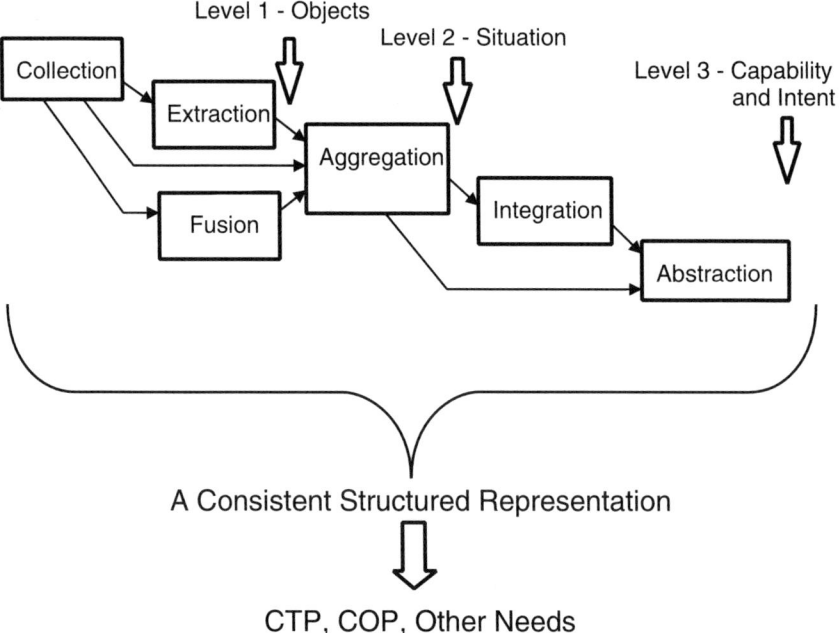

FIGURE 6.9 Phases of the information exploitation process. COP, common operational picture; CTP, common tactical picture.

the output of each step can also contribute to higher-level representations or abstract views of the battlespace.

Collection refers to all forms of data or information acquisition. From a set of raw collected data it may be necessary to extract a specific piece of information. Extraction often involves substantial processing, as when doing model-based recognition of an object in an image. Similarly, data objects of different origin may be fused to form a more authentic representation of that same object. Such extracted and fused data can then be aggregated with other unprocessed data or information to form a more composite representation. Aggregation is exemplified in the building up of an air or undersea defense picture. The final phase is abstraction, which is the replacement of a number of individual, not necessarily similar elements with a single, higher-level representation that spares the viewer excessive detail or clutter. That higher level may also involve information that has been integrated to form an insight that individual data elements might not reveal. This string of steps can be recursive as information is passed up the chain of command. Lastly, when information is exploited successfully, the result is a consistent representation of the battlespace situation that includes specific views, such as a common tactical picture (CTP) or, at the CINC level, the common operational picture (COP).

6.2.5.1.2 The Importance of Context or Metadata in Fusion and Abstraction

A key need in the fusion, aggregation, and abstraction of data is context. Context can be defined as supplemental information or metadata, providing the basis for more readily understanding a discourse, and it must be enlarged until the communicating parties are clear on what is being exchanged. All such information-bearing elements introduced in a discourse need a context for them to be meaningful.

So, what seems necessary, even critical to the emergence of a general battlespace picture, is the presence of two contexts. First, any information object, regardless of its size, must be attended by an explicit supplemental vector that describes what it is and the circumstances of its origin, including its veracity. Without such metadata, the information itself is difficult to integrate or use directly. Also, if there is ever to be the hope of automatic fusion, this background vector must be present. Ideally, the vector itself, and the definition of its elements, would be universally agreed to and accepted. That is much easier for basic or atomic units of information but probably more difficult as the scope of the information gets larger. For example, all sensor data must be attended by metadata that are equivalent to a camera model that enables an interpreter to merge that sensor output with the output of other sensors. Having the metadata and data together lets the couplet be an object with much broader utility in the information space.

A second critical context is one in which the information element gets interpreted or used. Just as in the case of the information element, this second or integration context can be defined more easily when the information elements to be aggregated are few and not complex. Because this second context is, by definition, one step higher in information abstraction, it may be harder to define and less adaptable to standardization. Yet it will still have common descriptors such as time stamps, perishability, scale or scope of resources (i.e., level of command), red or blue order of battle, veracity, and its ultimate name or use. At the highest level of aggregation or abstraction are the developing terms such as "common operational picture" and "image of the battlespace." Their tailored nature may also make them too varied to be universally interpreted. Predefined templates, widely used as a point of departure, may help. "The right information to the right person at the right time" is a maxim that rolls easily off the tongue but ignores the fact that all three "rights" have no generally applicable meaning. Such maxims become useful only when defined in the context of each specific situation.

Clearly, the context of aggregation at one level becomes the context of an information element for use at the next point of integration or abstraction. Further, it is important to maintain the distinction between content and context. This emphasis on explicit and usable context not only draws attention to needed supplementation but also avoids the pitfall of having to define data and knowledge, terms that are mostly in the eye of the beholder.

6.2.5.2 Near-term Assessment

The process of information exploitation is practiced in many military and intelligence contexts today. Within the limited scope of this report it is not possible to comment on those applications in any detail. However, it is fair to say that as one proceeds to the right through the steps in Figure 6.9, the process becomes less automated. Indeed, the bulk of research efforts have been devoted to matters of extraction (e.g., target recognition) and fusion. Generally, one can say that automatic extraction is widespread, automated fusion has shown some success, and automated aggregation has been demonstrated in situations involving like data and a relatively fixed operational environment.

The COP/CTP stands as one particular and important example. The tracks of moving targets detected by an individual sensor can be determined automatically (extraction). However, significant manual intervention is required to process out redundancies and uncertainties among the tracks from multiple sensors to form the composite picture (aggregation). Part of the difficulty in forming the composite picture is the lack of metadata, or context, as described above.

Conceptually, the COP should mean a consistent and not a common operational picture because, ideally, it is derived from a single, consistent, structured representation of the battlespace. Under the present COP formulation, information is collected at all levels, with the lower levels percolating upward what they believe is important. Thus, information in that distributed process becomes integrated with the perspectives and judgments of each level as the story moves on up. That is both good news and bad, for while insight, hopefully mostly useful, may be resident in that stream, it is more difficult to assure consistency. In other words, in the process by which the CTP/COP is now formed, consistency becomes subordinate to collective insight.

The process of exploitation is a difficult and long-studied one. While there do not appear to be major advances in the near term, evolutionary progress should be possible. One recommended avenue of pursuit is to be more systematic about the definition, capture, and use of context as described above. SPAWAR is very active in COP/CTP developmental efforts, and it should consider such context-based approaches. A worthwhile objective might be to find some type of template system that would permit tailoring tactical presentations while still retaining a consistent, perhaps standard ontology for the information used.

6.2.5.3 Future Capabilities

Information exploitation has a number of long-term requirements:

- The ability to automatically extract common objects from different, non-orthonormal imagery, without manual intervention;
- The fusion of information from disparate sources having different extraction methods (e.g., imagery, multi- or hyperspectral data, IR, and human intelligence);
- General algorithms that enable the automatic building of abstract representations that convey their salient features to the level of command involved;
- Rapid and synchronized interpretation of all-source data;
- Automated extraction support to improve responsiveness and help compensate for the decreased availability of skilled and informed analysts; and
- Harnessing the potential of video and other time-sequenced imagery with automatic monitoring and processing (e.g., tracking and classification of targets).

While these needs argue strongly for advances in the automation of extraction, they should not be taken to imply that all information and insight will come from machines. The exploitation process should be subject to human oversight and involvement, where appropriate.

There remain some profound difficulties in the automatic exploitation of information. One is the automatic composition of an accurate representation of the battlespace from elements with multilayered (repeatedly abstracted) uncertainties. Another is the integration of disparate input data. The normal technical approach to such compounded uncertainties is Bayesian statistics. Difficulties are caused by the lack of independence of the various conditional probabilities in a multidimensional graph and the sheer concatenation of uncertain events, even if they are perfectly independent. As has been already mentioned, any aggregation of objects to be abstracted should have a resultant profile that includes its uncertainty. Humans have a propensity in handling such data or information to avoid deriving the uncertainty of a new, integrated object either because it cannot be computed or because the multiplication of independent probabilities leaves the person discouraged. At best, humans use judgment in choosing the confidence in what has been created. The danger is that unqualified computer output is too often accepted without question. In the design of information-centric systems, one must remember that the knowledge gained will rarely if ever be perfect. Appraisals will be best served if honest assessments are given as the information is aggregated and abstracted.

The needs indicated above are all important in the long-term realization of the NCII. Over the years, much research has been devoted to the problem of exploiting information, and much future research is necessary. DARPA, for example, has been active in the field: two examples of current work are the Dynamic Multiuser Information Fusion (DMIF) and Dynamic Data Base (DDB) programs. The naval services should track and participate in exploitation research and sponsor it in areas particularly germane to them, since such research is critical to establishing information products, such as the COP and CTP.

6.2.6 Information Request and Dissemination Management

Users have many sources of information; they, in turn, create more information using their own value-added processes. But in today's information-rich environment, the burden is on the user to find the means to locate the right information. This situation will become more complex. The future will bring increased data collection capabilities in terms of the nature and number of sensors available to commanders at every level. It will bring a manifold increase in repositories of information from producers who are globally dispersed but accessible through networks. Given all the information that will be available, including that from open sources, the user may reach a state of information overload. New capabilities are needed that can provide users with ready means to easily discover and acquire the information that is most relevant to them.

The function of information dissemination management (IDM) is about managing the flow of information from providers to consumers who are globally dispersed but connected via networks. It is about providing integrated capabilities for awareness of, access to, and delivery management of information to support the full spectrum of military operations for users at the strategic, operational, and tactical levels. These distinct environments result in user requirements that differ in such parameters as time lines and interfaces and introduce different constraints that must be accommodated.

6.2.6.1 Near-term Assessment

Today, the user is often unaware of what information is available, or there are inadequate methods to access it without knowing beforehand exactly where it is stored. For instance, a user needing a specific imagery product but not knowing where it is located would have to query many servers to find the product. The means to tailor information flows for individual user communities or for individual users in accordance with a commander's intent are limited. Dynamically adjusting information needs based on an emerging operational situation is very difficult. Such changes are achieved with manual processes or with workarounds, if at all.

Certain intermediate capabilities are emerging that promise to enhance significantly the accessibility and distribution of information to users. These early innovations are intended to provide end-to-end dissemination of information consistent with the commander's intent and give select users an awareness of certain information as it becomes available. This represents a significant shift in capability through a focus on the end-to-end management of information to the user from the producer. These emerging services are tailored to specific user needs and user communities. Most are initially directed toward the strategic and JTF-level needs and use the DII-COE infrastructure.

More specifically, these new capabilities focus on the core services shown in Figure 6.10. These core services are transitioning from the DARPA Battlespace Awareness and Data Dissemination program and the Bosnia Command and Control Augmentation system, in conjunction with the DISA Information Dissemination Management and Global Broadcast System programs.[15] The services, which will become part of the DII COE, comprise awareness, access, delivery, and support for information needs. They rely on user profiles, command policies for management of content and resources, and the use of metadata schemas to describe information needs and policies, coordinating information access and dissemination across a federated infrastructure of repositories and networks. Realization of the new capabilities also requires that the information producers adapt to the metadata schemas in their architectures, as is occurring, for example, in NIMA's U.S. Imagery and Geospatial System.[16]

A series of demonstrations, exercises, and experiments was used to evolve and test these interim core services by providing geospatial information, imagery, intelligence order of battle, and logistics data. A recent military assessment of the results noted both the advantages and the shortfalls in their implementation

---

[15]The capabilities so implemented are often called "idm" (little IDM) in contrast to "IDM" (big IDM), a program aimed at providing the longer-term capabilities.

[16]Planned capabilities in this program go beyond the management of information for dissemination. They include deliveries of libraries to the Services, commands, and agencies for storage and access to imagery products and geospatial information.

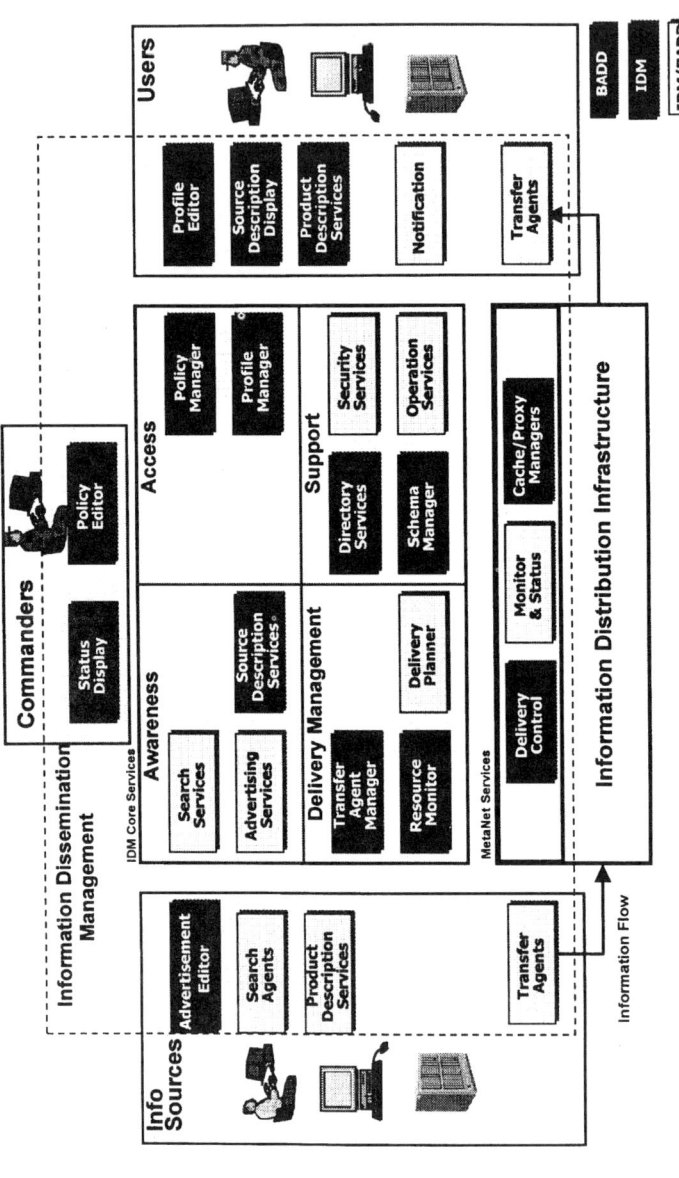

FIGURE 6.10 Information dissemination management core services. BADD, battlefield automated data distribution; IDM, information dissemination management.
SOURCE: Beaton, Robert, from a briefing to the System Architecture Panel, Committee on Network-Centric Naval Forces, April 16, 1999, Defense Advanced Research Projects Agency, Arlington, Va.

with respect to stability, performance, and ease of use. Some immaturities will need to be resolved by an evolving accommodation to joint warfighting requirements. An important issue at the moment is the absence of agreement on metadata standards for the publishers of information. The set of core services achievable in the near term is currently constrained to a subset of those needed by all the CINCs. They are also limited in the number of producers and consumers who can be linked, a constraint driven by the need for DII compliance, and the narrow set of producers who meet the metadata standards currently implemented (IDM, Link 16, and Intelink-S). In summary, the core services and associated developments are a significant step forward, but further effort is needed to achieve the longer term goal and set of needs.

6.2.6.2 Future Capabilities

Future IDM needs will be realized through a set of services that provides an information marketplace for users and functions in accordance with policies that may vary by the commander, by the operational region, and by the nature of the mission. The services require that all information producers have the means to advertise, publish, and distribute their information to a widely dispersed community. They require the ability to deliver published information to users over effective communication paths in a manner that is transparent to the user.

A program for longer-term IDM development has recently been established, with USJFCOM having the lead on developing the capstone requirements document[17] and the Air Force serving as the executive agent. Given the importance of information dissemination management to the NCII, the naval services should closely monitor and work with this program. Several important challenges face the program. It must ensure that a complete and robust set of core services is established. The services must be easy for the user to apply and must adapt to rapidly changing information needs in the face of evolving operational situations. They must also guarantee that information dissemination follows the commander's guidance. Furthermore, the program must ensure that adequate information assurance technology and practices are incorporated into the IDM functionality. Information dissemination poses critical vulnerabilities that must be protected against—e.g., denial of service, traffic analysis, and the insertion of false information.

One of the most critical challenges facing the long-term realization of the IDM paradigm is the scaling of the metadata standards that describe information needs, products, and policies. These standards must be satisfactory to and ac-

---

[17]A capstone requirements document (CRD) is a document that provides an overarching description of the goals or vision of its subject area. The capstone feature provides a shared vision of the top goals and objectives that guide the plans and activities of the individual supporting programs.

cepted by the entire user and producer community for the overarching IDM concept to work. Potential extension to allied and coalition partners must also be considered. Achieving such widespread agreement is a daunting task and quite possibly beyond the scope of an individual program office. It may well require concerted efforts at the senior levels of DOD and the intelligence community.

Even under the most optimistic conditions, one cannot assume that the IDM mechanisms will locate all relevant information. Thus, in implementing the NCII, additional search mechanisms should also be considered, such as those provided by software agents. DARPA's Control of Agent-Based Systems (CoABS) program focuses on the technologies of software agents to help manage information in an environment of heterogeneous systems. As such, it has utility in many applications, one of which is information acquisition. Among the potential applications in the program are managing sudden, irregular increases in bandwidth and optimizing resource allocations, brokering open sources of information for the user, and negotiating among disparate legacy systems to achieve interoperability. Agents can be mustered into a mobile team to search for information that is not "plugged into" the standard infrastructure. In the heterogeneous environment of coalition operations, connected with disparate networks, agents have great potential in facilitating the movement of information from providers to users. A powerful but simple example was demonstrated in Operation Allied Force, when software agents were used to direct imagery users to the right source with one access request.

CoABS will implement a prototype agent grid supporting diverse systems and using different types of agents for various services, such as brokering, searching, visualization, and translation. The results will be used to determine the best types of agent control, such as to provide quality of service and efficiently manage routing. Implementation will also allow exploring the best ways to codify intelligence in agents. The dynamic nature of information management, its changing run-time environment, and the changing way in which information is used, even ad hoc, offer special challenges in adaptation to the teams of mobile agents.

6.2.7 Information Presentation and Decision Support

6.2.7.1 Introduction—The Common Tactical and Operational Pictures

Just how much a substantial increase in the amount and timeliness of information can help a commander achieve his military objectives is to a large degree determined by how and when that information is presented. As used here, presentation means the pictures being displayed, the tools used to produce those pictures, and, finally, the display technology itself. The most "official" of the presentation pictures are the common tactical picture (CTP) and the common operational picture (COP), which are supported by the common tactical data set

(CTD). Because the CTP and COP are so prominent, they are now described briefly.[18]

The COP is an integrated view of CINC-level operations provided through the Global Command and Control System (GCCS). The COP is formed from the CTD and is a composite of the various CTPs associated with individual operations within the CINC's area of responsibility. Nominally, the joint task force and levels below that each develop their own CTP, but where in the hierarchy that ends is not clear. While the CINC is clearly responsible for creating and maintaining the COP in his area, Figure 6.11 shows that each Service also has its own COP, and other CINCs have theirs. Suffice it to say that the basic form of the COP is being defined and developed at the Joint Staff and DISA, and it is intended to become the means by which a high-level commander first obtains and then maintains situational awareness. For what follows, there is little need to distinguish between the various COPs or CTPs.

To increase its local utility the COP/CTP can be tailored by each of the joint force component commanders according to their mission and preferences. The COP/CTP is distributed horizontally to sister line and support units as an expression of the battlespace conditions that they have in common. Distributed vertically, the COP/CTP does three things: (1) it conveys to the next higher command level the substance of the subordinate commander's available information and perspective, (2) it is handed downward to provide context for interpreting that level's composed CTP, and (3) it helps distribute the commander's intent.

The embodiment of both pictures is a semiautomated situational map, a graphical depiction of the information available at that time to the command level preparing it. The map base can be overlaid with a number of reporting and tasking orders, and its information can be selected hierarchically with links to more in-depth information. The various CTPs follow the established chain of command in a specific area of operations, with each commander being responsible for maintaining the CTP depicting his area of responsibility. The COP/CTP system can also portray future conditions or situations such as the impacts of impending weather.

For a variety of reasons, much of the CTP is now manually prepared. What each display shows is agreed upon in only the most general terms. Much of the present COP/CTPs has to do with the sighting and tracks of red, blue, and neutral platforms in the sea and air or on the ground. Some intelligence products such as

---

[18]The responsibility for maintaining the CTP, COP, and CTD, their general composition, and the associated information flow and management are outlined in the Chairman of the Joint Chiefs of Staff Instruction CJCSI 3151.01, June 10, 1997, "Global Command and Control System Common Operational Picture Reporting Requirements," Washington, D.C. Further descriptive information is contained in Joint Chiefs of Staff, 1999, *Global Command and Control System (GCCS) Common Operational Picture (COP) Primer,* presented in a briefing to the committee, March 4, 1999, Washington, D.C.

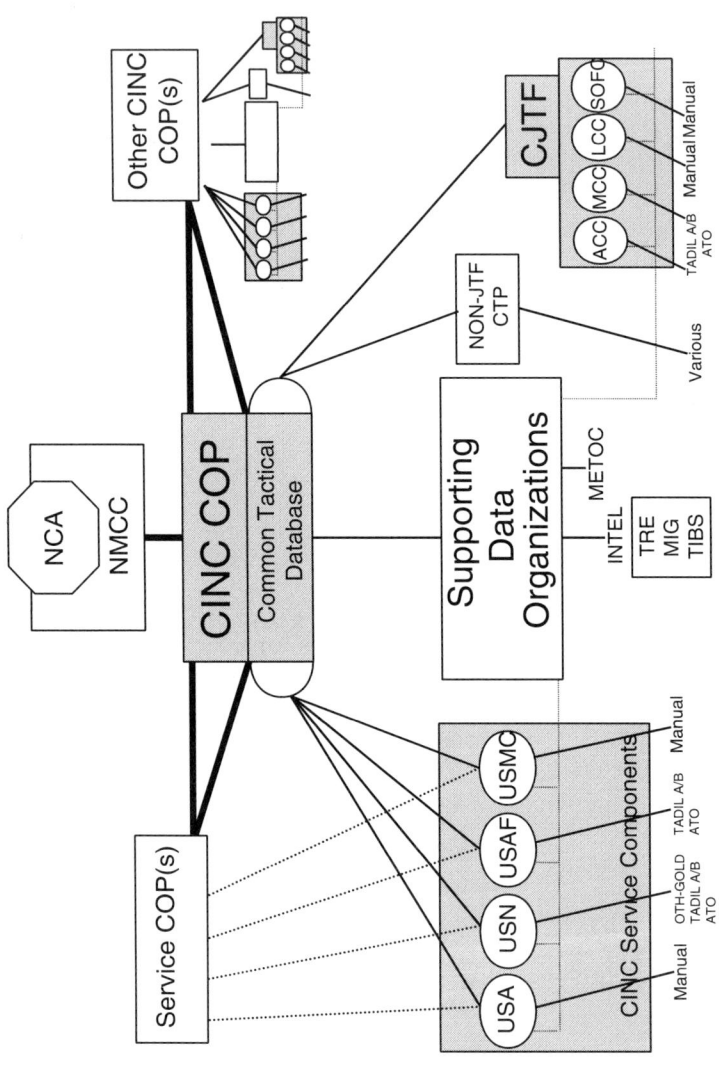

FIGURE 6.11 The common operational pictures (COPs) and their contributors as illustrated in the COP primer. Acronyms are defined in Appendix H. SOURCE: Joint Chiefs of Staff, 1999, *Global Command and Control System (GCCS) Common Operational Picture (COP) Primer*, p. 27; presented in a briefing to the committee, March 4, 1999, Washington, D.C.

electronic intelligence (ELINT) and Theater Intelligence Broadcast System (TIBS) observations are also integrated. The integration or fusion of information for the CTD is done mainly by so-called track managers.

6.2.7.2 Near-term Assessment

The implementation of the COP and CTP is widespread and serves as a basis for information presentation in the NCII. The COP is distributed through GCCS, which makes it available on major ships, while the CTP is distributed through GCCS-M, making it available on almost all ships. For the Marine Corps, a variant of the CTP is the principal battlespace picture in exercises such as Urban Warrior and the Extended Littoral Battlespace ACTD. The use of COP and CTP is a big advance over the previous practices that relied on separate display systems, each containing different information of relevance to the overall battlespace picture.

As noted above, the COP and CTP draw their information from several sources. This is a strength in that it allows input from many sources that gather input on the battlespace (although input on ground targets is currently very limited). But it is also a weakness since there can be inconsistencies among these sources (as discussed above under information exploitation). In addition, the bottom-up formulation of the COP and CTPs with significant manual intervention can lead to time delays in distribution. Furthermore, the information from the individual sources is displayed in the composite COP and CTP without an overall concept for just what information is required and should be displayed. Because the COP and CTP are relatively new products, no doctrine or guidance for how best to use them operationally has developed.

Advances in COP/CTP implementation and means of presenting information over the next few years are expected to be evolutionary, not revolutionary. One key item that should be addressed by the naval services and the broader joint community is the content required in the information displays and the methods for using this information. These requirements should be driven by the warfighters. The experimentation process is an important way to develop and refine these requirements, as has been already seen, for example, in a limited way in Urban Warrior and the Extended Littoral Battlespace ACTD. Questions to be addressed include what information is needed at each command level, how to best portray it, and how soon it should be distributed. Additional questions of interest at the tactical echelons are how to deal with the limited bandwidth available, especially for the lowest Marine echelons, and how to use the CTPs to synchronize operations. Furthermore, given the variability in missions, constituent forces, and the preferences of command, none of the COP/CTPs will be the same. Experimentation should be used to ensure that these differences do not introduce inconsistencies or incompatibilities in operations.

One other aspect of decision support should also be noted—conferencing, including video teleconferencing. This is an effective way for officers at physically separated locations to plan, exchange information, and reach decisions. The Navy has already explored this in IT-21. Over the next 5 years or so, high-quality, immersive, virtual roundtable conferencing is expected to become available in situations where wideband connectivity exists. The naval services should track these developments and consider their applicability to naval missions.

6.2.7.3 Future Capabilities

6.2.7.3.1 General Considerations

Research issues in information presentation that should be pursued to provide the flexible and responsive information management capabilities ultimately envisioned for the NCII include the following:

- New viewing paradigms other than just two-dimensional maps that include ways to quickly and intuitively grasp items of importance while deemphasizing less meaningful items;
- Automatic picture updates based on event-driven and temporal cues;
- Consistency over time and between command and functional levels in the portrayal of the battlespace and the displays on which it is depicted;
- Continuous planning methods that adapt to changing events or courses of action; and
- Methods that can, from the assembled information, suggest an enemy commander's intent or course of action.

While this is a daunting list of challenges, work is going on in the visualization area. For example, the DARPA Command Post of the Future (CPoF) program is concentrating on the types of visualizations that can increase the speed and quality of command decisions. CPoF will examine how pictures can be tailored and decision support tools adapted to changing conditions. It will build on the other programs at DARPA producing analysis, planning aids, and information management and will try to develop the following capabilities: decision-centered visualization, speech and gesture interaction, automatic generation of visualizations, and dialogue management. The Navy (SPAWAR Systems Center (SSC)) has been participating in CPoF and should continue. Hopefully, the Marines Corps can also participate in the present CPoF studies to examine how their particular tactical information presentation needs can be addressed.

6.2.7.3.2 Tool for Operational Architectures

One further topic deserves particular elaboration. The above near-term assessment notes that there is no overall concept for just what information is required and should be displayed in the COP and CTP and no doctrine or guidance for how best to use the COP and CTP operationally. This relates to the broader issue of the information needs and flow at the various levels of command to support an operation. As noted in Chapter 4, the operational architecture (OA) should lay this out. But, as also noted in that chapter, attempts at operational architectures have been too detailed and focused on the as-is situation to be useful generally.

What is needed is a tool (or tools) for developing OAs at the intermediate levels of abstraction, which in good measure is a research problem. This OA development tool would assist in defining the policies and rudiments of information flow with just enough precision that a system architecture could be defined for a specific operation. The tool should be easy to use and understand, have definable types of information and levels of detail, and be aware of the system resources available that carry and display the information needed. If developed, an OA tool would have broader applicability than just information presentation and decision support, but clearly it would be important to that functional area.

Since there is neither time nor need to start from scratch each time a naval force is assembled, an OA tool begins with templates. These can range from organizational charts and reporting relationships, to the most important information needs of specific units, to the available system resources and their characteristics that will eventually be represented in the associated system architecture. To the extent that the anticipated force structure is similar to one used in training and field exercises, templates derived from those areas will obviously expedite the task of defining a new operational architecture.

The OA development tool would be a computer-aided means to create operational architectures that are specific enough to define the information flow in general terms as well as to identify the system resources necessary to carry it out. Here are some features such a program ought to have:

- *A hierarchical structure.* Such a structure is able to define information flow starting with the highest level units and delving to the lowest but stopping where detail is sufficient to define a system architecture. A logical point of departure in forming such a hierarchy is the naval force's organizational chart with its joint and coalition linkages;
- *Interunit information relationships and descriptions of need organized by type and level of information.* Different levels in the hierarchy will have different aggregations of traffic, and where specific types of information have critical capacity or priority aspects, they will be identified. Factors considered include the following:

　　—Intra-Service or joint or coalition linkages,
　　—Nominal information requirements for each unit, including types of information or traffic flow and quality of service (capacity, availability, accuracy, delay, security/privacy);

- *Location of units.* This knowledge is useful when maintaining network equipment and for use by attached devices that do not carry such information themselves; and
- *System offerings.* A catalogue of presentation and network transport equipment and systems indexed by particular information needs helps define the system architecture.

An important guideline for designing such a system is that there is an iterative relationship between the operational and system architectures. The "intelligence" of such a system is an ability to recall previous iterations and solutions for a given mission and to stipulate by means of previously defined templates that may in part be rule-based, the following:

- All previously used information flows and their corresponding equipment;
- Equipment required for a specific information need;
- Needs not satisfied by equipment in the inventory, indigenous or leased; and
- Unit or platform locations not covered by elements in the system inventory.

This type of tool should fit easily under a DARPA program now under way called Active Templates. The program, which has interface shells that permit the building or use of templates in the context of interactive planning, allows the incorporation and use of recent experience and can employ automatic reasoning, including temporal. While the program is not yet addressing the building of an operational architecture, such a task seems well suited for the technical methods now under way.

6.2.8 Execution Management

The functional capabilities discussed in the preceding sections provide support for making command decisions. Once those decisions have been made, they must be conveyed to the appropriate operating elements and, in the face of rapidly changing events, modified if necessary. That is the purpose of the execution management function. One might argue that the preceding functions are all that is needed to convey and modify decisions. In a sense that is true, but the need for rapid adaptation is so central to network-centric operations that it would be best to explicitly identify a function that supports the rapid direction and redirection of force elements.

Four capabilities seem particularly necessary for execution management:

- Rapid, guaranteed delivery of command orders;

- Effective promulgation of commander's intent;
- Rapid feedback of battle effects (e.g., battle damage assessment (BDA)); and
- Rapid planning for force redirection.

In considering these capabilities, one should not think only in terms of a strictly hierarchical command model. That is, the initial orders could come from a senior echelon, but the rapid adaptation could involve decisions made only among the lower echelons.

6.2.8.1 Near-term Assessment

The delivery of command orders has long been a matter of high priority. While new means to improve communications are always being sought, especially in the face of limited bandwidth and jamming, effective means for communicating command orders have in fact been developed. The advances discussed earlier in communications and networking, information assurance, and system resource management should lead to further capabilities in this area.

The main method today of assuring that a commander's intentions are absorbed and executed by the available forces is the pre-mission command briefing, relayed down the chain of command until all members of the force are able to act in unison. These briefings convey the mission objectives, the enemy situation, distribution of responsibility across the participants, and the timing of the operation. Most of today's command briefings are based on two-dimensional map symbolism and the plan of execution is expected to hold until a new, similar briefing can be held. The question is whether more elaborate means are necessary to convey the commander's intent (as distinct from some of the more detailed aspects of the battle plan). The MCCDC has examined this issue, and its thinking is that more elaborate means are not necessary. Rather, what is required is that the purpose of the operation, as distinct from the specifics of movement or attack, be clearly stated. That way when there are failed assumptions or changing circumstances in the course of a battle, the forces have a rationale for how to adapt. Learning to convey *purpose* is largely a matter of officer training.

Dedicated narrowband voice channels enable rapid feedback of the most salient points about battle progress. More detailed feedback would come through such means as the CTP. As discussed above, while progress has been made in establishing CTPs, matters such as latency and consistency still need attention. Activities such as the Extended Littoral Battlespace ACTD are examining procedural and technical means for the real-time distribution of friendly and enemy force situation data. Another factor is BDA. Even if damage information is rapidly conveyed back to force planners, it is necessary to rapidly assess the effects of this damage in order to decide if forces can be directed elsewhere because of target destruction, or if additional forces must be applied to the original target. BDA is a difficult task, and means to carry it out much more rapidly are needed.

Traditional planning, for example, in assigning strike aircraft to their targets, is based on an air tasking order (ATO) prepared daily. Efforts to reduce ATO planning cycle time are under way and consideration is also being given to directing or redirecting aircraft in flight, based on recently gained information (e.g., the effectiveness of other sorties and the movement of enemy forces). One example of a planning system for such rapid redirection is the Real Time Targeting and Retargeting (RTR) program being carried out at SPAWAR. Other aspects of rapid force direction or redirection, to include the case of land forces, are being explored in the Navy fleet battle experiments and the Marine Corps Sea Dragon experiments.

In summary, effective realization of execution management is, in some important ways, dependent on those functional capabilities discussed in previous sections. Some new items raised in the above discussion require continuing attention: clear statement of purpose in the commander's intent (which may be largely a matter of training), faster BDA, and planning processes and tools that allow the rapid direction and redirection of forces. The planning capabilities are a matter of procedure as well as technology, so continued experimentation is critical to improvements in this area.

6.2.8.2 Future Capabilities

Ideally, the intent is to develop an integrated sensing, planning, and execution system that functions continuously, giving commanders timely situational reports and suggested options. Desirable features include tools that could be keyed by an operational plan to perform continuous assessments, that could enforce tightly synchronized action, and that could replan instantly as friendly or enemy assessments changed. Efforts to integrate sensor information are ongoing and are reflected in the discussion of functional capabilities in the previous sections. Efforts at rapid planning and replanning have begun with such activities as the SPAWAR RTR program, noted above, and the Joint Forces Air Component Commander (JFACC) program at DARPA. However, automatically generating battle options for typical situations is a difficult task and is likely to remain unrealized in the foreseeable future.

6.3 RECOMMENDATIONS

The committee's findings and recommendations, based on the foregoing discussion and assessment of progress toward realizing the functional capabilities needed in a common command and information infrastructure for the naval forces, are presented and discussed here.

Finding: The Department of the Navy has valuable ongoing initiatives (e.g., IT-21, GCCS-M) contributing to the functional capabilities necessary for an NCII. However, the ongoing developments do not provide a comprehensive approach to realizing the set of capabilities necessary for a common information infrastructure. IT-21, for example, is improving long-haul communications to all ships, and GCCS-M (including the COE) is also providing necessary functional capabilities (e.g., for information dissemination management and information presentation). The value of these enhancements is very significant in facilitating and improving the treatment of information in naval operations. But as can be seen from the numerous shortfalls discussed in the functional capabilities assessment, the capabilities are not being fully addressed. Furthermore, and perhaps more importantly, IT-21 and GCCS-M/COE do not offer a systematic framework for filling out the full set of functional capabilities. IT-21 focuses mostly on end-to-end connectivity (roughly the "lower layer" or supporting resource base as shown in Figure 4.2 in Chapter 4), which is of course very important, but it does not recognize that a broader assemblage of functional capabilities is necessary. Likewise, GCCS-M/COE does not offer a systematic framework.[19]

The committee's set of findings and recommendations drawn from its assessment of all the functional capabilities is given below. As is clearly seen, the set of recommended actions is large and, taken together, they make the point that much further technical advancement is required to realize the full range of functional capabilities required for the NCII.

6.3.1 Findings and Recommendations for Functional Capability Areas

6.3.1.1 Communications and Networking—General

Finding: Significantly increased in-theater SATCOM capacity is planned, but the Department of the Navy's stated SATCOM capacity requirements could be unrealistically low, especially considering increasing imagery demands. In addition, no comprehensive statement of requirements for direct communication links from in-theater sensors (e.g., U-2, JSTARS, UAVs) to ships could be found by the committee.

Recommendation: The Department of the Navy should conduct a comprehensive analysis of communication capacity requirements and projected availability, and identify remedial actions if significant shortfalls exist. The analysis should include long-haul communications and tactical data links, including direct links from in-theater sensors.

---

[19]The COE effort has described as a layered software architecture, but that is different than a systematic presentation of the functional capabilities (which largely correspond to the common support applications in COE terms).

Finding: Communications interoperability is increasing with the forces of other Services and joint elements but is still very limited with allied and coalition forces.

Recommendation: The Department of the Defense should explore with allies the means for improved communications interoperability, in particular those based on common commercial technologies.

Finding: Rapidly advancing and potentially revolutionary commercial satellite communications developments (e.g., wideband LEO satellites) are anticipated.

Recommendation: The Department of the Navy should make maximum feasible use of emerging commercial satellite communications infrastructure and technology.

6.3.1.2 Communications and Networking—Wireless

6.3.1.2.1 Waveform Interoperability

Finding: Programmable modular radios are achievable for most communications waveforms. The technical problems in handling the JTIDS waveform in a modular radio can be overcome with a relatively modest investment. Any perceived competition with the MIDS program can be defused by pointing out that modular radios are considered just another way of implementing a JTIDS radio.

Recommendation: The Department of the Navy should give preference to modular radio programs whose individual modules can switch dynamically among multiple waveforms. All modular radio programs should include modules capable of processing the JTIDS waveform.

Finding: To take full advantage of the potential value of programmable modular radios, an experimental program is needed to explore how this new capability can best be used.

Recommendation: Using joint combat information terminals, the Marines should experiment with simultaneous interoperation with the Navy, Army ground units, and Army airborne units.

Finding: A strategy is needed to ensure future compatibility and to prevent developers from introducing new waveforms that transfer costs to the information infrastructure.

Recommendation: Acquisition agencies contemplating the introduction or further purchase of radios whose waveforms are not emulated by existing programmable modular radios should be required, absent rarely granted waivers, to develop the PMR software that permits the emulation of these waveforms.

6.3.1.2.2 Antennas

Finding: The Department of the Navy is correct in continuing to give priority to the search for multifrequency, self-stabilizing, multibeam, electronically steerable shipboard and aircraft antennas. However, developmental antenna systems may not be affordable unless requirements are tailored or a breakthrough technology appears.

Recommendation: While continuing to push available technology in programs like the Advanced Multifunction Radio Frequency System, the Department of the Navy should also seek to validate potential breakthrough technologies and should attempt to adapt its transport architectures to the use of future low-cost electronically steered antennas developed for commercial applications.

Finding: The submarine will always be at a disadvantage in terms of maximum communications rate unless its antenna aperture can be made comparable to that of a surface ship, but that would be a very expensive undertaking. Two-way communication to a submerged submarine would be possible through the use of towed buoys or an acoustically linked autonomous vehicle.

Recommendation: The Department of the Navy should perform system engineering to quantify the effect of an improved communications rate, for both periscope depth and deeply submerged submarines, on the effectiveness of the entire network in relation to the cost involved. Based on those results, the Department of the Navy should invest as appropriate in improved submarine antennas.

Finding: The committee found no Department of the Navy program dedicated to developing architecture and apparatus to permit dismounted troops to interoperate well with other component systems, although multiple technology and position location identification (PLI) programs exist.

Recommendation: The Department of the Navy should obtain agreement between MCCDC and TRADOC on the characteristics of terminals for dismounted troops and on an architecture that will permit interoperability in communications and PLI, experiment with hub-and-spoke implementations of this architecture, and procure appropriate terminal equipment jointly.

6.3.1.3 System Resource Management

Finding: Existing means for system resource management will be significantly enhanced by the quality of service (QOS) features available in the most recent and emerging internet protocols (e.g., IPv6, RSVP). DARPA programs (e.g., Quorum) are also promising significant QOS advances in the near future.

Recommendation: The Department of the Navy should track and apply advances in QOS-related commercial technologies, and work with the developers of emerging standards to address military needs. The Department of the Navy should apply DARPA advances in system resource management technology (broadening what is now being done with Quorum and the DD-21).

Finding: Knowing the attributes of the end devices connected to a network will provide useful status information on those devices (including authentication) and allow information feeds to them to be tailored to their capabilities. Very little end-device information is made available now.

Recommendation: The Department of the Navy should promote research to define and make feasible the disclosure of end-device attributes to authorized network entities.

6.3.1.4 Collection Management

Finding: Collection management systems are now largely associated with individual sensor systems and associated tasking often involves a hierarchical, manual process. This results in lack of timeliness, integrated collection planning, and cross-cueing, which could be exacerbated as assets for collecting data increase in the future.

Recommendation: The Department of the Navy should support both planned evolutionary advances (e.g., NIMA tasking, processing, exploitation, and dissemination baseline/modernization plan) and potential revolutionary advances (e.g., DARPA Advanced ISR Management program) for data collection management.

6.3.1.5 Information Exploitation

Finding: The common operational and tactical pictures are primary means for representing the battlespace situation. Automated extraction of individual targets is accomplished, but much manual intervention is required to build a consistent representation of the overall battlespace in the COP and CTP.

Recommendation: In COP and CTP development, the Department of the Navy should apply more systematic techniques for the definition, capture, and use of context (metadata) for individual target data, to facilitate establishing consistent overall battlespace representations.

6.3.1.6 Information Request and Dissemination Management

Finding: Significantly increased ability for users to locate and transparently access information is promised by the information dissemination management (IDM) capabilities currently being deployed. Realization of a wide-scale IDM capability requires a more complete set of IDM services and, in particular, agreement across the producer community (defense and intelligence) on metadata standards for information products.

Recommendation: The Department of the Navy should work with the USJFCOM requirements developer and the USAF executive agent for the next-generation IDM program to achieve a widespread IDM capability. Agreement on metadata standards across the whole producer community could require concerted efforts at senior levels of DOD and the intelligence community.

Finding: While IDM could offer very widespread information search capability, not all information can be assumed to be "plugged into" its standard information products base, so complementary search capabilities are also needed.

Recommendation: The Department of the Navy should explore the use of software agent technology (e.g., in the DARPA CoABS program) as a means to provide users a rapid and transparent information search capability, and should incorporate it in more formal development programs as the technology matures.

6.3.1.7 Information Presentation and Decision Support

Finding: The COP and CTP represent important advances in combining information from many sources, but there is no overall concept for what information is required and how it should be displayed.

Recommendation: The Department of the Navy should continue to refine the development of information presentation through experiments. Warfighter input should drive information presentation development.

Recommendation: The Department of the Navy should develop a computer-aided tool to aid in the construction of operational architectures. In helping to elaborate information flows and needs, this tool will have broad utility, including

COP and CTP construction. A key aspect requiring research is the ability to specify the information flows and needs at intermediate levels of abstraction.

Finding: Near-term COP and CTP development is based on the use of tradition two-dimensional map-based displays. While such representations are certainly useful, they are limited in their ability to convey information.

Recommendation: The Department of the Navy should continue and expand participation in visualization research efforts (e.g., the DARPA CPoF program) and, as the technology matures, incorporate it in more formal development programs.

Finding: Conferencing, to include video teleconferencing, has proven valuable to naval forces in planning, exchanging information, and decision making.

Recommendation: The Department of the Navy should explore and incorporate as feasible the advances in conferencing capability (e.g., immersive, virtual roundtables) expected to be available through commercial technology in the next several years.

6.3.1.8 Execution Management

Finding: Conveying the commander's intent is central to execution management, and a clear statement of an operation's purpose is essential to expressing the intent.

Recommendation: The Department of the Navy, through training and experiments, should ensure that purpose is always clearly conveyed in statements of the commander's intent.

Finding: Execution management, especially at the increasingly fast pace anticipated for operations, requires the ability to dynamically assign or reassign targets to forces (e.g., in aircraft strike missions).

Recommendation: The Department of the Navy should continue to pursue further development of planning processes and tools (e.g., the SPAWAR RTR tool) to allow rapid direction and redirection of forces and should continue refining the use and development of these processes and tools in military experiments.

6.3.2 General Cross-cutting Recommendations

Each recommendation above is worthy of consideration; however, since the assessment for each functional capability area is already given in the chapter, these individual findings and recommendations are not discussed further here. Rather, the focus is on a general recommendation that builds on observations that cut across the assessments and will aid in realization of the individual recommendations.

Recommendation: The Secretary of the Navy, the CNO, and the CMC should develop a comprehensive and balanced transition plan to aid realization of the functional capabilities necessary for the NCII. Individual elements with which to begin building this plan are given in the individual recommendations above. General principles for use in developing the plan include the following:

- *Achieve balance within and across all the functional capability areas.* Improvements should be made in proportion to the extent of the shortfalls noted in each area, the relative importance of each area, and the feasibility of making progress in that area. Furthermore, to ensure that a balanced approach is being taken to needs for a given functional capability, a general structure of the following sort might be considered. Associated with each functional area are both an operational process that must be carried out and technology (i.e., a hardware/software) to support it. Furthermore, each function supports the warfighter, who will need direct access to it (recall the discussion of Section 4.1.2 in Chapter 4). Likewise, each function must also have certain capabilities in it to support the technical specialists who ensure its operations. Thus, there is a "2 × 2 matrix" (process, technology) × (warfighter, technical operator), and for each of the four elements of the matrix there should be a specified set of needs. Balanced planning for a given functional capability means that all these needs are defined and addressed.
- *Participate with the other Services, defense agencies, and the joint and intelligence communities in developing the functional capabilities.* Many of the functional capabilities are not under the direct control of the Department of the Navy, as would occur in a traditional program management situation. For example, SATCOM assets are shared, collection management occurs partly in the intelligence community, and next-generation information dissemination management is being developed by a USAF executive agent and will most likely be maintained by DISA. The naval services must track and encourage such developments, and ensure that naval needs are being addressed in them, providing funding where necessary to make that happen. While staff-level working groups are important in this process, naval involvement cannot stop there. Senior-level naval officials must be aware of progress in cross-community activities and, where necessary, step in to facilitate them and to ensure that naval needs are being met.
- *Take full advantage of research products.* For example, DARPA has (or has had) programs relating to every functional capability, and ONR/NRL has important programs in communications and networking and information

assurance. Such research offers the potential for significant advancement. Interaction with these research programs is a "two-way street"—both absorbing the technology and also influencing its direction. While there is some naval involvement in such programs, the committee observed a reluctance on the part of naval program managers. Incorporating research products into acquisition programs requires that the research products be matured ("hardened"). The naval services would have to allocate funds for this, which are perhaps best kept separate from the acquisition programs so they will not be absorbed for other purposes. Furthermore, explicit efforts to assess research programs to identify "low-hanging fruit" should be carried out.[20]

- *Utilize commercial technology as much as possible.* The rapid advances in commercial communications and computing technology and their potential for reducing costs in military developments have been widely discussed. The individual findings and recommendations for the functional capabilities presented in Sections 6.3.1 and 6.3.2 noted the use of commercial technology several times. And even in cases where it is not explicitly noted, an examination of the functional capability shows wide use of commercial components in the makeup of the overall capability. Development of functional capabilities should give first priority to use of commercial technology, although it is recognized that there are situations where it is not able to meet the needs.

---

[20]Some activities of this sort do occur under the Chief of Naval Research, and the recently established Chief Technology Officer under the ASN (Research, Development, and Acquisition) could also be involved.

7
Adjusting Department of the Navy Organization and Management to Achieve Network-Centric Capabilities

7.1 KEY DECISION SUPPORT PROCESSES AND THEIR INTERRELATIONSHIPS

The Navy-Marine Corps team takes pride in giving the United States the means to implement national policy unconstrained by national boundaries. The core of the naval capability is an integrated forward-deployed battle force of Navy and Marine Corps combat units. When joined by a robust network-centric command and control (C2) system, the battle force will be adaptable to a wide variety of situations across the whole spectrum of operations from peacetime presence and training to full-scale war. Unfortunately, the committee found that the integrated battle force concept is not reflected in integrated decision making for four key management processes that are basic to better implementing the concept of network-centric naval forces for more effective operations. These four key decision support processes include:

1. *Requirements generation:* clearly stating operators' mission needs;
2. *Mission analyses (assessments) and resource allocation:* aligning program and budget resources to meet mission needs;
3. *Systems engineering, acquisition management, and program execution:* integrating, acquiring, and deploying for interoperability; and
4. *Personnel management:* acquiring personnel and managing careers to meet network-centric needs.

The first three are key for determining who has responsibility for what missions and functions both within the Department of the Navy and across the

Department of Defense (DOD). The fourth, personnel management, concerns acquiring and retaining high-quality, trained individuals to execute all of the Navy Department's missions.

The objective in the integrated management of all of these decision support processes is to field the best mix of forces, materiel, and support to accomplish national security objectives and strategy within applicable funding constraints. This objective applies for the DOD in total as well as the Department of the Navy. The effective implementation of network-centric operations (NCO) will require the cooperative actions of all the military departments; none can reach maximum effectiveness within its own boundaries of responsibilities and resources. Jointness, interoperability (i.e., the ability for systems to work together), and the sharing of information across all boundaries within the DOD are essential.

In the same vein, the actions within the three major decision support processes are interrelated and so have to be mutually supporting and well integrated. Effectiveness in the three, individually and collectively, is central to rigorous assessment of important issues and informed decision making. To provide the leadership required for a successful transition to network-centric operations, the Department of the Navy will have to adjust its thinking and key processes from a platform focus to a network-centric orientation.

As illustrated in Figure 7.1, each of the three key decision support processes serves different elements of the Navy Department leadership. This differentiation is a result of the functions assigned in law to the Secretaries of the military

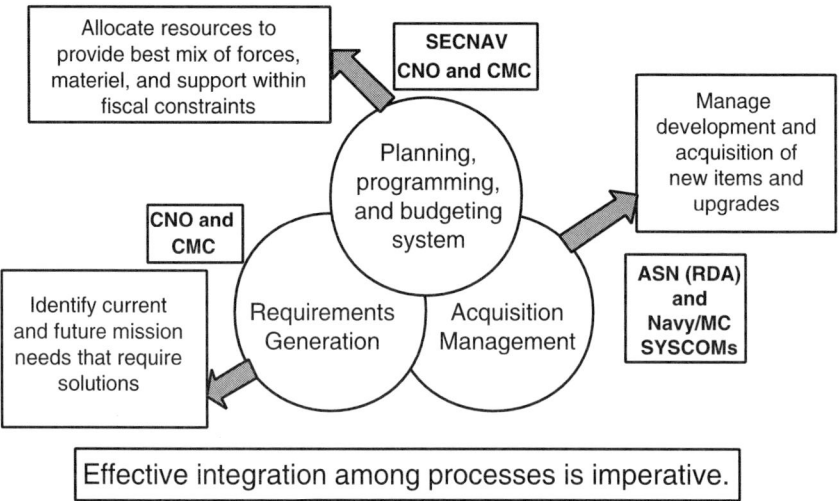

FIGURE 7.1 Major decision support processes in the Navy. Acronyms are defined in Appendix H.

departments, the Chairman of the Joint Chiefs of Staff, the Service Chiefs, and the military departments' Assistant Secretaries for Research, Development, and Acquisition. Box 7.1 gives details regarding the division of support for requirements generation and for acquisition management.

Effective integration among the processes is essential for achieving interoperability and will be even more important as the Department of the Navy shifts to a network-centric focus. This shift will require the cooperation of the combined military and civilian leadership in the Department of the Navy and the DOD to gain the full benefits of network-centric operations.

Transformations from one method of operation to another, such as from a platform-centric to a network-centric naval force, do not succeed in large organizations without strong support from the top. An organization's leaders are responsible for ensuring that those involved in change are meeting goals and objectives and that they persist in making progress. An important related aspect in transforming the forces is to develop *concrete measures of output* for the forces— i.e., measures of the ability to accomplish assigned military missions.

The first four of the following six sections cover the key decision support processes related to implementing the network-centric concept. Each covers the process as currently implemented, some weaknesses as they relate to implementing good practices for improving network-centric operations, and findings. The fifth section addresses current organizational responsibilities (and weaknesses) for implementing more effective network-centric operations. Suggestions and specific recommendations for improving the individual processes and, more importantly, for promoting their integration to achieve the larger goal of implementing more effective network-centric capabilities are offered in the last section.

7.2 REQUIREMENTS GENERATION: CLEARLY STATING OPERATORS' MISSION NEEDS

7.2.1 The Requirements Generation Process

A requirement can be defined as "an established need justifying the timely allocation of resources to achieve a capability to accomplish approved military objectives, missions or tasks."[1] Preparing a requirement that has a reasonable chance of successful development and acquisition depends on the balancing of capability, resources, and timeliness among the operational, technical, and financial communities. An intense and continuing dialogue among the three communities is required to avoid a failed development, i.e., one leading to a system that

---

[1] Melich, Michael, and Michael G. Sovereign. 1985. *The Requirements Process in DOD*, a report submitted to the President's Blue Ribbon Commission on Defense Management, informally known as the Packard Commission, Naval Postgraduate School, Monterey, Calif.

> **Box 7.1 Division of Responsibility for Requirements Generation and Acquisition Management**
>
> Requirements generation supports the Chairman of the Joint Chiefs of Staff (CJCS) across the DOD and supports the Chief of Naval Operations (CNO) and the Commandant of the Marine Corps (CMC) (beginning with the operating force commanders) within the Department of the Navy. Following passage of the Goldwater-Nichols Act of 1986, which strengthened the role of the CJCS, the Services still are responsible for the functions of manning, training, and equipping the forces to be provided to the joint regional and functional commanders-in-chief (CINCs). The Joint Requirements Oversight Council (JROC), chaired by the Vice Chairman of the Joint Chiefs of Staff (VCJCS), is responsible for reviewing and approving all requirements that have "joint interest,"[1] but the Service Chiefs retain the responsibility to develop requirements for the forces that they provide to the joint CINCs, who command U.S. military operations. The joint CINCs participate in the JROC process through their advice to the CJCS and the VCJCS (and also by stating their views on needs to their Service component commanders).
>
> Acquisition management supports the Under Secretary of Defense for Acquisition and Technology (USD (AT)) across the DOD and the Assistant Secretary of the Navy for Research, Development, and Acquisition (and the Systems Commands (SYSCOMs)) within the Department of the Navy. Program managers for major new acquisition programs, whose primary function is managing the development and procurement of new systems, are responsible to the USD (AT) and, under law, are to report to the USD (AT) through no more than two intermediate levels in their management chain. The acquisition management process falls under the purview of the civilian (Service Secretariat) side of the military departments, and the Assistant Secretary of the Navy for Research, Development, and Acquisition (ASN (RDA)), is the designated Service acquisition executive.
>
> The Planning, Programming, and Budgeting System (PPBS) supports the Secretary of Defense across the DOD and both the Secretary of the Navy and the two Service Chiefs within the Department of the Navy. The Service Secretary is responsible for submitting program and budget proposals to the Secretary of Defense for approval, and the individual Service Chiefs (and their staffs) retain responsibility for developing the proposed programs for their Service.
>
> The requirements generation and the acquisition management processes are event based (decision points occur based on readiness to proceed). They are intended to support the time-driven PPBS, which assists the military departments and the Secretary of Defense in preparing budgets presented to the Congress on specific dates. The fact that two of the systems are event based and the third, the most dominant, is driven by external calendar constraints (and also more recently by externally imposed funding constraints) has created tension rather than cooperation among the staffs responsible for managing the processes.
>
> Moreover, in recent years other circumstances have also proved detrimental to good integration. Some major programmatic decisions have been delayed until late in the budget preparation process; some long-term modernization plans have been destabilized during annual budget development cycles; and worldwide contingencies have required larger-than-expected operating funds.
>
> ---
>
> [1]The VCJCS can use the joint warfare capabilities assessment (JWCA) process to assess military capabilities and needs in different warfare areas. The JWCA process is managed by the Joint Staff and participated in by representatives from all four Services.

will be obsolete, ineffective in the field, or too expensive. This is particularly true for developments involving rapidly changing technology.

The rapid evolution of information technology, the inherent jointness of information networks, and the loss of technical leadership to the commercial market have made the generation of requirements for sensors and information systems difficult in all the Services. The implementation of the relatively new joint processes, the Service and platform orientation of the acquisition systems, and the need for Navy personnel to acquire new skills and understanding have contributed to the difficulty of the challenge. However, achieving the necessary dialogue is essential, whether by a linear or a spiral process.

Currently, a formal process is not in place to produce the integrated requirements for more effective NCO. Also, there does not exist a Navy (or DOD) element whose position, longevity, and interests are adequate to control the evolution of capabilities to support NCO. Historical precedent can be found in the Navy's General Board of the 1920s and 1930s. In that period of great uncertainty about the future, the General Board was invaluable in maintaining a map of what had been decided and why, and then in enforcing consistent, synchronized allocation of resources. However, the board may have been less innovative than what is needed today to lead the current transformation to more effective network-centric operations.

A formal requirement in the Navy is the responsibility of the Chief of Naval Operations (CNO), who uses the Office of Naval Operations (OPNAV) to provide top-down guidance and oversight of what is a very diffuse and diverse requirements generation process. In this process a formal requirement is staffed by OPNAV and approved by the CNO or Vice Chief of Naval Operations (VCNO) or their designated authorities. It is then sometimes approved by others within the DOD. Upon final approval it then may become the basis for an acquisition program, to be approved by Congress (if it is large enough to require that specific attention). Generally, platform sponsors dominate the early formulation of Navy requirements, while the requirements process in the Marine Corps rests with the Commandant of the Marine Corps (CMC) staffed by the Marine Corps Combat Development Command (MCCDC).

Occasionally requirements are discovered or generated by top-down analysis. Much more frequently they flow up informally from the operational community, which complains about deficiencies in existing systems, or from the technical community, which sees new opportunities in emerging technologies. Rarely is a formal requirement developed for a new system; most requirements are for modifications to existing systems.

Fixing deficiencies and embedding new technologies in existing systems are most often funded through repair and modernization efforts. In most cases no formal requirement is written, and OPNAV's visibility is limited.

7.2.1.1 Requirements Through Analysis

Over the last four decades requirements have been developed principally through operational analysis or, in the absence of such analysis, by assertion.[2] This process generally has been linear, with the desired end result specified in terms of performance parameters (sometimes with technical specifications written in as part of the "requirement"). However, for areas where the technology is changing rapidly, this approach is weak if the development and procurement of a required military system cannot be achieved within a reasonable time.

The committee believes that a new paradigm is needed to develop requirements for rapidly changing technology areas in the future network-centric world.

7.2.1.2 Requirements Through Experimentation

Experimentation provides a means to explore alternative doctrine, operational concepts, and tactics that are enabled by new technologies or required by new situations. That is, new technologies or situations may call for different ways of conducting operations. But without actual operational experience in using new technologies or in using existing technologies in new situations, experiments are the next best thing. For making informed decisions on future doctrine and requirements, experiments provide a better basis than does reliance on analytical studies and/or simulations.[3]

Experimentation can be performed on different scales, in different echelons, within different mission types, and with different operational communities. Experiments should complement modeling and simulation activities. Although they can fail in their ability to find the right solution, experiments should always provide knowledge about the ramifications of new ideas and technologies, to assist those who write requirements by reducing the likelihood that they will specify requirements for too much (something that cannot be achieved within reasonable bounds) or too little (improvement insufficient to justify development).

7.2.1.3 The Spiral Process

The spiral process, also called evolutionary development of requirements and systems, is an innovative method for fielding a system quickly by using

---

[2]Operational analysis involves the determination of functions to be performed, and the order and manner in which they should be performed, to carry out military missions or operations. Systems analysis involves the further step of evaluating the relative cost-effectiveness (and other benefits and disadvantages) of alternative means of accomplishing the same military mission or operation.

[3]Computer Science and Telecommunications Board, National Research Council. 1999. *Realizing the Potential of C4I: Fundamental Challenges*. National Academy Press, Washington, D.C.

commercial and government off-the-shelf equipment, with maximum user involvement throughout the process. The first spiral is usually regarded as the first development cycle of a system. Subsequent spirals allow technology insertion, the addition of new mission capabilities and upgrades, and enhancement of interoperability and integration, all in an environment of continuous user feedback.

The spiral process characteristically partitions the more traditional requirements generation and acquisition cycle into shorter, incremental cycles in which operators get hands-on access to the evolving system in each cycle and provide their feedback and modified requirements to a development team that is prepared to respond with adjustments. In so doing, the operators may modify their own operational processes and concepts based on use of the emerging capability. As such, the spiral process can also support reengineering of the operational concepts and doctrine. Each spiral has its own defined activities, performance objectives, schedule, and cost; and each spiral concludes with a user decision to field the system based on the requirements developed to date, continue with evolution, or stop.

The spiral process is a powerful alternative to the traditional requirements generation and acquisition processes. One of its advantages is that it offers a sound replacement in areas in which technology is changing rapidly and cycle times in the commercial sector are short compared to the traditional DOD requirements and acquisition processes. A major key to success is the involvement of operators, requirement generators, and technical personnel with appropriate resource allocation support from the financial sector. The spiral process also accelerates fielding of innovative operational processes and systems. Some of the successful innovations carried out by the military services are described in Chapter 2.

7.2.2 Requirements for Interoperability in Joint Operations

The DOD requirements generation process for joint operations is hampered by Title 10 language that assigns the Services the primary responsibility for equipping, manning, and training the Service component forces. The Joint Requirements Oversight Council is responsible for overseeing and prioritizing DOD requirements affecting joint operations and can review any requirement proposed by a Service. In addition, the Secretary of Defense and other elements of the Executive Branch have established separate entities to procure and operate major C4ISR systems (e.g., DISA, DIA, NRO, NSA).[4] Unfortunately, this ununified approach can further exacerbate the separation between users and

---

[4]DISA, Defense Information Systems Agency; DIA, Defense Intelligence Agency; NRO, National Reconnaissance Office; NSA, National Security Agency.

producers, hinder their dialogue, and result in "stovepipes" of communication and information.

Although it specifies that the Services define operational and system architectures, the Assistant Secretary of Defense for Command, Control, Communications, and Intelligence (ASD (C3I)) does not spell out the missions or any other taxonomy to guide building a comprehensive set of interlocking architectures across Services or stovepipes. As a result, responsibility for definition of systems' architectures has fallen largely to each individual Service, and lateral connectivity on the battlefield among Services has suffered. Even the Joint Tactical Information Distribution System is implemented differently within the various Services, thus limiting interoperability. A positive action has been the recent effort by the Joint Staff Joint Theater Air and Missile Defense Organization (JTAMDO) to identify fixes for existing problems and to anticipate problems with regard to tactical ballistic missile defense.

In addition, the Office of the Secretary of Defense has begun to require a command, control, communications, computers, intelligence, surveillance, and reconnaissance (C4ISR) supportability annex for every new major platform program. This annex is supposed to define the platform's interface requirements for information support. A great deal of effort has been spent on these annexes for the next generation surface combatant for the Navy (DD-21) and the joint strike fighter. However, refinement of the process is needed so that not every new acquisition program is forced to create its own similar infrastructure and architecture, which may not be consistent with those already developed.

Commanders-in-chief (CINCs), who provide input to the Joint Requirements Oversight Council (JROC) through the Vice Chairman, Joint Chiefs of Staff, often focus on their immediate needs rather than on longer-term systems. However, the CINC's specification of the Navy's numbered fleets as potential joint task force commanders is a positive step that has increased the numbered fleets' sensitivity to the requirements for interoperability and to the need for development of updated capabilities—for example, those aboard the USS *Coronado*.

All requirements for potential acquisitions with "joint interest" as defined by the Joint Staff are subject to JROC review, which provides for some examination of joint interoperability and commonality. However, the Joint Staff lacks the significant technical capability required to ensure meaningful scrutiny of each system. Smaller acquisition programs not designated as having joint interest are not reviewed. The JROC does use the C4ISR Decision Support Center, which is staffed almost entirely by contractors, to identify potentially important interface and connectivity requirements.

A recent joint event of some portent is the emergence of the U.S. Joint Forces Command (USJFCOM), which has begun to have an impact on the requirements generation process. USJFCOM, with its recently acquired Joint Battle Center and others, now has a significant operational experience base and an intense if not always friendly training relationship with the other regional CINCs. It may become an important agent for generation of joint interoperability requirements including hardware, software, and tactics, techniques, and procedures (TTPs).

USJFCOM is currently preparing the Capstone Requirement for Joint Tactical Communications, Command, Control, and Computers. This draft document identifies generic deficiencies and establishes top-down requirements, such as, "Survival information must be delivered within a threshold of 6 seconds and planning information within 30 seconds from when the information is initially processed and ready for transmission within the Joint area of operation." The effort is an interesting first step toward a top-down approach to specifying architectural constraints on all C4ISR systems, but the connection to specific Service system contract specifications has not yet been made, nor is the relationship to the Joint Interoperability Testing Center clear. Capstone requirements for NCO also may be a useful concept within the Navy, the leader in NCO among the Services.

A first step in this direction is the Capstone Battle Force Requirements Document being developed under the NAVSEA Battle Group Systems Interoperability Testing. The objective is to solve problems with regard to common time reference, data registration, combat identification, navigation, correlation algorithms, and metrics identified repeatedly in the All Service Combat Identification Evaluation Team exercises. The capstone document has a decidedly joint focus—Joint Integrated Air Defense System's Interoperability Working Group and the joint interface control officer are participating in its development.

USJFCOM is also the executive agent for the DOD experimentation program to support the evolution of Joint Vision 2010. This effort, which is just beginning, could become an important step in the spiral development of the NCO concept and of requirements for C4ISR systems.

In summary, the OSD, the Joint Staff, and defense agencies provide sporadic, conflicting, and sometimes onerous guidance with regard to C4ISR requirements rather than a well-structured, consistent, and testable process for elicitation and validation of cross-Services information needs. Promising exceptions are emerging, but a stronger mechanism for ensuring joint interoperability is required within the Services, even if only to supply points of contact for the joint efforts.

7.2.3 Tools for Developing Interoperability and Related Requirements

Rapidly evolving information technologies present serious challenges to but also opportunities for dealing with the evolution to NCO. New, commercially driven technologies are becoming available to the Department of the Navy far faster than would be possible with military research and development (R&D) only.

The refreshment of existing information systems at 2- to 3-year intervals may be possible and economically feasible because of reduced costs for procurement and operation. The new joint Global Command and Control System (GCCS) was built at a significantly lower cost than had been budgeted for continuation of its predecessor, the Worldwide Military Command and Control System. If the military systems incorporating commercial building blocks can adjust to a stepped-up pace for embedding new components in operational platforms as well as addressing related training, supportability, and security and other concerns, it should be possible to break out of the traditional 15-year acquisition cycle. Commercial communications services may take over large parts of the Joint Planning Network, for example, relieving the Navy of operating its own extensive communications utilities and applications.

Accomplishing a rapid evolution will place demands for change on all Navy Department processes, including the requirements generation process. Synchronization of the changing requirements across a battle force is a primary concern. Merely keeping track of the changes and ensuring interoperability on deployment have already become difficult.[5] New communications technology has made it possible to realistically link hardware-in-the-loop simulators of the major sensor, communications, and combat systems of the battle group. This distributed simulation capability, called the distributed engineering plan, can serve as a testbed for interoperability across the battle force. In addition, distributed simulations less expensive than the DEP can drive experiments leading to the development of new operational concepts, subsequent new or revised doctrine, and requirements for system modifications and new systems and components.

The combination of operational analysis capabilities and distributed simulation tools, the war gaming and fleet battle experiment capabilities at the Naval War College, and the doctrine development and experimentation program of the NWC's Navy Warfare Development Command (NWDC) could be extremely useful in defining future mission interdependencies and requirements. Inclusion of the U.S. Marine Corps in such an effort could be an excellent step toward closer synchronization of littoral requirements. An example of such a combined effort is that of the Air Forces' joint expeditionary force experiments (JEFXs) conducted under the auspices of the Air Combat Command's Aerospace Command and Control, Intelligence, Surveillance, and Reconnaissance Center (AC2ISRC). As a related effort, the AC2ISRC is procuring the software to support the aerospace expeditionary force (AEF), the new form of the deployed U.S. Air Force. The Electronic Systems Command at Hanscom Air Force Base has a new system program office for all command and control systems that is handling the acquisition for the AC2ISRC. These agencies are working closely together to iterate the development of the software through the EFXs and intermediate testing with users at intervals of about 1 year.

The Navy's requirements generation process should change to include interactive participation. Use of the spiral process for evolutionary acquisition has

---

[5]In one case, the result was the loss of the services of two cruisers for a considerable time. The NAVSEA BGSIT group has been designated to take on this problem.

been mandated for Air Force C2 systems, and the process is being used by the Air Force to develop the C2 system for its AEF. In general, the initial requirement in the spiral process should be a statement of functional performance at an easily accomplishable level, in order to provide a prototype rapidly and give the operational community a testable item with limited financial risk. After testing, the requirement may become tighter, evolve in another direction, or disappear entirely. In the spiral process, the operational community and the requirements writers are continually involved in an integrated products team (IPT)-like setting with the acquisition team.

New distributed simulation technology has given the Department of the Navy tools for improving the requirements generation process to respond to rapid technological change. The Department of the Navy should capitalize on the technology to reap the benefits of NCO.

7.2.4 Requirements for Synchronization Among Cross-Platform Capabilities

In addition to shortfalls in the requirements generation process, the committee believes that there are deficiencies in the resource prioritization and acquisition processes for support of NCO by the Navy (addressed in subsequent sections). In all three areas, the lack of focus, inflexibility, and lack of a capability to plan and deliver across platforms, and in conjunction with other Services, are key deficiencies. Because the platform communities place higher priority on the number and performance of their platforms than on the performance of the total networked system, cross-platform interoperability, including networking requirements, cannot be defended well and is often sacrificed when funding is scarce. To put it another way, the Navy provides resources for its platforms, not its battle forces.

Consensus is needed on operational architectures and synchronization of migration away from legacy systems as systems change. Furthermore, all parties must stick to the schedule if the transition to network-centric operations is to be accomplished as rapidly as is technically and economically possible. One holdout can delay the transition for the whole fleet.

The requirements generation process must have a firm basis in the operational community, which provides expertise for new ideas on how existing systems can be improved in terms of readiness, efficiency, and effectiveness for current missions; how existing systems can be adapted to emerging missions; and when such systems are no longer suitable or supportable. Some organization in, or with strong ties to, the operational community will have to be in charge of developing the requirements for the interfaces that join various different platforms and operational organizations. Priorities for changes should also come from the operational community. Because the NCII will become a matter of life and death to the operational forces, it is imperative to have a single commander who will be responsible for establishing requirements for all operational network services and improvements. Unfortunately, there is no one community within the Department of the Navy with operational credibility and authority for establishing requirements for the systems supporting NCO.

7.2.5 Requirements Generation Process—Summary of Findings

As discussed above, the committee believes that weaknesses in the requirements generation process are currently inhibiting or slowing the Navy's transition to network-centric operations.

Finding: The Office of the Secretary of Defense, the Joint Staff, and the defense agencies provide sporadic, conflicting, and sometimes onerous guidance with regard to C4ISR requirements. There is no well-structured, consistent, and testable process for elicitation and validation of cross-Service requirements for information. Although the ASD (C3I) is trying to help the Services remedy this problem, the efforts as yet lack a unifying structure across Services and stovepipes, and the C4ISR supportability annex process needs refinement to eliminate repetitive efforts.

Finding: The Department of the Navy lacks an architectural transition plan tied to a defined program of experimentation. Lack of a plan precludes designing experiments that will permit the Navy to evolve mature NCO concepts and develop related requirements.

Finding: The Department of the Navy permits discretionary rather than directed implementation of the results of current experimentation. As a result successes are not systematically incorporated to enhance and evolve a common information infrastructure (the NCII) architecture. Currently no one organization is accountable for ensuring the implementation of results Service-wide.

Finding: The spiral process offers increased opportunities for fast-track acquisition to accelerate implementation of system architectures in the field.

Finding: New distributed simulation technology offers the Department of the Navy tools for improving the requirements generation process to respond to rapid technological change in information systems.

Finding: Successful development of more effective naval network-centric operations will require that some organization within, or with strong ties to, the operational community be in charge of developing the requirements for the interfaces for different platforms and operational organizations. No one community within the Navy has the operational credibility and authority for preparing requirements devoted to the systems supporting NCO.

Finding: Inter-Service and joint efforts to improve interoperability have to be expanded. This could include better leveraging of emerging advanced capabilities of other Services and defense agencies. For example, capabilities being provided or developed by DARPA, DISA, NIMA, and NSA[6] are essential for the NCII and require incorporation into the NCII as they mature.

7.3 MISSION ANALYSES AND RESOURCE ALLOCATION: ALIGNING PROGRAM AND BUDGET RESOURCES TO MEET MISSION NEEDS

The Navy uses its integrated warfare architecture (IWAR) process led by N81 to develop assessments for each of its missions and supporting areas (see Figure 1.2 in Chapter 1). The purpose of the IWAR process is to provide the following:

- A current road map for warfare and support areas;
- A focus on capabilities vice components or subsystems;
- Cost-constrained coverage (i.e., costs constrained to 100 percent of total obligational authority;
- A linkage across the Navy's strategic vision, threat assessment, and programs;
- A translation from vision to guidance for the acquisition community;
- A foundation for allocation of resource; and
- An integrated product team approach.

The IWAR process, guided by CNO/4-Stars, is envisioned to do the following:

- Analyze end-to-end capabilities;
- Achieve total force capability with synchronized pieces;
- Accomplish battle force integration across platforms;
- Improve rigor and discipline;
- Prioritize capability areas inside the Navy total obligational authority;
- Tie together program execution, budget, programming, and out-years; and
- Provide early vision and stability for sponsors, claimants, program executive officers, and vendors to achieve efficiency.

---

[6]DARPA, Defense Advanced Research Projects Agency; DISA, Defense System Information Agency; NIMA, National Imagery and Mapping Agency; NSA, National Security Agency.

7.3.1 Mission Analysis: The IWAR Challenge

The committee believes that network-centric operations demand the integration of component systems into a coherent system, and progress toward NCO will surely involve some evolutionary improvements that integrate legacy systems. Ultimately the full power of NCO will be realized only if the network of sensors, weapons, and information is planned and developed as a coherent system. To emphasize the point, the notional example of a future power projection operation, as presented in Chapter 3 and shown in Figure 3.2, is referred to here. This scenario illustrates the complexity of the interactions among systems, the tight time lines associated with these interactions, and the difficulty of assessing and integrating the contributions of component systems to network-centric operations within one particular mission area.

Assessing the contributions of components and subsystems to the combination of the Navy's four major missions (see Figure 1.2) is even more difficult. Moreover, as is discussed below, the Navy currently lacks good measures of output for operational analyses and systems analyses of many of its existing missions and functions. Thus, assessing the contributions of network components to mission performance first requires developing preferred metrics and adjusting (or developing) analytic tools to evaluate them.

7.3.1.1 Depth and Continuity of Assessments

There are several shortcomings in the current IWAR assessment process. First, the process has limited resources available (both personnel and time). For example, only about 60 personnel are available in the N81 staff to lead and manage the entire IWAR process containing the 12 major analytic areas, each with multiple subdivisions.[7,8] Although other parts of OPNAV and the Department of the Navy provide analytic support, the committee believes that overall support by the IWAR process is inadequate to provide the CNO the needed decision-support information. Currently most of the IWAR assessments are qualitative and judgmental rather than quantitative (e.g., stoplight charts indicating green, yellow, or red status in particular areas), in part because supporting operational analysis and systems analysis capabilities in different functional areas lack adequate measures of output and models to support good trade-offs within or across missions. The key point is that there is a need to significantly improve quantitative results provided to decision makers.

---

[7] The entire N81 staff contains about 90 personnel, about two-thirds of whom are involved in the IWAR process.

[8] The 12 major analytic areas are maritime dominance, deterrence, information superiority and sensors, power projection, air dominance, sustainment, infrastructure, manpower and personnel, readiness, training and education, technology, and force structure (see Figure 1.2 in Chapter 1).

Second, the IWAR process is used primarily to make recommendations for the CNO's programming guidance to the resource sponsors and has limited impact outside the CNO's program objective memorandum (POM) development process. To be fully effective, the IWAR process must continue through the budget and execution processes to ensure that conclusions and decisions derived in the POM process are not overtaken by events. Ideally, the assessments should be based on analytic tools and assessment methods that the individual resource sponsors and claimants understand and use; be used throughout the planning and execution processes; and be available to the CNO and the Secretary of the Navy whenever resource decisions are required. However, the IWAR process and supporting operational and systems analyses capabilities do not currently enable continuous assessments from requirements generation through programming, budgeting, and execution.

7.3.1.2 Measures of Effectiveness and Performance

A major reason for the lack of depth in the IWAR assessment process (including its supporting elements in the resource sponsor and claimant communities) is that there are no good, analytical, objective measures of effectiveness (MOEs) and/or measures of performance (MOPs) for C4ISR, particularly for mission C2 systems. And, although an IWAR analysis may provide an assessment of how well a current or projected battle force performs in some network-centric operations, it does not provide the means for making system trade-offs. That is, would an investment in the NCII provide more or better warfighting capability than an equal investment of resources in another warfighting component (e.g., such as an improved weapon or sensor)? This situation exists because there are insufficient means to measure the warfighting value of systems for a battle force; there is as yet not a good integrated campaign effectiveness model for naval forces. Examples of what exists and what is missing are provided in the following list:

- Combat system operational requirements documents (ORDs) specify detection ranges, engagement envelopes, and raid kill probability;
- C4I ORDs specify throughput, net cycle time, and probability of corruption;
- Given a target, detection range, and weapons envelope, raid kill probability within a time window is determined by reaction time and weapon/delivery subsystem reliability;
- More time to react allows more depth of fire and supports higher probability of kill;
- The composition of battle force component systems predetermines a set of detection ranges, engagement envelopes, weapons reliability, and communications network;

- However, the reaction time of the battle force as a whole is not specified, and no good measures of battle force campaign effectiveness are provided because naval battle force dynamics are especially difficult to model.

Currently, the Navy lacks good MOEs and MOPs for evaluating network-centric operational capabilities and the contributions of different components to the larger goal. A means of developing better metrics for battle force mission capabilities lies in the design reference mission (DRM) concept. A DRM is a set of one or more mission scenarios describing what needs to be accomplished in a particular mission. It indicates the range of capabilities of the adversary, the environment, rules of engagement, the range of capabilities of friendly forces, and any other factors that bear on the outcome of the mission. It must be comprehensive enough to cover the full range of mission possibilities, must have agreement from the warfighters that it is comprehensive, and must be kept up-to-date.

Battle force DRMs are being developed by NAVSEA, and DRMs are in place for theater ballistic missile defense and DD-21. DRMs are in progress for theater air defense (TAD) battle management command, control, communications, computers, and intelligence (C4I), TAD overland cruise missile defense, next-generation aircraft carrier (CVX), and the nuclear-powered aircraft carrier, CVN-77. However, DRMs for other warfare areas are lacking. If other DRMs were completed, it would be possible to determine, in quantifiable terms, such attributes (effective MOEs) as battle force reaction time for a range of sensors, weapons, and NCII capabilities. This in turn would allow for analysis of overall NCO capabilities to determine the "right" mix of components across missions and across platforms.

Another major deficiency in the IWAR assessment process is the treatment of new versus legacy components. Legacy components are generally inadequate to some degree because they were developed years ago for a different purpose and have evolved to where they are today. However, all have not evolved in the same way and, because they are closed subsystems, they are difficult to upgrade and standardize. On the other hand, a new component subsystem can incorporate the latest technological advances. All things being equal, operators would want the new component subsystem, but it takes significant resources and years to develop and field a new component subsystem. The fielding part is often overlooked in making component and component subsystem trade-offs.

Because developing and fielding a new component take years, forces will be required to continue operating with the legacy components, and during this time there is some likelihood that the component will be required for combat. To the extent that the legacy component is inadequate for new or revised concepts of operation, there is a risk that it, and the mission, will fail. Therefore, some resources will be required to extend the life of the legacy component or to improve its performance. Since resources are a zero-sum trade, the challenge is to determine how much to devote to maintaining and improving legacy components at the risk of lengthening the development and fielding time of new components. However, the Navy lacks an agreed-upon methodology for assessing the risk of failure of a military operation (i.e., a methodology that consistently arrives at the same answer for a given set of circumstances).

To make the choices about how much to spend on maintaining and improving legacy components versus buying new ones requires knowing exactly what is inadequate about the legacy component. Each year the All Service Combat Identification Evaluation Team conducts a joint Service exercise that is highly instrumented and provides the raw data for an engineering analysis. NAVSEA has been working on root-cause analysis from exercises and has amassed a considerable data bank. The distributed engineering plant will allow for hypotheses to be tested in a controlled and repeatable environment, thus allowing some degree of confidence in the success of proposed modifications to legacy systems. Given DRMs, it is possible to measure how much warfighting improvement can be obtained for a given investment in modifying legacy components,[9] as well as to measure the probability that a component will fail in an operational setting. However, the risk associated with warfighting failure must continue to be a judgment call on the part of Navy leadership.

7.3.2 Resource Allocation

There are three pieces to the resource allocation process in the Department of the Navy: mission analysis (assessment), programming, and budgeting. The current responsibilities for these portions of the process are as follows:

1. Mission analysis (assessment), discussed in the previous section, is currently chaired by N81 using the IWAR assessment process, with NAVSEA providing root-cause analysis and development of DRMs.

2. In the programming phase the resource sponsors are generally (a) N6 for Navy-wide telecommunications, computer centers, and the higher-level command applications and (b) the platform sponsors (N86, N87, and N88) for the mission/combat component subsystems for domain-specific applications and dedicated tactical data links.

3. Budgeting and execution are handled by the system commands in parallel with their OPNAV sponsors in the programming process.

In the programming phase, the Department of the Navy uses program ele-

---

[9] However, the Navy must also look at what the other Services are doing. If the Navy fixes its own components and the Army or Air Force does not fix its, the Navy Department will be no better off, unless it is the long pole in the tent. The Navy and Marine Corps can lobby the OSD, the Joint Chiefs of Staff, and organizations such as JTAMDO, but in the final analysis, such trade-offs are Service decisions.

ments (PEs) in the Navy portion of the Future Year Defense Program (FYDP)[10] to allocate resources to resource sponsors, who then propose changes to their programs to accommodate changing requirements and fiscal constraints. The PEs correspond primarily to platforms (e.g., cruisers, aircraft carriers, F/A-18 squadrons) and include all the resources associated with that platform (R&D, procurement, operations, manpower, and so on). Unlike the Army, which has PEs for divisions, the Department of the Navy has no PE for battle forces, and so Navy programming tends to be more platform-centric. Any oversight in resource allocations for battle forces must be identified based on separate analysis and is handled primarily on a case-by-case basis.

Since DRMs are lacking for many mission areas, comprehensive oversight of resource allocation for naval battle forces is spotty at best. There is a significant potential for resource mismatches between and lack of coordination among programs (e.g., the ES-3 program was discontinued by N88 but the shipboard data links developed for the ES-3 continued to be funded by N6—and are now used for the U-2).

In short, because utilities (i.e., information infrastructure) and applications (which ride on the infrastructure) are handled by different sponsors and systems commands whose jurisdiction depends on arbitrary definitions in vogue at the time, there is inadequate oversight for network-centric operations as a whole. There are often seams,[11] particularly between closed legacy system components, with deficiencies (and mission needs) in both the utilities and the applications, and there is no systematic means of ensuring that the seams are consistently covered. While the support for resourcing of individual programs is usually well addressed, the resourcing of the seams for the battle force as an entity is not. Unfortunately, the capabilities of the whole force may be less than the sum of the individual parts if seams and integration are not dealt with sufficiently. For example, when parts of this committee were reviewing land-attack targeting within force projection, they found, among other things, the following:

• Inadequate targeting for naval surface fire, including lack of an agreed-upon method for transmitting target coordinates from a deep-inland forward observer to an over-the-horizon firing ship; and

• Inadequate capability to detect, identify, track, and engage moving targets.

---

[10] Cohen, William S., Secretary of Defense. 2000. "Chapter 17: The FY 2001 Defense Budget and Future Year Defense Program," *Annual Report to the President and the Congress*, U.S. Department of Defense, Washington, D.C. Available online at <http://www.dtic.mil/execsec/adr2000/chap17.html>. Secretary of the Navy. 2000. *Budget of the Department of the Navy*, U.S. Department of the Navy, Washington, D.C., February. Available online at <http://navweb.secnav.navy.mil/pubbud/o1pres/db_u.htm>.

[11] The term "seams" refers to the interfaces in which compatibility and interoperability must exist between assets.

Although power projection is one of the four major Department of the Navy and IWAR missions, the seams among the deputy N8 platform resource sponsors for carrying out naval missions appear to be receiving less attention in resource allocation than the major platforms themselves. This situation is critical for major mission areas such as power projection (or even major portions of power projection such as striking land targets). These areas involve many parts of the Navy and Marine Corps team that in the resource allocation process are represented by different parts of the OPNAV staff (both in and beyond the N8 organization). This arrangement makes it particularly difficult to address the system component trade-offs for such mission areas.

In the budgeting and execution phases, too, more resources are provided for platforms than for the seams within and around battle forces. With the exception of the Navy's interoperability initiative in NAVSEA, programs are handled individually and the seams continue to exist. NAVSEA does look at the entire battle force but has limited or no authority to effect changes in individual programs. A corollary problem is that network-centric programs in particular are viewed as Navy, not joint—there is no formal joint advocate in either OPNAV or the systems commands.

The net result is an unfulfilled need to provide comprehensive oversight of network-centric operations in the programming, budgeting, and execution processes. Such oversight is the only way to ensure that the seams are adequately addressed and that there is joint advocacy in all three phases of the broader resource allocation.

A second problem is that the allocation of resources is insufficient for spiral acquisition. In an environment of tight fiscal constraints, the tendency is to specify an end product that can be fully justified to the Office of the Secretary of Defense and Congress. Spiral acquisition is often viewed as counter to this approach and as a process that puts Navy Department resources unnecessarily at risk. Such an outcome can be mitigated to some degree by specifying phased end products (e.g., baselines) that are the result of implementing spiral development in a homogeneous product line. However, current Department of the Navy and DOD resource allocation procedures discourage rather than encourage the good commercial practice of spiral acquisition for systems and applications dominated by rapidly changing technologies.

7.3.3 Mission Analyses and Resource Allocation Process— Summary of Findings

As discussed above, the committee believes that the following findings are important elements for the mission analyses and resource allocation process needed to better implement capabilities for network-centric operations.

Finding: The IWAR assessment process is not adequately staffed to provide good quantitative assessments of mission area capabilities and deficiencies.

Finding: The IWAR assessment process and supporting operational and systems analyses capabilities do not currently enable continuous assessments from requirements generation through programming, budgeting, and execution.

Finding: The Navy lacks good measures of effectiveness and measures of performance for evaluating network-centric operational capabilities and the contributions of different systems to the larger goal.

Finding: A comprehensive set of defense reference missions does not exist across all mission areas. A comprehensive set would provide the basis for developing good metrics for battle force mission capabilities.

Finding: The distributed engineering plan could provide considerable information that would be useful in testing potential modifications to legacy systems.

Finding: The Navy has no program elements that specifically identify the resources for funding the "seams" among individual programs within battle forces.

Finding: No one office in the Navy has the responsibility and authority to address the trade-offs among system components for major Navy and naval force missions (or even major portions thereof). The existing capabilities are more focused on trade-offs among components for major platforms.

Finding: Comprehensive oversight of network-centric operations as a whole within the programming, budgeting, and execution processes is needed to ensure an adequate consideration of "seams" and jointness.

Finding: Current Navy and DOD resource allocation procedures discourage the good commercial practice of spiral acquisition for systems and applications dominated by rapidly changing technologies.

7.4 SYSTEM ENGINEERING, ACQUISITION MANAGEMENT, AND PROGRAM EXECUTION

The process for acquiring networking capabilities is not at all straightforward, nor are there "silver bullet" management solutions that, once adopted, will address the management challenges for all time.

Previous sections stress that information superiority will require information systems interoperability (at the technical and data levels) that cannot be realized without eliminating or mitigating several major obstacles. This section examines the system engineering, acquisition, and program execution processes whereby the needs and requirements for new network-related capabilities can be managed within the mechanisms embodied in Goldwater-Nichols reforms and DOD acquisition directives.[12]

7.4.1 Acquisition Challenges to Achieving Operational Interoperability

Different military organizations approach their missions differently and implement mission-supporting doctrine and systems accordingly. For example, normal use for the Army Patriot Air Defense System was assumed to mean no friendly aircraft in the zone of fire; consequently, identification friend or foe capabilities were initially lacking. With Patriot now envisioned as a forward-deployed weapon system, such major shortcomings need to be eliminated. Lessons from operations in Grenada (1983) through Desert Storm (1991) have shown that joint operations as well as ad hoc assemblages of units from a single Service face serious obstacles in collaborating to achieve a common objective.

To avoid such difficulties in the future, system components that will be coupled for NCO must fit into a joint operational architecture. To ensure success, that architecture must be developed and maintained by operators, analysts, and system engineers throughout the development and acquisition programs.

7.4.2 Acquisition Challenges to Achieving Technical Interoperability

To succeed in making modern computer and communication systems interoperate, academic, industrial, nonprofit, and government organizations have contributed to a set of system and subsystem architectures and interface standards designed to reduce the barriers to technical interoperability. Although the Navy and the DOD should use these commercial standards whenever practical, the committee could not find comparable efforts for tactical networks. Successful automation of networked processes for tactical portions of the NCO system will require adopting standards and organizational discipline at the data, data definition, data structure, and processing algorithm level.

The Navy has taken some recent steps to enhance the systems engineering process within the SYSCOMs (i.e., NAVSEA) and within the Office of the Assistant Secretary of the Navy for Research, Development, and Acquisition (ASN (RDA)) (i.e., the appointment of a Chief Engineer). However, the systems engineering discipline is still insufficient for integration and interoperability of cross-platform and cross-SYSCOM system components within the Navy and

---

[12]Goldwater-Nichols Department of Defense Reorganization Act of 1986, United States Statutes at Large 100 (1986): 992-1075b.

Marine Corps. In addition, no single Navy organization below the CNO is in charge of each of the operational, systems, and technical architectures for network-centric operations. One or two responsible agents will be essential to enforce the necessary discipline across multiple acquisition programs.

7.4.3 Acquisition Challenges to Acquiring Information Networks

7.4.3.1 Legacy Component Constraints

In the past, communication and data processing design methodologies did not distinguish adequately between "closed" and "open" architectures, and specific application-dependent features were often embedded in generic, general-purpose computer and communications transport mechanisms. Consequently, changes on either side of the "interface" were difficult. Existing shipboard combat direction components, aircraft mission components, and tactical data links reflect these design approaches, making replacement an unlikely near-term option. More recently the Department of the Navy has taken steps toward open systems architectures, which should alleviate the problem of upgrades in the future. However, there are no specific Department of the Navy or DOD acquisition provisions or regulations that encourage or enforce the separation of utilities (transmission media and control processes for their management) from domain-specific applications (those that directly serve warfare mission areas and command functions) in the development of new system components.

7.4.3.2 Subsystem Acquisition by Independent Offices

Within individual program management offices and elements of the Department of the Navy military and civilian staffs, there is undue emphasis on accountability for the performance, cost, and schedule of the individual program, and limited oversight of and coordination among related programs. This management approach often leads to stovepiped programs with little regard for interfaces, interoperability requirements, or synchronization of procurement, installation, and training schedules. Often each program manager is encouraged to optimize his solution to his own program's "requirement" or some other directive.

Within the Department of the Navy acquisition community there is no one below the ASN (RDA) with the responsibility to oversee all aspects of battle force system interoperability and integration for new programs and to coordinate program execution across the SYSCOMs to ensure synchronization in development, production, and installation of systems important to network-centric operations. This makes it particularly difficult to ensure interoperability of components for major naval force missions that may involve five or more program executive officers (PEOs). The committee is concerned about the ability of the

Navy acquisition community, working under the existing structure, to achieve good system engineering results for NCO systems. For example, when the committee looked at the Department of the Navy's management approach for the power projection mission, it was concerned about the ability to achieve good system trades-offs with five or more PEOs having responsibilities for some portions of the power projection mission.

7.4.3.3 Managing Backward Compatibility versus New Technology Offerings

Legacy component constraints have often forced postponement of both potentially important improvements and new capabilities. A possible solution for the Navy and Marine Corps is to develop new capabilities that have two attributes: (1) interoperability through component partitioning, thus permitting insertion of new technology while enforcing compatibility with older generations, and (2) a preplanned technology refreshment cycle that will make legacy components obsolete in a coordinated and synchronized way.

7.4.3.4 Heterogeneously Equipped Units and Platforms

Typically, system component improvements on ships, submarines, or aircraft can be made only when platforms are not in use. Battle forces comprise diverse units, which are first brought together for an extensive period of workup, then deployed for several months, and then dispersed. The likelihood of having the same networked facilities aboard all platforms in a battle force is low unless such a capability becomes a top-command matter. The most likely acceptable solution, given the naval forces' operating environment, will be to continue to upgrade individual battle forces as the ships go through overhaul or prepare for a new deployment cycle. More importantly, there is no set of efficient practical procedures, or organization below the Secretary of the Navy, the CNO, and the CMC, with the authority for making postprogram and postbudget adjustments. Such adjustments are sometimes needed to accommodate exigencies occurring during development, production, and fielding of systems that could introduce asynchrony in component subsystems that must be interoperable in the battle force. Some of the constraints placed on the Navy are external (i.e., reprogramming limits), but others could be alleviated by improved Department of the Navy procedures and enhanced cooperation toward meeting top management goals.

Examples follow of how battle force programs can get out of phase. When a budget is approved, it should be internally consistent (e.g., what NAVAIR is doing with the E-2 should be on track with what NAVSEA is doing with Aegis). However, if Congress directs undistributed reductions to acquisition appropriations that each SYSCOM then may take differently, high-priority programs may have cost overruns that cause internal reprogramming; or a crisis such as in

Kosovo may cause the Administration or the Congress to redirect portions of the budget, and each SYSCOM might implement changes differently. Such events, which might appear to be solely financial problems, involve technical issues, and financial personnel do not necessarily have the expertise to recognize technical disconnects. What is needed is real-time budget execution oversight to advise top leadership when programs are getting out of phase, plus a mechanism to manage the consequences. This oversight should help to get the battle force components to work together for deployment. Getting the money to integrate across programs is a part, but not all, of the process.

7.4.4 The Acquisition Process

The committee believes that the DOD requirements process, including the JROC, and the DOD 5000 series of acquisition directives, offer sufficient flexibility to acquire and upgrade NCO system components. Success lies in intelligent and disciplined application by senior officers and civilians who understand how to build and operate information infrastructures. The same leaders also must understand funding so that critical applications and elements of the information infrastructure evolve across sensors, weapons, and platforms. The Department of the Navy and the DOD could benefit from applying good commercial practices based on lessons learned in the rapidly changing commercial information businesses.

7.4.4.1 Commercial Sector Lessons in Acquiring Network-Centric Capabilities[13]

Commercial enterprises have successfully developed and used network-centric capabilities with applications involving heterogeneous systems and complex interconnectivity. They have learned how to manage several essential development and operating processes. The Navy can tailor its acquisition and system engineering policies and practices accordingly. Several key process-oriented activities should provide a mechanism to separate development of applications and domain-specific capabilities from those processes associated with the development and evolution of the common information infrastructure. Seven such process-oriented activities are discussed below.

1. *Design and development methodology.* Both spiral development and the contrasting "waterfall," or sequential, approach involve competent planning,

---

[13]Anderson Consulting (Mark Goodyear, Hugh W. Ryan, Scott R. Sargent, Stanton J. Taylor, Timothy M. Boudreau, Yannis S. Arvanitis, Richard A. Chang, John K. Kaltenmark, Nancy K. Mullen, Shari L. Dove, Michael C. Davis, John C. Clark, Craig Mindrum). 1999. *Netcentric and Client/Server Computing: A Practical Guide.* Auerbach Publications, CRC Press LLC, Boca Raton, Fla.

analysis, design, construction, and test phases. Successful projects always involve incremental phased development and some degree of iteration regardless of the name given to the methodology. Successful projects often provide prototype solutions and obtain feedback as a result of extensive user involvement. A good acquisition methodology will incorporate processes that (a) encourage and focus activity on definitive objectives, (b) are complete and verifiable for correctness, (c) produce deliverables that can be measured and easily used in the next phase of the activity, (d) include the operational user throughout, and (e) result in useful increments of capabilities in 18- to 24-month cycles.

2. *Program directives and organization.* The commercial program directive is the contract requiring all participants to agree to participate materially in the decisions to be made during the life of the program. A good program directive should state in some detail the goals, objectives, measurements to be used, roles of users and acquirers, and other relevant factors including interoperability.

The accuracy of the time and resource estimates will depend on the quality and scope of analysis and preliminary design undertaken as part of the program definition phase. Normally, networked system component developments will support more than one mission or application area. Each of these application areas must have a manager responsible for ensuring that the mission evaluation criteria are met within the time, budget, and quality standards established. Managing the technical components is a major undertaking involving architectures, networks, information standards, and infrastructure operations.

A systems engineer must be responsible for interoperability and interfaces, including those that extend to other systems beyond the scope of the program. Most successful network-centric information system developments have full-time user participation in the program office. Such interaction can ensure that user concerns are being heard and also that the user community is developing the necessary changes in its own practices and procedures in anticipation of a phased delivery of capabilities.

3. *Architectures—Operational, Systems, and Technical.* Development of information support for network-centric system operations must respect three architectural requirements:

- *Operational architecture:* a model that shows the relationships of all the stimuli flowing into the mission area and all the responses flowing back across the boundary into the outside world;
- *Systems and technical architectures:* the components that provide automation and communication support for an application. Technical architectures employ the set of standard building blocks, hardware, and software involved in an application; and
- *Legacy component linkages:* legacy components' expected contributions to meeting the required objectives, empowerment of the system engineers of all affected legacy components to resolve compatibility and transition issues, and resources for accomplishing these purposes.

4. *Information and data modeling.* An information model provides the detailed foundation for all data design decisions affecting all applications throughout a networked enterprise. Such a model includes entity-relationship diagrams, transaction and static volume estimates, attribute listings and properties, unambiguous definitions, state-transaction diagrams, and entity matrices. Lessons from the commercial world show that this kind of detail is critical to the success of networked applications.

5. *Network design and delivery.* Communication technology has evolved so rapidly that the term "networks" represents the convergence of computing and communication theory and practices. The network design and implementation segments of a program can be separated from the applications development segments, once the functional capabilities desired in the applications are known. Application specialists are not typically strong in networking, and communication specialists are not usually conversant with modern networking technology. Therefore, ensuring end-to-end consistency, performance, and fault management will require operational staffing that crosses mission, applications, and information system organizational boundaries.

6. *Applications design and implementation.* Systems, hardware, and software engineering methodologies need to be tailored to become part of a larger development and test environment. End-to-end testing of each component requires a development infrastructure capable of replicating realistic operational environments.

7. *Management of change.* Management of change is a relatively new management dimension. A major network-centric-oriented development team can itself be an agent of change, since users must accept and implement changes in the operating environment favored by the system solution. In an ideal setting, highly motivated, skilled users are involved from the beginning in developing the concept of operations and in interacting meaningfully in every phase of system evolution. User participation is the most effective means for enhancing communication, provided that information on status and other essential items is given to all higher levels of management and formal reviews with project sponsors are conducted. Managing expectations must become a key objective, especially for programs that take a long time to complete.

7.4.4.2 Acquiring Battle Force and Tactical Interoperability

Numerous deconfliction problems occurred in managing the air war in Desert Shield/Desert Storm. Furthermore, the Navy has been experiencing compatibility and interoperability problems in fielding new or upgraded combat direction system components for three classes of ships (Aegis, carriers, and non-Aegis combatants). These and other shortcomings have caused the DOD and Navy leadership to implement several organizational and process changes needed across a number of independent programs and platforms. Three of the most important are as follows:

- Establishment of JTAMDO (1998), intended to resolve conflicts dealing with concepts of operations, operational architectures, measures of effectiveness and assessments, and other interoperability issues;
- Chartering of the new Commander, Naval Sea Systems Command, to be responsible for battle management C4I/combat systems interoperability for deploying battle groups among all SYSCOMs and PEOs; and
- Establishment of the ASN (RDA) Chief Engineer to be responsible for the architecture, integration, and interoperability of current and future C4I/combat/weapon systems used by the Department of the Navy.

The committee endorses these DOD and Department of the Navy efforts to (1) align requirements, assessments, and the acquisition process with a network-centric philosophy, (2) make battle force and joint interoperability a system requirement, and (3) bring discipline to the system engineering process and provide benefits to meet the need for competent officers and civilians to have rewarding career paths.

However, the committee believes that more must be done to make the major cultural shift from platform-centric to network-centric naval operations. As a specific example, no office similar to JTAMDO exists to resolve conflicting demands associated with concepts of operations, operational architectures, and other interoperability issues for important operational areas such as land attack within the power projection mission. With sensors, aircraft, missiles, and other systems from all four Services working in the same airspace, interoperability across Service lines is essential. To be fully effective, network-centric forces must go beyond any one Service or military department.

7.4.5 Systems Acquisition and Program Execution Process— Summary of Findings

Based on the above discussion, the committee believes the following findings are important to making improvements in the acquisition and program execution processes and to achieving more effective network-centric operations.

Finding: System components that are tightly coupled for network-centric operations must fit into a joint operational architecture that is developed and maintained by operators, analysts, and system engineers over the life of the development and acquisition program.

Finding: While the Department of the Navy has taken some recent steps to enhance the systems engineering process within the SYSCOMs (i.e., NAVSEA) and within the ASN (RDA) (i.e., the appointment of a Chief Engineer), use of systems engineering methodology is insufficient for integration and interoperability of cross-platform and cross-SYSCOM systems.[14]

Finding: No single Navy organization below the CNO is in charge of each of the operational, systems, and technical architectures for network-centric operations.

Finding: No one below the ASN (RDA) has the responsibility for overseeing and coordinating all aspects of battle force interoperability and integration for new network-centric operations system components. Furthermore, no one below the ASN (RDA) coordinates program acquisition across the SYSCOMs to ensure synchronism in the development, production, and installation of new components important to network-centric operations for all naval missions.

Finding: There are no procedures, or an organization with authority below the Secretary of the Navy and CNO, for making postprogram and postbudget adjustments. Such adjustments are sometimes needed to accommodate exigencies that occur during development, production, and fielding of systems. As a result asynchrony might occur in the fielding of network-centric operations system components that must be interoperable in the battle force.

Finding: The Department of the Navy has not developed a policy or a set of procedures for deciding what is the most applicable design and development methodology for particular system developments. It has not adopted the spiral development approach for software and hardware applications for systems in which technology is changing at a very rapid rate.

Finding: The Department of the Navy has not yet found an effective and timely means to couple its experimentation process with the development portion of its acquisition process.

Finding: The Department of the Navy has not yet found a mechanism or management team approach for prioritizing the development, procurement, and installation of network-centric elements key to improving interoperability across platforms, or for managing the significant cultural change from platform-centric to network-centric operations.

Finding: No office similar to JTAMDO exists to resolve conflicting demands associated with concepts of operations, operational architectures, and other

---

[14]The need for system engineering is discussed in Section 2.3 of Chapter 2.

interoperability issues for important operational areas, such as land attack. With sensors, aircraft, missiles, and other systems from all four Services working in the same airspace, interoperability across Service lines is essential.

7.5 PERSONNEL MANAGEMENT: ACQUIRING PERSONNEL AND MANAGING CAREERS TO MEET NETWORK-CENTRIC NEEDS

7.5.1 Context and Introduction

". . . the people are still the most important part of our military—their quality and their training and their morale . . . ."
—President Clinton, news conference, July 21, 1999

Meeting the needs for network-centric operations will require people with different kinds of information and knowledge. Commanders will need the intellectual capabilities to apply information technology in complex and sustained operations and to encapsulate higher-level thinking in the context of information technology. Other individuals steeped in the fundamentals and concepts of information technology (e.g., computer scientists and software designers) will be needed to design architectures and networks to meet operational needs and to ensure that information dissemination equipment can be adequately installed and maintained. And future Navy, Marine Corps, and civilian personnel at all levels should have, and will need to maintain, computer skills sufficient to plug and play equipment components (e.g., sufficient skills to operate versus design and maintain essential terminals).

However, the potential gains from modern technology, and getting the knowledge from those who have it to those who need it, are not likely to be achieved with current DOD personnel management practices.

The current status of education and training for personnel in the information technology (IT) workforce (defined in the next section) is representative of specialty training in the U.S. Navy as a whole. This training supports multiple platforms and mission areas. The officers are given graduate education at the Naval Postgraduate School to meet billet subspecialty requirements, and the program is managed by the N6 organization on the CNO's staff. Officer training in IT-related functions, which are platform specific, is provided and managed by the platform or warfare community sponsor. There is limited and sporadic training/education of the officer corps at large.

Enlisted personnel are trained to meet equipment-specific requirements and in many cases gain their knowledge through on-the-job training while on ships or at shore stations. Their training is managed by the Chief of Naval Education and Training for quality and content but is controlled through the enlisted personnel detailers for training beyond entry-level skills. Community involvement is generally limited to the warfare specialties, with minimal oversight in the support ratings used throughout the Navy.

Civilian personnel receive the least training of any of the three groups of the IT workforce. A basic assumption in civilian personnel management is that civilians are "best qualified" when they are hired. Therefore it is assumed that they do not need training. This sizable element of the workforce generally receives no or minimal training. There is no career management or training preparation for civilian personnel during their career progression.

7.5.2 Identifying the Information Technology Workforce

Of the 826,000 people in the department of the Navy, 372,000 are active-duty Navy, 92,000 Navy selected reserves, 171,000 active-duty and reserve Marines, and 191,000 civil service personnel. Of the total, about 49,000 (6 percent) are considered to be in the information technology workforce even though almost everyone now uses computers on the job.

This apparent dichotomy results because the IT workforce is still identified by industrial age designators (see Table 7.1 for the IT workforce designators for civilians) and because the use of computers is becoming as common as the use of telephones. The military and the civil service IT job codes are both out-of-date and too rigid to accurately reflect the skills of the workforce. Enlisted IT specialists were called "radiomen" until October 1999. The civil service codes were last updated in 1985, long before the World Wide Web became an integral part of work. As a result, the Navy Department's "IT workforce" as interpreted today probably has few if any of the individuals described above as being needed for command positions (e.g., information-to-knowledge converters), many operational specialists (e.g., computer scientists), and only a small portion of computer users. Most likely the latter group includes only those individuals specifically assigned to jobs whose primary job requirement is computer skills.

The Office of Personnel Management (OPM) is updating the civil service codes, but it may be years before they are officially implemented. Tag codes do not exist to indicate the skills that have been accumulated by individuals in the workforce (i.e., to understand the existing skills, locations, and needs). In addition, there is no dated IT certification in personnel records to indicate how up-to-date an individual's certification is.

7.5.3 Understanding the Changing Environment

Information technology is changing more rapidly than any segment of warfighting doctrine or force implementation. It is changing the way we work and the way we live. Many of the new jobs in the expanding economy are in IT.

In 1998, the Defense Manpower Data System estimated there were 2.5 million information technology jobs in the United States, with another 1.3 million IT

TABLE 7.1 Department of the Navy Civilians Considered as Being in the Information Technology Workforce as of June 1999

Code	Title	Navy	Marine Corps	Total
334	Computer Specialist	7,553	471	8,024
335	Computer Clerk and Assistant	836	124	960
391	Telecommunications	774	50	824
850	Electrical Engineering	1,140	29	1,169
854	Computer Engineering	431	19	450
855	Electronics Engineering	9,703	45	9,748
856	Electronics Technician	4,208	177	4,385
1515	Operations Research	654	19	673
1520	Mathematics	744	0	744
1550	Computer Science	1,671	35	1,706
	Other categories	2,097	283	2,380
	Total	29,811	1,252	31,063

SOURCE: Compilation of data courtesy of Defense Manpower Data Center, Office of the Under Secretary of Defense for Personnel and Readiness, Arlington, Va., November 1999.

jobs to be added in the next 10 years.[15] The demand for IT workers is estimated to outstrip the supply by 50 percent. According to a study done in February 1998 by the Information Technology Association of America (ITAA), there is a 346,000-person shortage of IT workers in the United States, including a 20,000 to 25,000 shortage in Northern Virginia alone.[16]

The DOD believes that since 1992, it has lost 23 percent of its IT workforce through downsizing and attrition. In addition, 50 percent of federal civil service

---

[15]Data supplied by Defense Manpower Data Center, Office of the Under Secretary of Defense (Personnel and Readiness), Arlington, Va., November 1999.

[16]Information Technology Association of America and Virginia Polytechnic Institute and State University. 1998. *Help Wanted 1998: A Call for Collaborative Action for the New Millennium.* Virginia Polytechnic Institute and State University, Blacksburg, Va.

personnel will be eligible for retirement in 5 years. A further complicating concern is that most DOD workers in IT are supporting outdated legacy systems.

Young IT workers today are not looking for security. Most young workers do not expect to be in a particular job more than 3 to 5 years, and in the IT world it is less. Companies are stealing IT and knowledge manpower from one another by offering training, a better work environment, and higher pay. IT workers are looking for interesting work and high pay. In the DOD, starting pay for a GS-7 is $26,000, while the starting range in the private sector goes from the mid-$30,000s to the high $40,000s. This disparity makes it very difficult to recruit bright and competent IT personnel to the DOD. In 1999, for example, a Web-site manager was ranked the most desired job for the young worker.[17] In contrast, military careers ranked 74th for officers and 160th for enlisted personnel. Recruiting and hiring skilled DOD personnel have become daunting challenges, with a Navy recruiting shortfall in 1998 of 7,000 personnel, and a shortfall of 18,000 enlisted for shipboard manning. Although FY99 results were better for the Navy, the problem cries for near-term attention.

Since IT work in the military is changing rapidly, it is not known what skills will be needed for future work. The Department of the Navy (and all of the DOD) needs to evaluate who should do the work and what is the right balance among military members, government civilians, and contractors.

Competent personnel will be required to address the following critical areas:

- Information and knowledge management (extraction, presentation, application, and sharing);
- Technical design and sustainment (architectures, network design, and connectivity maintenance); and
- Applications (functional users).

All of these areas require markedly different kinds of education and training, some of which have not yet been addressed by the Department of the Navy or the DOD. Most important, all future Navy Department personnel will need some level of information technology knowledge.

7.5.4 Links and Interdependencies Among People, Technology, and Information

The Clinger-Cohen Act of 1996 established a Chief Information Officer (CIO) with the responsibility to ". . . assess and address hiring, training, classification and professional development needs of the Federal Government with respect to Information Resource Management." More specifically, the National

---

[17]Krantz, Les (ed.). 1999. *Jobs Rated Almanac*. St. Martin's Press, New York.

Defense Authorization Act of 1999[18] gives the CIO authority to do the following:

- Review budget requests for all information technology;
- Recommend IT budgets to the Secretary of Defense;
- Ensure interoperability and system standards within and between Services, DOD agencies, and national security agencies, and between other government agencies;
- Eliminate duplicate information within and between Services and DOD agencies; and
- Coordinate with the Joint Chiefs of Staff.

However, no one has the responsibility for the following:

- Understanding who can and should do the IT work—military active duty, reserve, civil service, and/or contractor;
- Developing career paths for personnel with significant IT capabilities;
- Developing the training plans to keep the Department of the Navy up-to-date on IT;
- Understanding the content of IT work and defining codes to keep classifications up-to-date in a rapidly changing world; and
- Integrating IT workforce considerations for both military and civil service personnel.

Present job skill codes are inadequate to provide the detail needed to truly understand the current work structure and desired manning skills. However, some progress is being made. In April 1999, the Space and Naval Warfare Systems Command (SPAWAR) initiated an analysis of the "technical" job codes used to identify needed information technology skills in the military.[19] The new job codes and specialty titles recently proposed by OPM are designed to reflect the major categories of IT work in the federal government. They will be available for implementation in 2000. OPM has also released a set of IT skills and competencies that will be used in recruiting and selecting job candidates. The competencies allow agencies and organizations much greater flexibility in identifying people who have the necessary skills to perform the job functions.

---

[18]Strom Thurmond National Defense Authorization Act for Fiscal Year 1999, United States Statutes at Large 112 (1998): 1967-1968.

[19]The ongoing study, entitled "Job Task Analysis for Computers, Information Systems and Networks," will result in a final report in the summer of 2000.

7.5.5 Training and Education—What Is Being Done?

The IT education and training needs for the Department of the Navy have been fragmented, not only for officers, but for enlisted personnel as well. The needs cross all warfare and community resource sponsor lines and encompasses numerous programs, (e.g., the Global Command and Control System-Maritime (GCCS-M), data management system). Training is further complicated and is a constantly moving target because technology and training demands change for successive generations of equipment.

Both N6 and SPAWAR are making an effort to address the education and training problems of the military. To date civilians are included in this plan only marginally.

The N6 training strategy (developed in 1998) is called Navy Communications, Information Systems and Networks (CISN) training to support C4ISR information operations (IO). However, there is not yet any consistent record with regard to training for implementation of NCO. An oversight board has been set up, jointly chaired by N1, N6, and N7. The N2, N8, the fleet CINCs, and SYSCOMs have representatives on the board. This is an essential first step for achieving the right mix and quality of personnel for NCO, but the benefits are not yet visible.

The board's job is to validate training requirements, identify and allocate resources, and implement training initiatives. The CISN C4ISR/IO training working group has representatives from all of the above organizations represented on the oversight board and the Chief of Naval Education and Training.

The training objectives are to (1) link all programs and systems, (2) identify training requirements and resources, (3) identify officer communities, enlisted ratings, and civilian positions, and (4) establish system acquisition and training development standards.

N6 wants to take the training to the sailor rather than have training only in the classroom. Training centers of excellence have been established to focus on quality education for the fleet.

In addition, there is a large push for "distance learning," and some training is being collocated with afloat technical support. But distance learning is a mode of delivery. If the courses are not well designed for the learner and learning, they can be a waste of time. Distance learning courses should be field tested before being provided broadly, and once available to all, a learner feedback system should be built. Feedback from students should address questions such as: How easily understandable was the material? How long did it take to finish? Would you recommend the course to others? Feedback should also be obtained from students' supervisors to get their assessment of the value of the courses.

7.5.6 Training and Education—What Do People Know Is Available?

Considerable effort appears to be under way to expand the training and education opportunities for Department of the Navy personnel, but finding them is a treasure hunt. Existing IT education and training opportunities are not well distributed and therefore not well understood within the Department of the Navy workforce. The Web sites that exist are not linked, and knowing the Web address is essential to finding the training opportunities. The productivity payoff from the Web comes from easy access to information that is entered only once and then linked for many to find. What is needed in the Department of the Navy is improved easy access to eliminate the scramble for information.

7.5.7 Career Paths

Career paths have been established for the enlisted IT specialist rating. However, there are no established career paths for civil service employees.

The national IT worker shortage could become a serious problem for the Navy. While the technical work to provide the Navy/Marine Corps intranet is projected to be provided by contractors, the application of this type of solution to other IT areas could present a problem with respect to sea/shore rotation assignments for experienced IT enlisted personnel. The Department of the Navy will never be able to compete with the commercial sector solely on a financial basis to attract IT workers, but it should consider special pay for critical skills in the fleet and take advantage of the signing bonus authority for recruiting skilled civilians. In addition, the Navy should take full advantage of term appointment authority for 3- to 5-year term appointments to attract civil service personnel with desired skills. Young people in IT fields may be attracted more by challenges and opportunities to work on critical problems than by longer-term job security. However, some clear delineation of career paths for both military members and civil employees with strong IT capabilities should provide significant promotion opportunities for those who want to stay with the Department of the Navy and, more generally, the DOD. Other productive actions could include training civilians along with military members when comparable skills are required and developing "career banding" for civilian personnel to provide for more interesting career paths.

Workforce planning should begin now to meet requirements for civilian employees and military members with significant IT skills. Timely planning is particularly important with respect to the civilian community, to take advantage of the potentially large number of retirements and the resulting opportunity to realign the IT workforce during the next 5 years.

7.5.8 Personnel Management Process—Summary of Findings

Based on the above discussion, the committee believes that the following findings are important for improving the personnel management process as it applies to achieving more effective network-centric operations.

Finding: All future Department of Navy personnel will need some level of information technology knowledge.

Finding: Civil service codes for IT workers, including those essential to network-centric operations, are out-of-date. The Office of Personnel Management is updating a limited set of IT civil service codes, but it may be years before they are officially implemented.

Finding: Existing IT education and training opportunities are not well distributed and, therefore, are not well understood within the Department of the Navy workforce.

Finding: There is a need to analyze the content of the desired IT work for both the military billet and civilian position structures.

Finding: There is a need to analyze existing job skills to assess the current functions being done and how they are carried out. Present job skill codes are not adequate to provide the detail needed to understand the present work structure and manning skills. The manning, technology, training, and resource requirements analysis done for both the military billets and civil service jobs was last updated in 1985.

7.6 ORGANIZATIONAL RESPONSIBILITIES FOR EFFECTIVE NETWORK-CENTRIC OPERATIONS INTEGRATION

The committee believes that successful network-centric operations will require high degrees of cooperation, trade-offs, and interaction among the stakeholders responsible for the management functions that must be integrated to implement network-centric operations, depicted in Figure 7.2.

Historically the functions have been carried out more or less sequentially, although there have been exceptions when threat or other circumstances were sufficient to cause some acceleration of events. Within the Department of the Navy, the lead responsibilities (oversimplified with emphasis on the Navy versus the Marine Corps) for the major functions are as shown in Table 7.2. However, all functional areas in the Navy/Marine Corps involve continuous relationships among representatives of the fleet commands, type commands, OPNAV and Marine Corps staff, and the SYSCOMs.

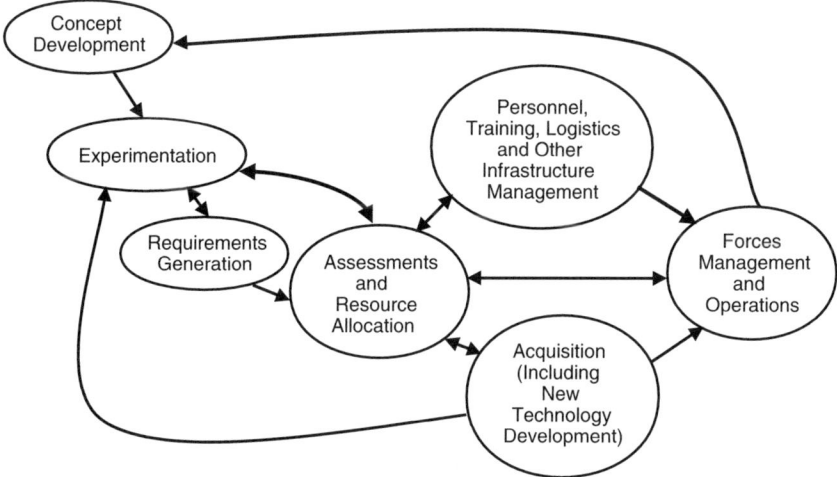

FIGURE 7.2 Functions involved in network-centric operations integration.

TABLE 7.2 Lead Responsibilities for Selected Navy and Marine Corps Functions

Function	Lead Responsibility
Concept development, experimentation, and requirements generation	CNO through NWDC and CMC through MCCDC
Assessments and resource allocation	CNO, CMC, and SECNAV with N8 staff lead
Acquisition including new technology development	ASN (RDA) and Navy/Marine Corps SYSCOMs
Personnel, training, logistics, and other infrastructure management	Navy and Marine Corps sponsors (e.g., N1, N7, N4) and numerous claimants
Forces management and operations	Fleet/fleet Marine force commanders and type commanders

NOTE: Acronyms are defined in Appendix H.

Within the Navy the concept of "clusters" as used below describes the force management and operations organization and how it functions in these portions of the key decision-making support processes. Figure 7.3 shows the basic, simplified relationships in the uniformed side of the Navy between the CNO, the CNO staff (OPNAV), and the principal Echelon 2 Commands. For completeness, the relationship of the SYSCOMs to the ASN (RDA) in the civilian side of the Department of the Navy is also shown because the SYSCOMs have three "customers" (the CNO, the fleet CINCs, and the Service acquisition executive).

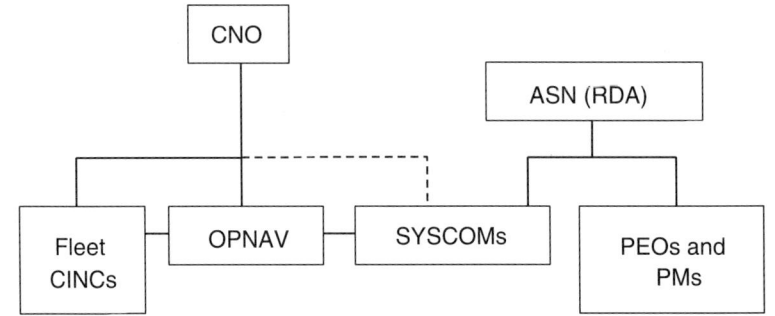

FIGURE 7.3 Basic structure of the uniformed side of the Navy. Acronyms are defined in Appendix H.

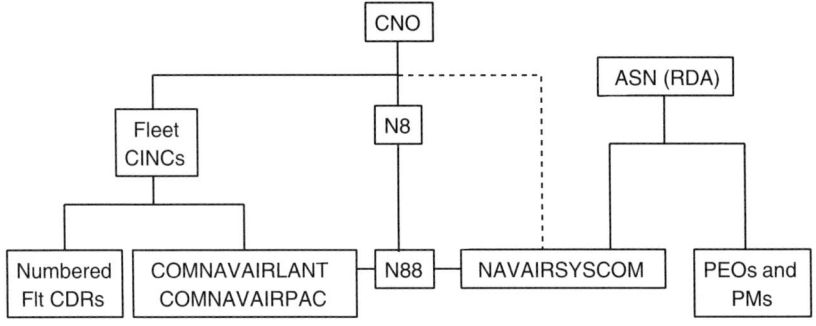

FIGURE 7.4 Air community cluster. Acronyms are defined in Appendix H.

In practice, because the various communities within the Navy have different requirements, the next echelon down is divided into clusters by platform communities, as shown in Figure 7.4 (for the air community). A similar structure exists for the submarine community, as shown in Figure 7.5.

Similar relationships exist for the surface and amphibious communities. However, these relationships are by no means exclusive—for example, the aircraft carrier desk in N88 deals with NAVSEA, and the Tomahawk desk in N86 (Surface Warfare) deals with NAVAIR. However, the primary relationships, such as in the aircraft and submarine communities, deal with fundamentally different operational, engineering, and programmatic needs. These organizational relationships allow for a continuous focused flow of information between the principal players throughout the requirements generation, allocation of resources, acquisition, in-service engineering, and execution processes. However,

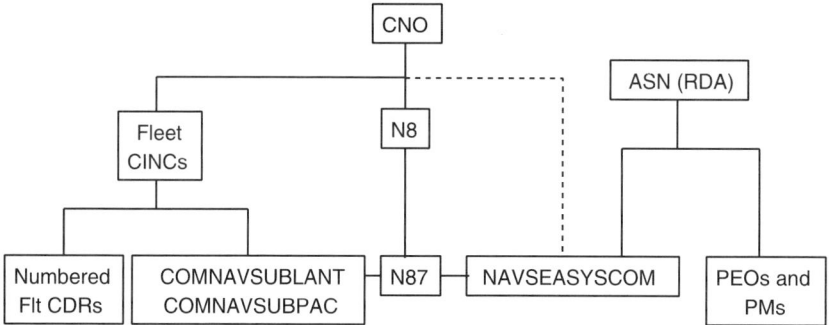

FIGURE 7.5 Submarine community cluster. Acronyms are defined in Appendix H.

the community cluster does create an artificial division between the various entities that will make up the battle force, and there is no formal mechanism or structure to relate the various functions to battle force requirements, allocation of resources, acquisition, in-service engineering, and execution processes. Simply, what is missing is an integrating function to ensure cross-mission and cross-platform consistency.

This lack of a suitable mechanism to promote network operations and interoperability is even more apparent when considering the C4I organizational relationships depicted in Figure 7.6.

1. Note the differences that exist between those responsible for the information infrastructure and those responsible for platform communities. There is no one like a type commander for C4I reporting to the fleet commanders to represent the fleet operational view. While the fleet commanders' N6 have significant expertise, they do not have the resources or responsibilities of a type commander.

2. Whereas the operational units (type commands, numbered fleets) are in the fleet commander's chain of command, the C4I supporting activities (e.g., Naval Computer and Telecommunications Command, Naval Security Group) are Echelon 2 commands under the CNO and other supporting activities (e.g., the Navy Network Design Facility is a field activity under SPAWAR).

3. There is no direct linkage between requirements for individual platforms and the requirements of the battle force as an entity, particularly as it relates to C4I.

This lack of a formal, institutionalized operational focus makes the integration of C4I (and network-centric operations) difficult across the entire spectrum of requirements generation, resource allocation, acquisition, and in-service engineering for the battle force.

Another dimension of the issue can be seen in the relationships between the

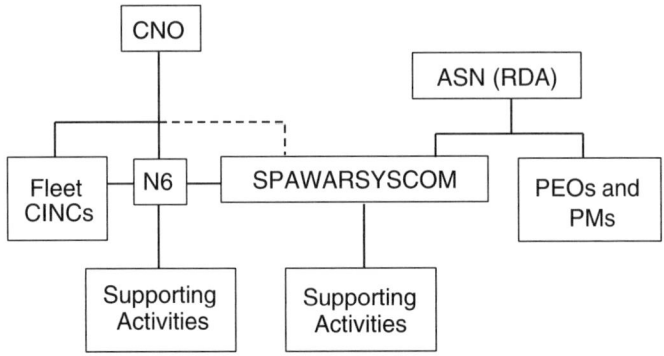

FIGURE 7.6 C4I community cluster. Acronyms are defined in Appendix H.

various offices in OPNAV and their counterparts in the SYSCOMs that are responsible for engineering support and portions of the acquisition function, as depicted in Figure 7.7 (the Secretariat and the Marine Corps are also shown in simplified form).

In Figure 7.7, NAVSEA is shown three times in support of the warfare community sponsors (N85, N86, and N87), but they all are different branches within NAVSEA. Note the Chief Engineer's (CHENG) office under ASN (RDA). This new position was established to overcome the obvious potential for seams between the various branches within NAVAIR, NAVSEA, and SPAWAR. The three systems commanders also agreed that NAVSEA would act as an integrator between them for interoperability,[20] and work closely with the CHENG. Although this arrangement appears to be working, there is no formal organizational structure that will sustain it beyond the tenure of the present incumbents. Furthermore, NAVSEA has no formal authority to enforce interoperability standards over other SYSCOM branches. Today it works because of the quality of the people in the job and the commitment of the three systems commanders to make it work. However, it is not clear that the "bully pulpit" method of management will continue to be effective when other priorities emerge.

As noted in Table 7.2, the lead responsibilities in the Department of the Navy for assessments and resource allocation are the CNO, CMC, and Secretary of the Navy with N8 staff lead, and for acquisition are the ASN (RDA) and Navy/

---

[20] In a prior generation, the Chief of Navy Material, a four-star admiral, was the superior in the chain of command between the CNO, the Secretary of the Navy (SECNAV), and the systems commands. This added layer provided some integration of effort, most notably in the Antisubmarine Warfare Directorate (PM-4), but was disestablished because it was top heavy and did not provide enough added value.

ADJUSTING NAVY ORGANIZATION AND MANAGEMENT 329

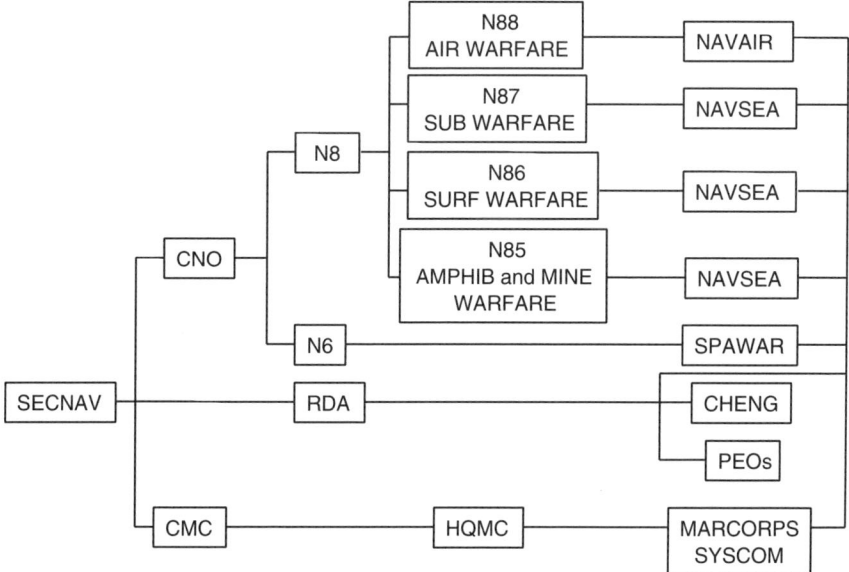

FIGURE 7.7 OPNAV offices and SYSCOM counterparts responsible for engineering support and some acquisition functions. Acronyms are defined in Appendix H.

Marine Corps SYSCOMs. The committee believes that these lead responsibilities are appropriate but it does have concerns regarding integration for NCO in both of these functions. When the committee looked at the major missions of the Navy/Marine Corps and the resource allocation and acquisition functions, it noted that there is still a strong platform versus mission orientation in the management organizations under the N8 and the ASN (RDA). For example, the committee observed the following for the power projection mission (one of the four major missions of the Navy/Marine Corps team):

• The network-centric approach is not being emphasized within the OPNAV resource allocation process (compared to the approach of emphasizing the contributions provided by individual platforms); and
• Within the PEO structure under the ASN (RDA), the management mission for key components of the force projection mission is distributed among six PEOs.

Although the committee does not have a specific recommended solution for improvement, it is concerned about the degree to which system trade-offs and systems engineering can be accomplished.

7.6.1 Organizational Responsibilities—Summary of Findings

Based on the above discussion, the committee believes that the following findings are relevant to achieving more effective network-centric operations.

Finding: The fleet OPNAV/SYSCOM organizational relationships, which are primarily platform-centric, have served the Navy well for many years and ensure a tightly integrated focus across the requirements generation, resource allocation, acquisition, engineering, and execution processes within each community. However, an integration function for cross-platform and cross-mission needs of the battle force in these processes is missing.

Finding: The lack of a functional type commander resource for C4I that can interact with the other platform type commanders exacerbates the cross-platform integration problem.

Finding: No formal organizational structure crosses platforms and missions within the systems commands (including the Marine Corps Systems Command) to enforce C4I interoperability standards.

Finding: There is uncertainty about the extent to which the CNO staff and the Navy acquisition structure are suited to end-to-end acquisition of NCO-oriented subsystems in the system context, especially with reference to the power projection mission.

During the committee's extensive discussion of the organization of the Navy structure for coordinating NCO system requirements, some members of the committee expressed the belief that, owing to the legacy of earlier maritime strategies, the Navy has not put sufficient emphasis on the power projection mission in the N8 organization and in the PEO structure. The N8 organization reflects submarine warfare, surface warfare, and air warfare, with power projection a part of each office but not the focus of any. Meanwhile, air dominance is well served by the focus of the office of surface warfare, and strategic deterrence is served by the office of submarine warfare. It appears that power projection lacks a true advocate in N8. The same may be true of sea dominance, although this issue was not examined in as much detail by the committee. In the PEO structure, air dominance is the focus of PEO (Theater Surface Combatant (TSC)). At least five PEOs strongly relevant to power projection are primarily product oriented, the products being platforms and weapons in many cases. Therefore, management of end-to-end system designs and acquisitions as such is considered to be problematic. The same may be true for such system designs in other areas, although both the N8 and the PEO structures have been successfully adapted to the need in areas such as antisubmarine warfare (ASW), cooperative engagement capability (CEC), and in the growing theater missile defense (TMD) effort. The ASN (RDA) has recently announced the redesignation of the Program Executive Office for DD-21 as PEO (Surface Strike), assigning it responsibility for NAVSEA Program Manager, PMS 429's Naval Surface Fire Support, including the Advanced Land-Attack Missile program, as well as the DD-21. This represents a major step in the direction of concentrating attention on power projection systems as a whole, in parallel with the concerns the committee expressed in this area. The committee's recommendations also pertain to making targeting an integral part of the strike system, to strike warfare from the air, and to the relationship between and coordination of naval surface warfare and air strike warfare. The committee commends the entire power projection area to further scrutiny of the kind that led to this most recent PEO reorganization, in both the PEO and the N8 contexts.

Within the context of this study, other members of the committee addressed and argued against making recommendations on these two issues; they favored what they regarded as more pragmatic recommendations to improve implementation of network-centric operations. Among other things they believe that recommendations on the issues above will deflect Navy attention from recommendations made in more important network-centric challenge areas—i.e., the recommendations focused on (1) improving integration within and across all decision support processes and (2) developing improved output measures and mission/system component trade-off analyses and assessments. Given these divergent views and the uncertainty they reflect about the true management situation applicable to overall network-centric operations system planning and acquisition, the committee concluded that recommendations to the Navy Department and the CNO would be in order, to review the N8 and the PEO structures and adjust them *if necessary* and *as appropriate* to accommodate end-to-end system designs for NCO subsystems, including especially those relevant to the power projection mission. These recommendations are included with the others that follow.

7.7 RECOMMENDATIONS

7.7.1 Committee's Approach

In the "Findings" sections and related text, the committee identifies more than 30 problem areas in management functions and processes that are key to achieving fully effective network-centric operations. These shortfalls are summarized in Box 7.2, where they are organized by management function. However, the benefits of network-centric operations will not be achieved if these areas are addressed individually and in isolation from each other. Consequently, the committee did not develop individual recommendations for each finding. An effective dialogue and integration among the different decision support processes are needed for successful implementation of the network-centric operations con-

Box 7.2 Gaps in Management Functions That Have to Be Addressed for Effective Implementation of Network-Centric Operations

Requirements generation
- No advocate for intra- or inter-Service interoperability as it pertains to information
- Insufficient integrating of requirements that cross platforms and missions in the naval service
- Need for integration, particularly in the seams
- Inadequate coupling among concepts, fleet experiments, and fleet operations for network-centric operations
- Lack of operationally oriented information requirements; no continuity of discipline for synchronization and prioritization of requirements

Mission analyses
- Inadequate output measures of effectiveness/measures of performance for network-centric operations
- No systems trade-off analyses
- Particularly lacking cross-platform and cross-mission analyses and assessments
- No reference point to measure against
- Assessment process not continuous—stops after the program objective memorandum
- Assessments not made of impact of acquisition and execution decisions as they occur

Resource allocation
- Network-centric operations not treated as an entity
- Various components funded separately
- Not treated as a coherent system of systems
- Lack of a cross-platform/cross-mission sponsor
- Insufficient funding provision for spiral development
- Insufficient allowance for technology "push"

System engineering
- Insufficient system engineering in making design trade-offs
- Insufficient system engineering input to requirements and assessment processes
- Insufficient application of system engineering to battle force
- Each component/system engineered separately
- Need for system engineering for system of systems

Acquisition
- No single office responsible for each of operational, systems, and technical architectures
- Lack of spiral development in system design(s)
- Need to avoid merging of utilities and domain-specific applications, such merging leading to closed, inflexible legacy systems

Program execution
- No single point of contact for operations information for network-centric operations
- Fleet support provided by multiple offices and often uncoordinated
- Lack of resource reallocation oversight for network-centric operations as a whole
- No assessments of how decisions made about one system affect all other components
- Inadequate coupling among concepts, fleet experiments, and fleet operations for network-centric operations

Operations
- No single provider of services for network-centric experiments
- No central point for development of tactics, techniques, and procedures (TTP)
- No central TTP development for battle force network-centric operations
- No single provider of trained personnel and network services

Personnel management and training
- Need for some information technology understanding and skills for all naval personnel
- Inadequate database of personnel qualifications and skills
- Inadequate career development for personnel with excellent information technology skills
- Recruiting and retention of qualified personnel difficult in competitive environment
- No common training requirements/standards for network-centric operations and operations information
- Education in network-centric operations lacking or provided unevenly
- Documented job descriptions and codes for military and civilian billets not reflecting the transition in work being performed

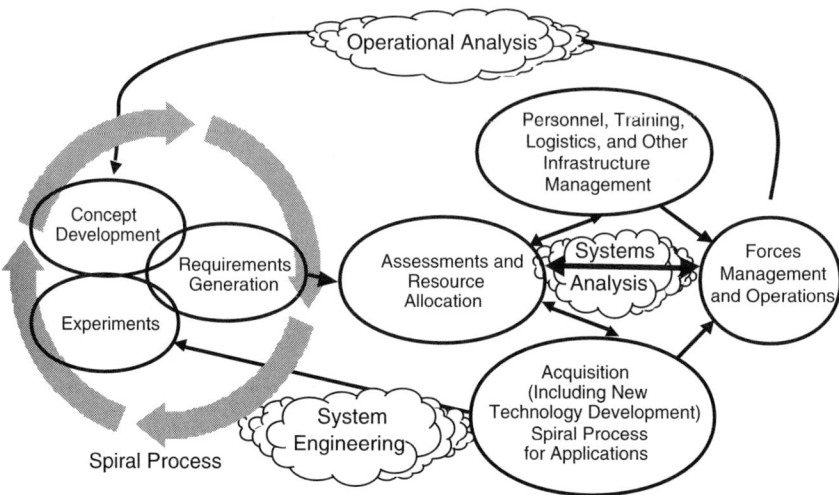

FIGURE 7.8 Functions and their relationships for effective integration of network-centric operations.

cept. The committee also believes that any set of recommendations to be implemented should be viewed as an integrated whole. To achieve this integration the committee believes that the Department of the Navy should build on its existing organizations with some changes in emphasis, rather than attempt to totally restructure the department or create a new additional stovepipe for all network-centric responsibilities. The difficulty with even attempting to create a new entity to be responsible for all or a major portion of network-centric operations is that such operations span almost all Navy and Marine Corps activities. Therefore, the committee strove to develop a pragmatic approach, taking into consideration the restrictions that exist within the DOD and the Department of the Navy as a result of laws and culture developed over many decades.

The goal is to create a better-integrated set of the basic functions needed for effective NCO and to achieve shorter cycle times for applications involving rapidly changing technology. Compared to the more or less sequential set of functions shown in Figure 7.2, the committee believes it has developed specific recommendations to make the processes work more as depicted in Figure 7.8.

7.7.2 Specific Recommendations

The committee's integrated set of recommendations, summarized in Figure 7.9, contains a few major organizational changes described below in this section. Because some of the organizational recommendations have applicability to multiple functions and processes, additional clarification is appropriate.

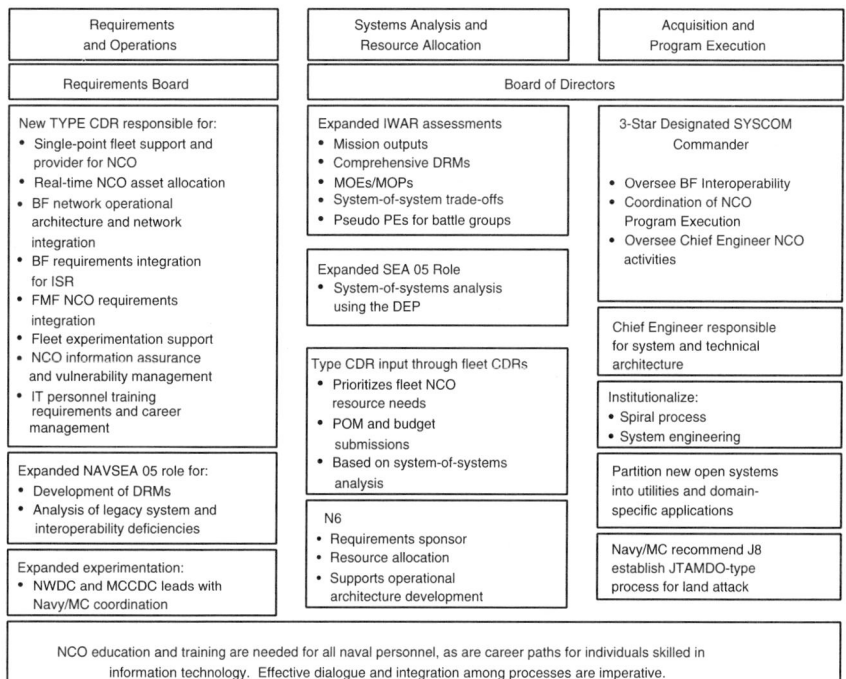

FIGURE 7.9 Key recommendations for managing network-centric operations. BF, battle force; DEP, distributed engineering plant; DRM, design reference mission; FMF, fleet Marine force; ISR, intelligence, surveillance, and reconnaissance; IWAR, integrated warfare architecture; MOE, measure of effectiveness; MOP, measure of performance; PE, program element; POM, program objective memorandum; TYPE CDR, functional type commander.

In reviewing these recommendations, note that only one new position is recommended: the creation of a functional type commander for Operations Information and Space (shown in Figure 7.9 under the requirements and operations functions). Two new boards (with individuals in existing Navy Department positions as members) are also recommended. The first is proposed to improve the overall implementation of the integrated network-centric operations concept in the Department of the Navy (a "board of directors"); the second is proposed to improve the integration of network-centric requirements ("requirements board") in the Navy. The committee recommends that one of the Navy SYSCOM commanders be double-hatted as a Navy deputy to the ASN (RDA) to integrate and oversee a number of Navy network-centric acquisition-related activities. The committee further recommends that the Navy/Marine Corps push J8 to set up an organization, similar to JTAMDO, for land attack to deal with cross-Service integration issues with that aspect of NCO. One last broadly based recommendation is to expand the type and quality of systems analyses and systems trade-off evaluations to be performed in the decision support processes. These evaluations should assist those leaders making the important resource allocation decisions affecting the mission capabilities and outputs of Navy and Marine Corps forces. Although this last type of recommendation does not affect any organizational alignments or responsibilities, its implementation will require some changes in emphasis in the duties of the OPNAV staff.

The following sections describe the first three of the above recommendations (those for the functional type commander and the two boards affecting multiple functions) in some detail and then pull together in one place the full set of recommendations made in this chapter and summarized by title only in Figure 7.9.

7.7.2.1 Three Major Recommendations Crossing Current Functional Lines of Responsibility

Recommendation

1. *Create a functional type commander for a new Operations Information and Space Command.* Navy combat operations in the information age currently suffer from the fragmentation of responsibility for operations information and information warfare. This lack of focus is the natural outgrowth of stovepipe product lines, the hallmark of successful procurement, fielding, and training of separate mission-specific platforms. The essential idea of information sharing and weblike interactive processes has not yet permeated the development and fielding of C4ISR systems to the level needed to ensure open-architecture connectivity.

There is currently no focal point for information aspects of naval warfare below the level of the VCNO. The deputy chiefs of naval operations (DCNOs) for Intelligence (N2), for Plans, Policy, and Operations (N3/5), for Space and Electronic Warfare (N6), and for Resources, Warfare Requirements, and Assessments (N8) all have important roles in this critical area of naval warfare, but no one has the primary responsibility. In addition, multiple organizations have uncoordinated responsibilities for fleet operations and support and for manpower and budgeting, as well as for operational, system, and technical architectures for information systems.

From the battle group and battle force level there is no single point of contact for integration of operational architectures and networks, or for integration of requirements for intelligence, surveillance, and reconnaissance (ISR) and for the Marine Corps fleet Marine force into the battle group or battle force. Information assurance, vulnerability management, and computer network defense have no coherent visibility at the fleet level. Finally, with the evolving career paths for officers, enlisted, and civilian personnel, there is no focused community management that will ensure viable career paths for individuals with critical skills.

The time is propitious for making information operations a warfighting mission with a fleet role and with responsibility comparable to that represented by the current type commanders. Such a warfighting mission with the appropriate organization (see Appendix F) not only would give a single point of contact for fleet support and a single provider of network services but also would provide the requisite warfighting focus and emphasis at the fleet level. Creating a new organization, a functional type commander for Information Operations and Space, reporting to all fleet commanders, would provide the integrating mechanism to enable the proper fielding and testing of new network-centric systems, hardware, and TTPs.

In arriving at this recommendation, the committee considered various alternate approaches to carrying out the necessary functions, and the likely problems and benefits that would attend the creation of the new position. One alternative was leaving the organizational situation as it is now, with a lower-ranking officer functioning with each fleet to deal with its information network matters. This arrangement would not provide adequately for the broad and fundamental nature of the change needed to fully implement network-centric operations in the fleets. The committee also considered a recommendation for creating multiple flag positions for each fleet, but this approach did not appear to resolve the problems of achieving consistency of equipment, planning, and operational techniques in the operational forces throughout the Navy. Only a single individual could achieve that.

After considering the pros and cons of various alternatives, the committee concluded that the need to achieve assured consistency and interoperability warrants having the functions be the responsibility of a single individual with a high enough rank such as the functional type commander suggested below.

This new functional type commander would be responsible for the following areas, many of which are covered in a piecemeal manner within the CNO's staff:

a. Be (i) the single point for information support to the fleets, (ii) the information provider for network-centric operations, (iii) keeper of the fleet portions of NCII, and (iv) the Navy operator for space assets.

b. Through an integrated information warfare operations center, perform real-time network-centric information technology asset allocation, conduct information assurance and vulnerability management, provide trained personnel for fleet support, and manage and deploy red teams for vulnerability management.

c. Be responsible for the network operational architecture (including the integration of that architecture with the systems architecture developed under the cognizance of the ASN (RDA)) and battle force integration.

d. Perform battle force requirements integration for ISR and prepare the fleet's priority list for network-centric information operations POM submissions and for short-term resource reallocation.

   e. Be responsible in the Navy for integration of Marine Corps fleet Marine force requirements with fleet NCO requirements and fleet experimentation support.

   f. Be the community personnel manager for all IT personnel in the Navy—officers, civilians, and enlisted personnel—including career field planning, rotation assignments, and training requirements.

   g. Serve as the Navy component commander to CINC Space, which has been assigned responsibility for DOD network assurance.

   h. Provide operational support to the fleet experimentation program.

This functional type commander should report only to the three fleet commanders and represent them in the POM and budget process for information operations issues. When this role is combined with the committee's two other recommendations that provide a mechanism to reallocate resources during a given budget cycle, this functional type commander should be able to respond to emerging fleet requirements, to maintain cutting-edge technology, and to support the development and fielding of new systems. This commander's role in providing operational advice and support to the fleet experimentation program should enhance the quality of experiments and add value to battle group and battle force operations.

The recommended Commander, Operations Information and Space Command, should be supported by the Commander, Naval Security Group Command, the Commander, Navy Space Command, and the Commander, Naval Computer and Telecommunications Command in their entirety and by the fleet operational support function of the Space and Naval Warfare Systems Command. The new commander should assume command of a major portion of the DCNO for Space and Electronic Warfare (N6), except for the policy function. An organizational view of the recommended command is provided in Appendix F.

Recommendation

2. *Create a requirements board.* The current requirements generation process is not sufficiently responsive to the demands of information technology development. Computer technology and related processes can become outmoded before the POM process can even react to the fielding of the results of the prior requirements generation process. In addition, multiple organizations have uncoordinated responsibilities for operations and fleet support, requirements generation, manpower planning, and budgeting. These disparate, and sometimes overlapping, activities are not effective and clearly do not optimize the Navy's efforts in future systems development. With the Navy on the threshold of conducting warfare in the information age, bold new initiatives are necessary.

A requirements board should be established to deal with information operations and to integrate the various competing requirements as presented by the fleets for rapid improvement of complex at-sea operations. This board should be responsible for the following areas:

a. Coordinate and rank requirements for the NCO decisions across all naval platforms with a view to the future. The board would control the evolution of connectivity and capacity requirements to mesh with platform networking capabilities. Through the VCNO, it would ensure that the Navy speaks with one voice at the joint level.

b. Encourage the full requirements dialogue among the communities and ensure that requirements trade-offs have been examined and validated by the best possible means, including use of the battle group engineering plan. The requirements board would make sure that the requirements remain flexible within the spiral development process in acquisition and improve the visibility into the repair and modernization programs.

c. Validate the requirements for all NCO systems.

The requirements board members would have four broad functions:

a. Develop policy and implement strategy for conducting information operations;

b. Advise the CNO on the strategy and doctrine, personnel, education, training, technology, and resource requirements for moving the Navy from platform-centric to network-centric warfare;

c. Establish the linkage to the Navy After Next[21] from this new level of warfare operations; and

d. Rank emerging NCO requirements based on fleet commanders' recommendations and the results of fleet experimentation.

The requirements board should be chaired by the VCNO and have the N6 as the executive director (until the recommended Information Operations and Space Command is established). The membership of the board should consist of the deputy fleet commanders, the president of the Naval War College, the DCNO for Plans, Policy and Operations (N3/5), and the DCNO for Resources, Warfare Requirements, and Assessments (N8).

The Board should specifically address four broad areas of information operations: (1) battle force operations, (2) NCO requirements, (3) IT manpower, and (4) NCO budgets. The area of operations should be much broader than current

---

[21]The Navy After Next is the Navy beyond that defined by the time horizon (5+ years) of the program objective memorandum (POM) process.

forums and should focus on the warfighting impact of new and emerging developments. Since the fleets will all be properly represented, progress should be possible in areas such as sensor-to-shooter priorities, long-range land-attack targeting, mobile targeting, and friendly fire. In all of these areas the board should seek solutions for the Navy After Next.

The board's responsibilities should cut across the current stovepipes, both by platform and by mission. It would be small enough and senior enough to make the necessary trade-off decisions and implement priorities in the requirements area, with concurrent fleet input. This forum for the prioritization of requirements could be a forceful lever to compel attention and achieve progress in the cross-platform areas.

The board should also approve a general education program for all officers, enlisted personnel, and civilians with continuity throughout each career path to ensure familiarity with the basic language, thought processes, and skills required to perform at a high level in their specialty.

The budget is an important area for this group, which should provide recommendations on priorities for funding, particularly across platforms and stovepiped areas. Again, since the fleets and the requirements leadership are adequately represented, this could be a very effective forum to ensure that information operations and its specific programs receive the visibility and priority they require.

Recommendation

3. *Create a board of directors.* The above two organizational changes are designed to focus the operational development of network-centric naval operations and to integrate the many priorities within the requirements sphere. There is also a need to provide focus for the acquisition and program execution portions of information operations. This review, oversight, and prioritization of the acquisition, installation, integration, and program execution portions of network-centric operations should be conducted by a new board of directors consisting of individuals with the authority to make adjustments in different areas.

The recommended board of directors should have the Under Secretary of the Navy as the chairman, and the VCNO and Assistant Commandant of the Marine Corps (ACMC) as members. Other members should be the three Navy system commanders of NAVSEA, NAVAIR, and SPAWAR, and the Marine Corps Systems Command, along with the ASN (RDA) (who should serve as the executive director). The DCNO for Plans, Policy, and Operations (N3/5) and the DCNO for Resources, Warfare Requirements, and Assessments (N8), as well as the Assistant Chief of Staff (ACOS) for Plans, Policy, and Operations and the ACOS for Programs and Resources of the Marine Corps staffs, should also be members. Requirements sponsors (N2, N4, N6, N85, N86, N87, and N88) should be advisory members to be consulted concerning the operational impact of potential program adjustments.

The 12-member board of directors should replace or augment the informal arrangements that now exist between the systems commanders and could oversee the coordination of the scheduling and funding of all cross-platform systems. The board should also ensure that the technical standards and components for interoperability being developed for new systems will be installed in an integrated manner to provide the necessary battle group and battle force compatibility, including that with units of the fleet Marine force. The board should be responsible for establishing battle group and battle force system response times. It also should recommend priorities among fleet modernization efforts and new system developments from a technical and operational standpoint.

Emergent and in-year funding requirements for deploying critical systems should also be dealt with by this group, through reprogramming, if necessary. This board of directors should function as a Navy Department resource board for information operations and for the supporting systems and R&D programs. In turn, this establishment would provide a network-centric operational focus, at the appropriate level, for the entire acquisition and program execution effort (including installations in battle groups) for both new systems and upgrades to existing systems. The results should include improved judgments regarding priorities and trade-offs among new systems and fixes to legacy systems that are more responsive to integrated fleet needs.

7.7.2.2 Requirements Generation Recommendations

The committee's recommendations for the requirements generation area are as follows:

1. *Provide the resources for and expand the role of the Navy Warfare Development Command in generating concepts of operations and operational architectures, and designing and analyzing the results of operational experiments. NWDC and MCCDC should coordinate Navy and Marine Corps experimentation efforts.* Also, involve the NWDC in the IWAR process.

2. *Create a requirements board.* This board of Navy senior leadership headed by the VCNO should act as the institutional memory and keeper of the integrity of the requirements generation process. (Details are described above.)

3. *Create an Information Operations and Space Command, a functional type command, to provide a focal point for NCO activities, consolidate responsibility for information support, and prepare requirements for NCO systems.* (Details are described above.)

4. *Propose the establishment of a joint office, similar in scope to JTAMDO, for joint fire and land attack.* JTAMDO has expertise from all the Services and the major agencies. It is defining the operational architecture and the inter-

operability requirements for air and missile defense. Thus, an analogous effort for joint fire could be a step toward jointly identifying the operational architecture and information requirements for land-attack, strike, and time-critical targets. A continuing joint fire testbed would be a step forward from the biannual Roving Sands exercises,[22] which have demonstrated continuing difficulty in integrating intelligence and command and control systems to pinpoint time-critical targets.

7.7.2.3 Mission Analyses and Resource Allocation Recommendations

The committee's recommendations for the mission analysis area are the following:

1. *Staff and provide resources for the IWAR process to enable continuous assessments from requirements generation through programming, budgeting, and execution.*

2. *Make developing of output-oriented MOEs and MOPs for network-centric operations a high priority.*

3. *Put more emphasis on and provide sufficient resources for developing a comprehensive set of DRMs across all mission areas and keep them current. Up-do-date DRMs are the single biggest improvement that can be made to mission analysis.*

4. *Make more use of the distributed engineering plan in testing potential life extensions of and improvements to legacy systems.*

5. *Develop a methodology that everyone agrees to for assessing the risk of failure of a military operation. Recognizing that it will be a subjective assessment, strive to achieve a methodology that consistently arrives at the same answer for a given set of circumstances.*

The committee's recommendations in the resource allocation area are as follows:

1. *Create a board of directors to add comprehensive oversight of network-centric operations as a whole to all three major decision support processes (requirements generation, resource allocation, and acquisition and program execution).* The oversight should also ensure that network-centric operations are always considered in the joint context. In addition, the CNO/CMC should review

---

[22]Roving Sands is a field training exercise that is the world's largest joint theater air and missile defense activity sponsored by the U.S. Joint Forces Command. Information is available online at <http://www.af.mil/news/May1996/n19960523_960496.html>.

how system trade-offs and resource allocation balances are addressed in the Navy/Marine Corps staffs to assist them in achieving maximum mission effectiveness for all naval force missions. This is particularly important for the power projection mission, which may be the most stressful to the Navy/Marine Corps team in terms of achieving the full benefits of network-centric operations. (Also see details on the recommended board of directors as described above in Section 7.7.2.1.)

2. *Strengthen the assessment process along the lines discussed in the mission analysis section (e.g., develop better measures of effectiveness and measures of performance) and expand assessments so that they continue through budgeting and program execution.*

3. *Put more emphasis on and resources toward phased implementation of spiral acquisition in homogeneous product lines.* Include resources for backfit where appropriate.

Therefore, the committee recommends that

4. *The organization of the Navy's N8 office should be reviewed and adjusted as appropriate and necessary to increase emphasis on all aspects of the power projection mission, including strike and countermine warfare, amphibious and airborne assault, and fire support and logistics support of Marine Corps forces from the sea.*

7.7.2.4 Acquisition and Systems Engineering Recommendations

The committee's recommendations in the acquisition and systems area are as follows:

1. *Create a board of directors for NCO integration, acquisition, and program execution.* (Details are described above.)

2. *Establish a three-star deputy for Navy NCO integration to the ASN (RDA).* This deputy should be a designated Navy SYSCOM commander and be double-hatted into this role. The deputy should oversee all aspects of battle force system interoperability and integration and coordinate program execution across the SYSCOMs to ensure synchronism in development, production, and installation of systems that implement network centricity. This responsibility specifically includes these same functions for the NCII. The deputy should be the executive agent of the board of directors for NCO integration, acquisition, and program execution for Navy systems. The deputy also should oversee the activities of the Navy Chief Engineer and the NAVSEA battle force interoperability engineering function.

3. *Institutionalize system engineering in the acquisition of Navy systems and in the execution of programs.* Under the recommended deputy for NCO integration, strengthen the role of the Navy Chief Engineer in institutionalizing system engineering as a methodology for achieving the integration and interoperability of cross-platform and cross-SYSCOM systems. The Navy Chief Engineer should oversee a system engineering cadre drawn from the three Navy SYSCOMs for this purpose. The SYSCOMs should receive the resources and staff to support this activity. In addition, the ASN (RDA) should look at the best means to address system engineering for the entire system and not just the functional parts. (This latter portion of the recommendation applies to all naval missions and particularly to the power projection mission, which affects so many elements of the Navy/Marine Corps team. All elements of the combined joint team must work as an integrated whole to achieve maximum NCO effectiveness. This result cannot be achieved if system trade-offs and system engineering efforts are addressed at the component level.)

4. *Assign the Navy Chief Engineer the responsibility and authority to develop and maintain the system and technical architectures (consistent with the Chief Information Officer standards and policies) and the interface, interoperability, and other standards required for compatibility of network-centric systems.* This responsibility should include coverage of the same functions for the development of the NCII.

5. *Institutionalize a process of spiral acquisition for network-centric systems, focusing on domain-specific applications that serve individual mission domains.* Network-centric systems should be funded and planned for continual technology refreshment and functional evolution both during development and throughout the full life cycle. Provision should be made for continuous operator involvement in the development and evolutionary cycle, preferably by assignment of operational user representatives from the operating forces to the program management offices. Budgetary allocations should be included in these programs for technology refreshment and functional evolution. The recommended Commander, Operations Information and Space Command, should assist the fleet in providing user feedback and choosing priorities for system upgrades overseen by the general board. The board of directors (for NCO integration, acquisition, and program execution) should recommend program adjustments necessary for expediting the spiral acquisition process.

6. *In keeping with the committee's findings above regarding mission balance in the management of requirements generation review and system acquisition, the ASN (RDA) should seek the best means to address the design and engineering of NCO systems to eliminate as nearly as possible any distortion of the overall NCO perspective through undue emphasis on any single naval force mission or any one platform.* In particular, the Navy Department PEO structure

should be reviewed and provision made, as is found appropriate and necessary, for management of the acquisition and the oversight of mission-oriented, networked major subsystems of the overall NCO system. In doing this, special attention should be given to end-to-end (surveillance and targeting through effectiveness assessment), fleet-based, land-attack (strike and fire support) subsystems for Navy, joint, and coalition missions.

7. *Institutionalize the fleet experimentation process with increased rigor, and establish a mechanism for injecting its products into acquisition.* Designate the Navy Warfare Development Command as the Navy Fleet Experiment Commander and provide resources to NWDC for this role. Charter the Navy Chief Engineer to provide system engineering support for the design of experiments and for the definition of MOPs and success criteria, and provide resources to the SYSCOMs for this role. Charter the recommended Commander, Operations Information and Space Command, to represent the fleet in providing operator insight in the design and evaluation of experiments. The recommended Navy requirements board should rank the requirements of successful products for acquisition. The recommended board of directors for NCO integration, acquisition, and program execution should recommend budgetary adjustments for the transition to network-centric operations.

8. *Propose the establishment of a joint office, similar in scope to JTAMDO, for joint fire and land attack.* Until such an office is set up, the Navy and Marine Corps should participate more actively in the "Attack Operations" pillar (one of four) in JTAMDO in which a working integrated process team is looking at the targeting of time-critical targets, such as mobile missile launchers.

7.7.2.5 Personnel Management Recommendations

The committee's recommendations in the personnel and career management area are as follows:

1. *Institute network-centric operations education and training for all naval personnel at all levels within the Navy and Marine Corps.* All elements of the Navy need to be more appreciative of the principles, benefits, and risks associated with network-centric operations.

2. *Develop career paths for both military and civil service employees with significant information technology expertise.* Provide opportunities for those personnel with significant NCO critical IT expertise who wish to make a career in the Navy/Marine Corps military or civilian team to achieve very high (if not the highest) positions in the Department of the Navy. Ensure that all Navy/Marine Corps line and field-grade officers have expertise in IT and are capable of effectively using operational information in network-centric operations. Train mili-

tary personnel and civilians together when the IT learning requirements are shore based to better integrate the military and civilian parts of the combined team.

3. *Analyze the current and projected IT work so that more informed decisions can be made about who should do the work—active duty personnel, reserve military personnel, civil service employees, or contractors.* Assess how functions can be expected to be performed in 5 years and 10 years. Evaluate manning, technology, training, and resource alternatives for those periods.

4. *Update military and civilian IT job codes to match the desired IT specialty work.* Analyze present job skills to assess current functions and how they are carried out (present job skill codes are not adequate to provide the detail needed to understand the existing work structure and manning skills).

5. *Distribute IT education and training opportunities well within the Department of the Navy workforce and make them readily accessible.*

7.7.3 Recommendations Summary

This chapter sets forth the collective judgment of the committee on the keys to implementing improved network-centric capabilities. The chapter shows how to modify the decision support and personnel management processes to achieve network-centric capabilities as major enablers in the conduct of naval operations. The key management processes (i.e., defining what is needed, allocating resources, acquiring systems, staffing critical billets, and operating a global information infrastructure) are described. In Figure 7.10, the committee's major recommendations that affect the processes necessary for effective NCO integration are shown below the functions that would be most affected by the specific recommendations. The committee believes that the changes recommended could be implemented by the Navy, without revisions to law or DOD directives, and that their implementation will make a significant improvement in the success of future naval operations.

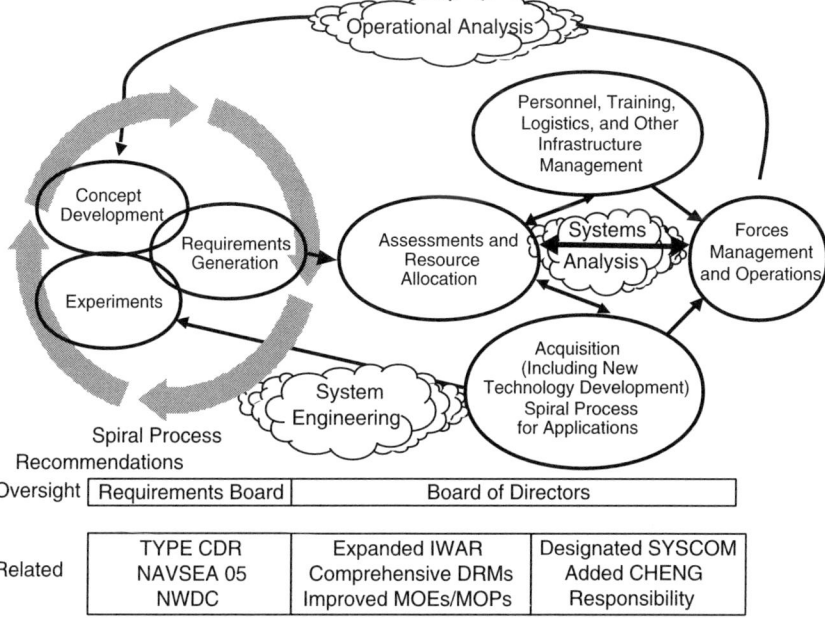

FIGURE 7.10 Functions for effective integration of network-centric operations shown in relation to major recommendations made in this report. CHENG, Chief Engineer of the Navy; DRM, design reference mission; IWAR, integrated warfare architecture; MOE, measure of effectiveness; MOP, measure of performance; NAVSEA, Naval Sea Systems Command; NWDC, Navy Warfare Development Command; SYSCOM, Systems Command; TYPE CDR, functional type commander.

Appendixes

A

Admiral Johnson's Letter of Request

 CHIEF OF NAVAL OPERATIONS

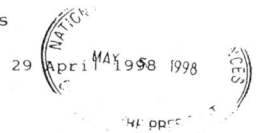

29 April 1998

Dear Dr. Alberts,

On 31 March, I was briefed on the results of the Naval Studies Board report, "Technology for the United States Navy and Marine Corps, 2000-2035." I was greatly impressed by the exceptional thoughtfulness and relevance of your work. I look forward to using the study recommendations in Navy's planning and programming process as we move toward the twenty-first century.

As a follow-on to that superb work, I would like you to consider undertaking a new study, "Transition to Network-Centric Naval Forces." I believe that realization of the full potential of network-centric warfare will prove critical to the operational success of future naval forces. It is a concept which I find very exciting and I request your assistance in helping that concept become reality. Should you agree to undertake such an analysis, my staff will develop detailed terms of reference for the study in consultation with the Chairman and Director of the Naval Studies Board.

As always, thank you for your support. I truly value our close working relationship with the National Academy of Sciences and look forward to working with you on this important new study.

Sincerely,

JAY L. JOHNSON
Admiral, U.S. Navy

Dr. Bruce M. Alberts
President, National Academy of Sciences
2101 Constitution Avenue, N.W.
Washington, DC 20418

B

Current Sensor Capabilities and Future Potential

To clarify the advantages and limitations of individual sensor characteristics, as well as the requirements that these characteristics may place on the performance of a tactical network-centric sensor grid, this appendix discusses each major sensor class in some detail, and it describes current state-of-the-art, likely paths for future growth. To contain the discussion, the sensors participating fully in the grid, that is, those used for surveillance, reconnaissance, and sensor-to-shooter targeting, are emphasized over those dedicated to a single weapon such as a gun control radar. Actually the only major difference between the two types of sensors lies in the relatively short-range and subsecond response requirements of the weapon's terminal phase, which contrasts with the long-range, seconds-to-minutes response requirements for surveillance, reconnaissance, and targeting. The fundamental technology characteristics are the same for all applications—only the detailed design parameters differ.

B.1 RADAR

Microwave sensors represent the dominant class of air/land battlespace sensors. Inevitably orders of magnitude physically larger than optical sensors with equivalent resolution, microwave radar easily compensates for this size disadvantage with its long-range, all-weather, "imaging" capabilities. As a result, every Navy platform has several radars: for search, navigation, missile fire control, gun fire control, target illumination, and so on.

Although there are still mechanically scanned, microwave tube-powered radars on naval platforms, modern radar implementations, both surface-based and airborne, are uniformly configured as electronic-scan, phased-array architectures with all-solid-state microwave power generation. The SPY-1 Aegis radar, the workhorse of the cooperative engagement capability (CEC), represents a pioneering naval implementation of a phased-array radar, albeit with a conventional centralized microwave tube power source. With four fixed-array faces, its electronic scan provides 360° coverage for ship self-defense via search, track, and weapons control.

Today, the Navy is considering the development of at least four new radars—the multifunction radar (MFR), an X-band radar for short-range ship defense; the volume search radar (VSR), an L-band radar for medium-range search and cueing to replace the SPS-49; the high-power discriminator (HPD; an X-band radar for ship-based theater missile defense); and a possible future long-range multifunction C/S-band replacement for the current Aegis radar. All are envisioned to be electronic-scan, phased-array architectures with active monolithic microwave integrated circuit (MMIC) solid-state transmitter/receiver (T/R) modules.

These surface platform-based sensors are capable of producing "images" of the surrounding air and sea surface space in the traditional radar sense of a georeferenced "map" of the estimated locations of significant observed returns. For an isolated radar, operating in a platform-centric mode, the precision with which the locations of these target reports are defined is of mixed quality—for although a radar usually can provide high-precision range and Doppler measurements, the angular precision is generally poor because of the large radar wavelengths and the dimensions of practically sized radar antennas. Beam widths measured in degrees or finite fractions of degrees are not uncommon. At the range of the target, even for modest ranges of 10 or 20 km, the positional uncertainty perpendicular to the radar beam could be measured in tens to hundreds of meters, whereas the range uncertainty along the beam direction could be less than 1 m. Exploiting the combined measurements of a number of dispersed radars immediately provides a quantum jump in radar-imaging capability without any change in the participating radars' operational characteristics.

Synthetic aperture radar (SAR) accomplishes much the same thing with a single radar sensor, in a different way—by moving it, with a crucial difference in the point of view—looking down rather than up. When an airborne radar is moved along a linear path and appropriate sequential measurements are made from a number of different spatial positions, the accumulated data can be combined in such a way as to duplicate the performance of a virtual antenna equivalent in size to the distance the platform flew during the collection of the data. The resulting radar images of the ground are of "optical quality," with uniform meter to submeter resolution in all dimensions, and can be obtained over large surface areas, at ranges up to hundreds of kilometers, through almost any kind of weather. SAR sensors are currently available in the battlefield on joint resources such as the Joint Surveillance and Target Attack Radar System (JSTARS) (APY-3), the

> **Box B.1 Hot Topics in Radar Today**
>
> - Digital radar for increased performance, microelectronic compactness, and reduced costs—digital waveform generation, digital receivers, digital true time delay, digital beam forming, and so on
> - Integrated array architectures—bricks versus tile approaches for lower cost per module, for instance, ~$50 per module, from today's ~$2,000 per module
> - Increased use of real-time synthetic aperture radar (SAR)/ground moving target indicator for ground surveillance and targeting—in-sensor processing enabled by the growth in computer capabilities described by Moore's law
> - Higher-power T/R modules via wide-bandgap semiconductor technology, e.g., silicon carbide, gallium nitride, and so on
> - Distributed radar similar to the cooperative engagement capability
> - Antistealth capabilities via distributed, bistatic configurations, and, perhaps, low frequencies
> - Ultrahigh-frequency SAR for foliage penetration combining low frequency with very large virtual apertures

U-2 platform, and the Global Hawk unmanned aerial vehicle (UAV), with future plans for space platforms (e.g., Discover II).

Real-time SAR requires enormous amounts of computational power, from gigaflops to teraflops. In the past, computers with this kind of throughput were large and deployable only on the ground. Early air- or space-based SARs were forced to transmit the raw data to dedicated ground stations for rapid, but not necessarily real-time, availability of the images. Fortunately, computer technology today has advanced to the point where the necessary computational throughput can be provided, in sufficiently small size and weight and low enough costs, to be deployable directly on the UAV or aircraft for real-time SAR.

Current thrusts in radar technology are described in Box B.1.

B.1.1 Radar Performance

B.1.1.1 Resolution

Because radar frequencies support large bandwidth signals and long pulse duration, the range measurement capabilities of radar can be quite good—tens of meters to meters or centimeters, if desired. If classification is of interest, hundreds of megahertz of bandwidth might be employed to get a resolution of a few centimeters for automatic target recognition (ATR), whereas in air traffic control (ATC), where only detection and general location of targets are necessary, range resolution might be relaxed to as much as 100 m, leading to a narrow-bandwidth, much cheaper radar. Similarly, ground moving-target indicator (GMTI) radar, using Doppler techniques, can detect radial motion as slow as 1 or 2 km/h.

In angle, radar beams are typically on the order of a degree or so, which corresponds to antenna dimensions of 60 wavelengths or larger. For X-band (i.e., 10 GHz), a 1° beam width antenna would be 1.8 m across, whereas at L-band (i.e., 1 GHz), it would be 18 m. Of course, bearing estimates are not limited to these dimensions, for with a large enough signal-to-noise ratio, beam-splitting interferometric techniques (e.g., monopulse) can estimate directions to small fractions of the beam width.

B.1.1.2 Field of View and Field of Regard

Although the generation of multiple simultaneous physical beams from a phased-array radar is possible, operational radars are never operated in this mode, but rather explore their environment one beam at a time.

In a conventional, mechanically scanned radar, the beam position is fixed relative to the antenna and thus provides a very restrictive instantaneous field of view (IFOV) equal to the beam width, which for typical radars is measured in fractions of a degree to several or even several tens of degrees. The field of regard (FOR), on the other hand, represents the angular portion of space over which the radar may be pointed by steering the antenna. In a sense, this is a design parameter, and mechanically gimbaled radars, which can rotate a complete 360° in azimuth and tilt up from horizontal to 60° or 70° or even 90° (the zenith), have been built.

Phased-array radars, on the other hand, can electronically scan their beams over a wide IFOV, without any mechanical assistance. In practice, deviation angles of up to ± 60°, in any direction off the axis of the array, are readily obtainable, with some compromise in performance at the larger angles because of beam-spreading "squint" effects. All four of the new phased-array radars now under consideration for missile defense will have an IFOV of this magnitude for each single face. For a fixed array, the IFOV and FOR coincide. To expand the FOR of a phased array, two approaches are commonly followed. One simply expands the radar to include individual fixed faces for each azimuthal quadrant and further tips each array face up so that both the horizon and the zenith directions fall within the IFOV capability of the face. This is the approach used by the current Aegis SPY-1 and will no doubt be the design adopted for the future MFR and VSR radars. An alternate approach, which may produce an advantageous cost/performance trade-off in some circumstances, uses a hybrid mechanical-electronic concept. Raytheon's HPD theater missile defense radar, for example, which is an evolution of the existing theater high-altitude area defense (THAAD) ground-based radar (GBR), has only a single array face. The array is mounted on a gimbaled platform that is mechanically steered to provide IFOV coverage of the full hemispherical FOR.

SAR radars address their fields of view and regard in quite a different manner because of the way the raw data are collected. Imagery is generated in a swath parallel to and offset from the flight path of the plane, and the resolution achieved can be varied from high to low by contracting or expanding the width of the swath. Or small, very-high-resolution "snapshots" can be taken anywhere from the minimum to maximum range that the radar can address transverse to the flight path—some SARs support a mode in which the transmitter beam is kept focused on a small region of interest as the plane flies past. The FOR of a SAR depends on the product of the maximum-to-minimum usable range of the sensor and the speed of the plane, whereas the IFOV is highly variable and can vary from low-resolution imagery of the whole swath to a very-high-resolution snapshot of a small portion of the FOR, typically the same number of pixels per second—a trade-off between resolution and search rate.

B.1.1.3 Range

The range capability of a radar is somewhat of a design parameter, as it varies with implementation parameters such as transmitter power, the microwave wavelength (e.g., S-, C-, X-band), antenna dimensions, detector sensitivity, and the like, as well as the scattering cross section of the intended target set and perhaps some environmental variables. Radars are thus deliberately designed to meet their mission requirements. Surface-based air search radars like the SPS-49 can detect targets 200 to 300 nautical miles away in any direction, and presumably the VSR will be designed with similar specifications. Airborne surveillance radars, such as the APS-145 early warning radar on the Hawkeye E-2C, reach out even farther, to as much as 600 nautical miles, although the JSTAR's SAR is capable of imaging areas at a range of up to 250 km (~140 nautical miles) transverse to the flight path.

B.1.1.4 Geopositioning Accuracy

A single traditional radar, whether phased array or not, has poor geolocation capabilities because, although its range measurement uncertainty can be very small if its signal bandwidth is large, e.g., centimeters to a few meters, its angular resolution is always poor in practice because of the limited aperture sizes available; e.g., $0.1°$ to $1°$ or $2°$ beam widths are typical. A $0.1°$ beam width at 10 km range gives a transverse target location uncertainty of ± 8.5 m, which grows to ± 85 m at 100 km range.

Combining two or more such radars, in a network-centric warfare (NCW) or CEC-like cooperative mode, immediately reduces the combined position location uncertainty to values on the order of the range resolution—degraded by the nonradar problems of determining the individual radar positions accurately via the Global Positioning System (GPS) or some other way. The ultimate limit on geolocational position accuracy is very likely to be dominated by the GPS accuracy of several meters, rather than the inherent capability of the individual radars.

The geopositional accuracy of SAR sensors is also controlled by the fundamental resolution of the imagery, which can be flexibly varied between broad swaths with tens of meter resolution to small snapshots with submeter resolution, as well as the GPS difficulties of determining the absolute location of the SAR platform at any given time. On the other hand, SAR ground imagery is of such quality that cross-correlation with highly accurate National[1] imagery may permit sensor resolution-limited performance to be achieved.

B.1.1.5 Area Coverage Rates

Search radars—whether mechanically scanned or phased array, ATC, or military—scan the full 360° upper hemisphere out to many hundreds of nautical miles in about 5 to 10 s. If a nominal 450 km range and a 6-s sweep interval, similar to that of the SPS-49, are chosen, the corresponding area coverage rate would be about 10^5 km^2/s—a very high rate of coverage—but the resolution is also quite low. Very typically, a primary search radar is designed to encounter and be able to detect and locate up to several thousand candidate targets during a single full azimuth sweep.

Surface-threat, self-defense radars, such as the MFR, try for a faster, approximately 1-s update rate and are horizon-limited to line-of-sight (LOS) ranges of a few tens of kilometers. Assuming a 20 km range and a 1-s sweep, the area coverage rate would be about 1.2×10^3 km^2/s—two orders of magnitude less than that of the long-range search radar, but no doubt done with much higher spatial resolution—more pixels per square kilometer.

Theater defense radars do not try to search large areas and so have minuscule area coverage rates. These radars are designed to detect, track with high accuracy, and classify incoming threats with their decoys and are cued to small IFOV baskets, within which the targets have been localized by other wide-area coverage sensors. Typically only a few tens of objects are expected to be found in the IFOV.

SAR sensors can generate low-resolution images of kilometer-wide swaths at the velocity of the airplane or trade this for a number of high-resolution snapshots using the same number of pixels generated per unit time. JSTARS according to the press is capable of mapping (at unspecified but low resolution, no doubt) 1 million km^2 in 8 hours, which translates to an area coverage rate of about 35 km^2/s, which is not high when compared with ordinary search radar performance. It is also claimed that the Global Hawk's SAR will be able to survey, in 1 day, with 1 m resolution, an area equivalent to the state of Illinois (40,000

---

[1]The term "National" refers to those systems, resources, and assets controlled by the United States government, but not limited to the Department of Defense.

square nautical miles), which translates into a fairly low rate of 1.6 km²/s, a rate compatible with high-resolution imaging. For example, if we hypothesize a platform velocity of 200 m/s and a 10 km swath to be imaged by a SAR at 1 m resolution, all of which sounds quite reasonable, the resulting area coverage rate would be 2 km²/s at a pixel rate of 2×10^6 pixels/s.

In contrast to Global Hawk, the Discover II program is targeting a much more capable, spaceborne SAR with a pixel rate of about 20×10^6 pixels/s.

B.1.1.6 Communication Data Rate Requirements

Building on the information above, it is possible to estimate the communication data rate loads implied by the different classes of radar sensors.

Non-SAR radars, as mentioned before, produce highly preprocessed images, with the information data rate heavily reduced through the simple expedient of reporting only "hits"—an elementary form of ATR. If sampling at a particular beam position (i.e., a dwell) finds no candidate target returns of significance, nothing is reported for that "pixel." A typical report will necessarily consist of a number of digital words describing target location parameters, such as bearing and range, or Kalman filter coefficient updates of information—altogether as many as twenty 32-bit words may be necessary for a worst-case total of 640 bits per report.

Thus a search radar, which may encounter as many as 2,000 targets on a single, 6-s, 360° scan, would require a maximum communication bit rate capability of about 200 kbps—although operating ATC radars often see no more than 500 targets at a time and often transfer the reports at 50 kbps over ordinary telephone lines. Horizon search radars, such as the MFR, with their horizon-limited range capabilities, expect to encounter only a few tens to a hundred or so candidate targets to deal with and so, with a 1-s update rate, can expect to need minimal capabilities, similar to the ATC example above—i.e., about 50 kbps.

But SAR, the true imaging radar sensor that generates data for every pixel, without exception, will require much higher communication bandwidth capability in order to participate in a network-centric sensor grid—but not nearly as much as is required by a capable modern electro-optical camera, as discussed in the section on electro-optical sensors (Section B.2). Practical SAR sensors produce pixel information at rates comparable to what is implied by the Global Hawk performance capability described above under "Area Coverage Rates" (Section B.1.1.5). Each second, an area of 1.6 km² is to be sampled at 1 m × 1 m resolution, leading to a pixel rate of 1.6×10^6 pixels/s, which is fairly typical of such systems. Assuming that the location information is implicit in the raster format by which the images are read out, each pixel will need no more than one 16-bit word (or even less) on average for an output reporting data rate of about 25.6 Mbps—which does indeed resemble the requirements of high-quality optical cameras, albeit at the low end of the requirements. Here again, it would be useful to be able to apply some automatic information extraction algorithms via local processing, so that only the compressed, salient information would have to be passed over the network-centric sensor grid communication infrastructure.

B.1.1.7 Spectral Issues

Different portions of the microwave spectrum are used by different classes of radars, not so much for acquiring additional target-background characteristics for ATR, as is the case with optical sensors, but more often to resolve implementation-application trade-offs. For example, an X-band radar at 10 GHz can achieve the same angular resolution as an L-band radar at 1 GHz with a 10 times smaller antenna. And so X-band is often preferred for high-accuracy applications or for missile seekers where aperture is at a premium. Similarly, the search rate capability of a radar is proportional to the product of the transmitted power and the area of the antenna. In addition, since low-frequency radars need large physical antenna in order to maintain even modest angular resolution and microwave power is much easier to generate at the lower frequencies—e.g., one can obtain T/R modules with hundreds of watts capability at 1 GHz of L-band whereas the current state of the art produces only about 10 W for an equivalent X-band module at 10 GHz and much less than 1 W for frequencies of 35 GHz and beyond—search radars are always L-band or lower.

B.1.1.8 Environmental Interactions

With few exceptions, radars are "all-weather," long-range, imaging sensors and for these capabilities they are highly valued. Radar frequencies are, in fact, absorbed and scattered to a minor extent by atmospheric constituents, but not nearly to the extent to which these same obstacles obstruct electro-optical systems. Rain certainly introduces attenuation, but ordinary radars, operating in the 1 to 35 GHz range, suffer very little performance degradation as a result. The effects are the same for fog, dust, and clouds.

Only as the frequencies move up into the millimeter range (i.e., 40 to 50 GHz or higher) do serious atmospheric absorption effects appear. Although there is interest in millimeter radars for short-range precision guidance applications, no broad situation awareness roles for them have yet been found, so these limitations are unlikely to have any impact on network-centric operations issues.

B.1.1.9 Susceptibility to Countermeasures

Clearly radars are susceptible to a variety of countermeasures. Jamming is effective, and all military radars are designed with this possibility in mind. As active sensors, radars inevitably emit radiation, thereby inviting physical attack by missiles like the high-speed antiradiation missile (HARM). For a single radar, operating in a platform-centric mode, often the only protection is to shut down, and the countermeasure has proven effective. On the other hand, with cooperating, distributed radars, as would be characteristic of a network-centric configuration, such a counter may be ineffective as only local portions of the network of sensors need be temporarily shut down, while the rest of the network continues to track the threats. Stealth techniques applied to aircraft have proven effective against many radars. However, very-low-frequency radar, because of size resonance effects, may have certain advantages in the detection of such targets, although perhaps with poor localization capabilities. More interestingly, since a large part of stealth technology depends on the exploitation of geometries that reflect the incident radar waves away from the transmitter, rather than scattering them back, bistatic approaches offer interesting counterstealth possibilities. Radiation scattered away from the transmitting radar may well be receivable by another radar receiver on the battlefield, and this kind of cooperative behavior is just what the network-centric sensor grid concept is going to encourage!

B.1.2 Technology Trends and Future Growth in Radar

B.1.2.1 Digital Radar

One of the most exciting developments in radar today is the vigorous push of digital techniques into many areas that have been traditionally analog (see Box B.1). The idea of a "digital radar" promises increased flexible performance, microelectronic compactness, and reduced costs.

Originally radar was based entirely on analog components and techniques—transmitter, antenna, receiver, and signal processing with the results output as analog inputs to a video display. Eventually, analog-to-digital conversion was introduced at the output of the receiver, and digital signal processing has now been a regular feature of high-performance radars for at least several decades. Exploiting the exponential explosion in computer technology, digital signal processing has permitted the continuing introduction of powerful advanced signal processing algorithms (e.g., for space-time adaptive processing) as well as the implementation in real time of complex tasks (e.g., SAR) previously doomed to off-line processing. Other tasks, such as digital beam forming of phased arrays, even at the subarray level, have remained impractical up to the present.

But today, as digital clock rates move into the gigahertz range and computational capabilities grow from gigaflops to teraflops, digital beam forming, particularly at the subarray level, now appears feasible. The term "digital radar," however, suggests much more than signal processing and beam forming. It now is possible to consider replacing many of the remaining analog radar components with much more compact, lower-cost digital equivalents—e.g., receivers, waveform generators, and so on. The idea is to move the analog-to-digital converter (ADC) away from the signal processor and as close to the front end of the radar as possible—in the extreme, an ADC at every T/R element in the phased array. The transmit waveform would be generated digitally and the signal transported in digital form over fiber-optic data lines to the individual T/R modules where it is digitally delayed to achieve true time-delay phase shifting, passed through a digital-to-analog converter, amplified by the solid-state T/R module, and transmitted. After some filtering and low-noise amplification, the received signal would be digitized directly at the microwave frequency or after a single stage of down conversion, digitally delayed as appropriate, and sent via fiber optics to the digital signal processor where pulse compression, beam forming, space-time adaptive processing (STAP), and so on will be carried out at real-time speeds.

Figure B.1 illustrates the performance of current state-of-the-art ADCs. To date, all available ADCs fall more or less below the diagonal line, which represents a form of jitter limitation. Successful implementation of digital radar concepts requires performance above the jitter-limit line. Currently the Defense Advanced Research Projects Agency (DARPA) and others are investing heavily

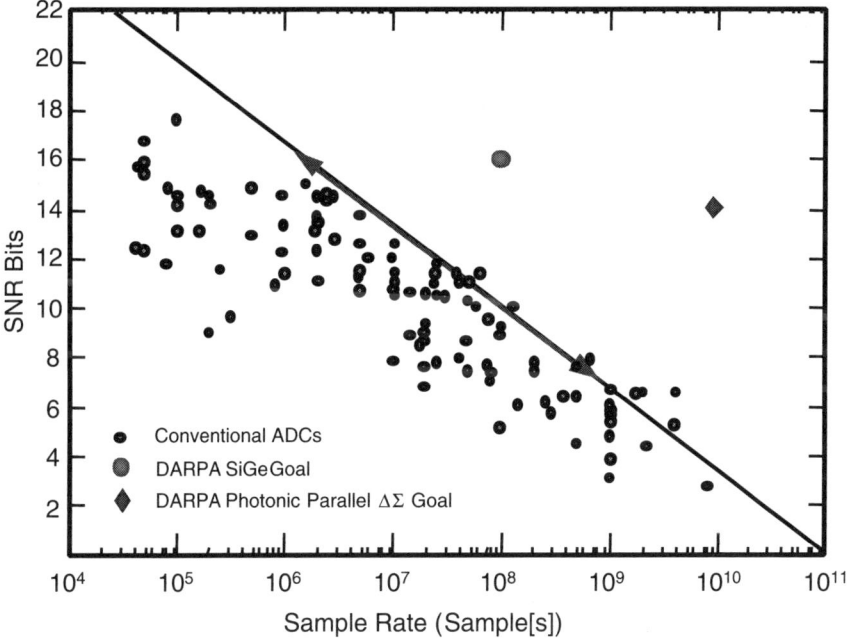

FIGURE B.1 State-of-the-art analog-to-digital converter (ADC) performance with current $\Delta\Sigma$ "breakthrough" targets indicated. SNR, signal-to-noise ratio. © Copyright 2000 Raytheon Company. All rights reserved. Reprinted with permission.

in finding ways to circumvent this apparent technology limit through the exploration of very-high-speed, so-called delta-sigma ($\Delta\Sigma$) 1-bit sampling techniques, which are common and successful in high-fidelity audio today at much lower sampling frequencies. A few more bits above the line will make digital radar a reality.

The increases in performance and decreases in size and weight can be enormous—one study indicated over a 100-fold decrease in the volume of the receiver hardware by going from analog to digital. Key to reaching these goals is the development of high-bit, gigahertz sample rate ADCs that are compact, low power, and inexpensive. DARPA, the Office of Naval Research (ONR), and others are currently supporting major thrusts in this much-needed ADC technology, and one can expect to find radars with digital receivers, and perhaps digital beam forming and digital true-time delay, deployed within the next 5 years. Many radars today already employ digital waveform generation.

B.1.2.2 Array Architectures—Low-cost Transmitter/Receiver Modules

Phased arrays are expensive. If the total cost of the antenna structure is divided by the number of elements (i.e., T/R modules) in the antenna, costs of $1,000 to $2,000 per element are the norm. With 10,000 to 20,000 elements in a typical high-performance radar, such as MFR or HPD, the antenna alone can cost tens of millions of dollars.

The fundamental building block of a phased-array radar—the T/R module—is a complex device containing multiple GaAs integrated microwave circuits (e.g., MMIC amplifiers, phase shifters, and so on), digital circuits (e.g., controllers), microwave, digital and power interconnects, radiating elements, mechanical support and cooling structures, and so on. For performance reasons, this module must be packaged to fit into an area on the face of the antenna of only one-half wavelength by one-half wavelength—at X-band, this is only 1.5 cm in each dimension.

The development of MMIC technology over the past several decades has greatly reduced the costs for these chip components, through the adoption and extension of techniques from silicon integrated circuit manufacturing. Most of the remaining costs lie in the packaging, interconnects, assembly, and testing, and these aspects can be minimized by the use of highly integrated modular array architectures. So-called brick architectures, which are common today, meet the packaging challenge by building back along the third dimension away from the face of the array. Generally a modular structure is created with 4, 8, or 16 T/R modules built into a single integrated unit with integral power supplies and cooling. Costs for these highly integrated designs are now dropping to about $500 per element.

An alternate architecture, the tile array, attempts to meet the wavelength constraints by building directly in the plane of the antenna with unpackaged ultraminiature components. A large subarray of 64 elements or more is created, not from individual T/R modules, but as a single monolithic structure in the form of a sandwich of functional layers (e.g., radiating elements, MMIC components, interlayer microwave, power and ground interconnects, distributed direct current (dc)-dc power converters, and so on)—generated largely by printed-circuit techniques.

These structures weigh very little and cost about $100 or less per element but are currently limited in power capabilities (e.g., a few watts per element) by the miniaturization requirements. It is hoped that progress in wide-bandgap semiconductors such as GaN can increase the power capabilities of the MMIC components to 5 or 10 W per element without increasing the size, making these architectures extremely promising and competitive, particularly for future airborne applications.

B.1.2.3 Real-time SAR

Air- or spaceborne synthetic aperture radar, with its high-resolution, all-weather, ground-imaging capabilities, complemented with a ground-moving target indication or GMTI mode, is unquestionably the premier land surveillance and targeting sensor on the battlefield. These capabilities are already available on the JSTARS, U-2, and Global Hawk and Predator UAVs; and further developments in hardware and software will permit more compact and versatile implementations, with efficient on-board real-time processing and extended capabilities for foliage penetration (FOPEN) and mine detection. In assessing its mix of organic versus joint sensors in the battlespace of the future, the Navy should carefully consider the merits of deploying its own SAR-equipped UAVs.

B.1.2.4 SiC and GaN High-power Devices

Although the modern high-performance phased arrays that are deployed or under development at present are uniformly based on GaAs or InP MMIC T/R technology, it has long been known that semiconductor materials with a larger bandgap than GaAs and InP are possible and would offer enormous benefits. Devices made from wide-bandgap materials could be operated at much higher temperatures and voltages than GaAs and InP and would be expected to show higher degrees of linearity. ONR has been in the forefront of this technology for years, and the community's effort has begun to bear fruit in the last several years. Both SiC and GaN, originally valued for their ability to produce the blue light characteristic of their large bandgap, have shown great potential for high power generation of microwave energy. Although GaAs and InP devices produce less than 1 W/mm of gate periphery, GaN, in particular, has demonstrated as much as 3 to 5 W/mm. In a few years, today's 10 W X-band T/R modules could be replaced by physically similar but much higher power GaN-based equivalents with 50 or more watts of power output. Introducing this technology in tile array configurations could greatly increase the attractiveness of these lightweight, inexpensive architectures.

B.1.2.5 Distributed Radar

The CEC is an innovative Navy program that has decisively confirmed the expected benefits of distributed networked configurations, demonstrating precise geolocation, robust tracking, and extended area coverage. CEC, which in its original implementation coupled only identical SPY-1 radars, is now experimenting with the incorporation of other types of radar, e.g., Patriot and Airborne Warning and Control System (AWACS), into its meta-radar. Network-centric operations will extend these concepts to more flexible mixes of different kinds of sensors (perhaps combinations of active and passive radar) on an opportunistic and adaptive basis, leading to hybrid distributed sensors with exceptional capabilities.

B.1.2.6 Antistealth

Future efforts should consider distributed configurations where some of the radars operate in a passive mode, perhaps in a time-varying, adaptive way, so that the antistealth benefits of bistatic configurations might be realized. The use of low frequencies (i.e., below L-band) would enhance the detectability of the stealth targets, whereas the distributed cooperation would greatly mitigate the poor angular resolution and permit practical-sized antennas to be utilized without compromising the overall performance of the networked radars.

B.1.2.7 Foliage Penetration

Penetrating foliage is one of the tasks that suffer from fundamental physics-based obstacles. Low frequencies are needed to penetrate—whereas high frequencies are required for good spatial resolution and imaging from reasonably sized antennas. With the rapidly advancing maturity of SAR, it is now possible to create, in a quite practical manner, a very large virtual (synthetic) antenna for a low-frequency radar that neatly sidesteps the physical limitations of real antennas and permits both penetration and good spatial resolution simultaneously. Such a radar is currently under development and will soon be available.

For urban environments, passive millimeter-wave imaging offers interesting possibilities as it can image through single nonmetallic walls of ordinary thickness with surprising effectiveness, and at the same time the equipment required can be reasonably compact.

B.2 ELECTRO-OPTICAL SENSORS

Optics and radar share a common physical basis—both exploit the propagation of electromagnetic waves. Thus the fundamental equations relating such parameters as aperture size and signal bandwidth to beam widths and measurement accuracy of the sensors are the same for both disciplines. However, the portions of the electromagnetic spectrum exploited are completely different. Whereas radar wavelengths are measured in millimeters to centimeters and meters, optical wavelengths are orders of magnitude smaller—in the neighborhood of a micrometer, that is, 10^{-6} m.

Because of this profound difference in characteristic dimensions, optical systems are always far smaller than radar systems of equivalent angular resolution. For a radar to match the beam width of even a modest-sized optical system, its antenna must be about four orders of magnitude larger—centimeters for the optical systems implies hundreds of meters for the radar!

In addition to this striking size/performance advantage, electro-optical systems are often significantly simpler than radar systems to implement. Optical systems make prolific use of simple mirrors and lenses of common materials that are transparent in the visible and infrared, conveniently supplying the electromagnetic phase shifts needed for precision beam control and focusing; optical imaging detectors are sensitive to the point of being able to detect single photons; and multipixel detector focal plane arrays for imaging can readily be implemented with microelectronic fabrication technology.

Because of the high quality of optical imagery, and the ease with which very narrow optical beams may be generated from small apertures, optical sensors are common on the battlefield. Most military platforms support one or more electro-optical sensors. The majority are imaging sensors, which include charge-coupled device (CCD) cameras, electronically amplified low-light-level, night-vision equipment that operates in the visible; forward-looking infrared (FLIR) cameras; infrared search-and-track (IRST) surveillance systems; and three-dimensional imaging ladar. Among the nonimaging sensors are laser range finders, target illuminators, and remote laser chemical/biological detection systems.

The difference in physics between radio frequency (RF) and electro-optics (EO) brings some disadvantages as well, as many environmental constituents, such as rain, clouds, fog, and dust, which are relatively transparent at microwave frequencies, are frequently opaque in the optical regime due to absorption or scattering. Environmental effects thus seriously limit the usefulness of optical systems in many scenarios.

As a result, for top-level situation awareness, air defense, and ground targeting, radar sensors that can "image" at very long ranges through all kinds of weather are preferred. Optical sensors are used more often as adjuncts for acquisition of scene details when the weather is fair or when the ranges involved are short and high-resolution imaging is required, as in endgame precision targeting.

> **Box B.2 Hot Topics in Electro-Optics Today**
>
> - Inexpensive, uncooled, ultralow-power, throw-away infrared cameras
> - Continued growth of focal plane arrays size— $\geq 10^6$ pixels
> - Pixel-aligned multiband and/or multipolarization focal plane arrays
> - Three-dimensional ladar image exploitation for automatic target recognition and automatic aim point selection
> - Hyperspectral imaging for improved detection and classification
> - Electronic beam steering via optical phased arrays to eliminate mechanical gimbals

Current thrusts in electro-optics technology are described in Box B.2.

B.2.1 Electro-Optical Sensor Performance

B.2.1.1 Resolution

Depending on the design details, i.e., the wavelength and the aperture diameter, passive imaging optical systems are characterized by pixel azimuth-elevation (az-el) dimensions from microradians to many milliradians. The corresponding pixel dimension on the objects to be imaged can vary from millimeters to tens of meters, depending on the range to the object and the angular characteristics of the optical system. Active laser-based EO sensors, both imaging and simple range finders, combine the excellent az-el two-dimensional resolution characteristic of all optical systems with the range accuracy of a radar. Three-dimensional resolutions of centimeters to meters, in all dimensions, are readily obtained over ranges of 1 to 10 km or more—weather permitting.

B.2.1.2 Field of View and Regard

Although optical configurations with a very large IFOV can be implemented, the optical design is challenging and the implementation hardware increases rapidly in complexity and cost as the IFOV increases. In addition, the finite size of the detector arrays permits only an equal number of scene pixels to be examined simultaneously. Thus to achieve the high-spatial-resolution performance for which optics is so valued, only a small IFOV can be examined at any instant. As a result of this kind of trade-off, typical fielded optical imaging sensors—both visible and infrared (IR)—are characterized by "narrow" fields of view (FOVs) of 1° or 2° in elevation and azimuth, whereas the "wide" FOVs are four or five times larger.

Traditionally, these rather small FOVs are extended to the desired much larger FOR by the simple expedient of mounting the sensors on a gimbaled mount. Gimbaled optical sensors with ± 90° in azimuth and/or in elevation, or even with a full 360° azimuth capability, are common.

B.2.1.3 Detection

Modern imaging optical sensors use highly integrated, monolithic arrays of semiconductor detectors known as focal plane arrays (FPAs). As semiconductors, FPAs participate fully in the inexorable growth of the electronics industry, with the numbers of detectors on a chip increasing exponentially year by year. Today, silicon-based CCD arrays for visible imaging are available, with more than 10^6 detector elements on a single chip along with all the required readout circuitry. Infrared focal plane arrays, fabricated from more exotic semiconductor systems (e.g., InSb for 3 to 5 µm mid-wave IR, or HgCdTe for 8 to 14 µm longwave), are not far behind, with 25,000-element (i.e., 512 × 512 pixels) FPAs already in such naval equipment as the Thermal Imaging Sensor System 3 to 5 mm surveillance sensor. The response times of all classes of FPAs are fast enough to permit signal readout at video rates (30 to 60 Hz).

B.2.1.4 Range

As pointed out above, practical optical sensors, whether passive or active, perform well only over relatively short ranges—typically 10 to 20 km at best. Line-of-sight requirements for high spatial resolution from small T/R apertures, as well as temporal variations in environmental obscurants, serve to limit the passive sensors. Active laser sensors share these same vulnerabilities and, in addition, suffer from the combination of relatively low available laser powers and small T/R apertures which, even in good weather, further limit range performance and severely restrict the sensors' search capabilities. For these reasons, active laser systems typically have to be cued to the target neighborhood by another wider FOV system, optical or radar, in order to achieve reasonable target acquisition times.

These short-range characteristics relegate most electro-optical sensors to secondary and special-purpose roles in the battlespace.

B.2.1.5 Geopositioning Accuracy

The importance of knowing the sensors' location to high accuracy—presumably by Global Positioning System (GPS) and/or an adequate Inertial Navigation System (INS) capability—has been discussed. To complete the picture, for the purpose of situational awareness a good, geographically consistent common operational picture (COP) must be generated by incorporating accurate location information estimates for features and objects observed and reported by the sensor. Optical sensors are ideally suited for this because of their inherent spatial resolution—particularly when combined with a laser range-to-target. With appropriate calibration, optical sensor measurements can be localized easily relative to the sensor to submeter accuracy. In addition, the high-resolution complex scene structure, generally obvious in optical images, is ideal for correlation of these images with reference imagery (e.g., satellite resources) to obtain precise determination of absolute image point geolocation.

B.2.1.6 Area Coverage Rate

The area coverage rate at which an imaging optical sensor can collect data within its FOR is determined by the competition between a number of factors—the IFOV of the sensor and the number of pixels on its FPA, the rate at which the FPA is sampled, the speed with which the gimbals can slew the IFOV across the FOR, and the motion of the platform on which the sensor is mounted. In many forms of optical imaging cameras, the update rate greatly exceeds the gimbal slew capability, and so redundant images are collected and the area coverage rate is determined solely by the telescope slewing characteristics and the platform motion. The IFOV projected on the ground is determined by the number of pixels in the detector array and the pixel resolution. For example, if the pixel image on the ground is 1×1 cm and the FPA is square with 10^6 pixels, the IFOV would cover an area of only 100 m^2. If the pixel size on the ground was 1 m^2, the same FPA would result in an IFOV on the ground of 1 km^2 or 10^6 m^2. So slewing could produce area coverage rates as low as a few hundred or as high as a few million square meters per second.

Some optical sensors, such as IRST sensors and various airborne three-dimensional imaging ladars, generate their images by scanning a one-dimensional linear array of detectors across the optical image of the scene through internal mirror motions. Others simply use the platform motion (e.g., a UAV) to sweep the image of the detector array over the region to be imaged—a so-called "push-broom" scan. Both techniques can collect nonredundant pixels at the update rate of the FPA, so the area coverage rate is easily calculated from the size of the FPA image on the ground and the angular speed of the internal scan or the linear velocity of the push-broom platform. For example, if the image of the FPA on the ground is 100 m and the UAV velocity is 200 m/s, the area coverage rate will be 2×10^4 m^2/s. If the linear FPA has, for instance, 2,000 elements (arrays as large as 10,000 elements are available and have been flown in high-resolution surveillance systems) and the frame update rate is 60 Hz, then 1.2×10^5 pixels are sampled per second with an average pixel area of 1/6 m^2—the pixel dimensions on the ground would be 5 cm along the array and 3.3 m along the flight path. This hypothetical situation does not result in as high-resolution imagery as what one might hope to get from an optical sensor—for better results, the platform should fly more slowly.

Nonimaging optical sensors, since they typically measure the equivalent of a single pixel per measurement, have very little area coverage rate capability and are strictly limited by the repetition rate of the laser. For conventional laser range finders (YAG and related systems), this is often only one to several tens of sample per second—many fielded laser range finders operate in this regime. With the development of solid-state, diode-pumped monolithic YAG, "microchip" lasers, and their extensions, repetition rates as high as 500 or 600 Hz have been demonstrated, permitting much higher pixel sampling rates, although these rates are still far lower than the capabilities of the passive imaging sensors.

B.2.1.7 Communication Data Rate Requirements

As has been suggested, sensors that generate images—which is just what electro-optical sensors do best—produce raw data at prodigious rates. Current generations of infrared FPAs can have as many as 250,000 detectors (i.e., 512×512 detectors), whereas visible silicon-based CCD cameras have a million (e.g., $1 \text{ K} \times 1 \text{ K} = 10^6$) or more. Given the dynamic range of typical IR and visible scenes, the output of each pixel is commonly quantisized to 12 bits. With video frame rates of 30 to 60 Hz (which is, in fact, typical of both visible and state-of-the-art IR sensors), it takes only a simple computation to discover that such imaging sensors generate raw data at rates between 90 and 720 Mbps. Attempting to transfer this data throughout the network-centric sensor grid by means of general purpose communication links would be disastrous.

Somehow the raw data have to be processed locally to extract only the salient information that may be of interest to a fusion node or decision maker, and only this minimal critical information communicated. The simplest form of preprocessing may be to apply lossy data compression techniques—of say, 40 to 1—which would immediately reduce the requirements to a more manageable 2 to 20 Mbps, while retaining much useful information about the scene observed. On the other hand, modern video image compression techniques transmit only the image changes from frame to frame, achieving large compression ratios with little information loss. Even better would be to apply a powerful ATR technique (perhaps still to be discovered) and reduce the useful information to only a few words of data associated with a limited number of candidate "targets" or other interesting aspects of the scene. As is reiterated in the discussion of radar sensors, this is precisely what an ordinary air defense or air traffic control radar accomplishes with its elementary form of ATR processing, leading to communication requirements that are low enough that ATC radar reports are customarily passed over ordinary telephone lines at 50 kbps or less.

IRST and other imaging sensors, which may employ push-broom linear array scan rather than two-dimensional framing, generate data at a much slower rate, as the one-dimensional linear arrays generally have far fewer detector elements than the two-dimensional arrays. A state-of-the-art linear array may consist of 10,000 or fewer detector elements. At 30 Hz sampling and 12 bits per sample, this kind of sensor generates raw data at more modest rates of 3 to 4 Mbps, which offers a more manageable communication burden, but which still may be further reduced by appropriate preprocessing.

Under many circumstances, sophisticated automatic information extraction techniques may not really be necessary, particularly for those two-dimensional FPA framing sensors that may be used for situational awareness, that is, for surveillance and reconnaissance, and that are of particular interest for integration into the network-centric sensor grid. This is because the sensor frame rates of 30 to 60 Hz clearly produce enormous amounts of redundant information which need not be transferred in total to other users in the grid. This high update rate is what is needed to refresh a display so the human eye detects no flicker, even if the image is completely static. The useful information in the scene, which may need to be communicated to another user, is determined by the dynamics of the scene itself—as, for example, through the motion of objects in the scene or changes in the lighting. In many scenarios, image update intervals of seconds, that is, frame sampling rates measured in fractions of hertz rather than tens of hertz, may be adequate, thereby reducing the raw data communication requirements by factors of 100 or more, even without sophisticated local processing.

Active imaging optical sensors, such as a rapid pulsed ladar that generates three-dimensional, range-to-pixel imagery, differ from the passive FPA imaging sensors just described in optical detector focal plane arrays, which are capable of simultaneous, independent, multiple time-of-flight measurements (i.e., need to define range-to-target at each pixel), do not yet exist—except perhaps as development items. In practice, it is not yet possible to collect such data in parallel. To date, such sensors have been operated sequentially, scanning the laser beam in some kind of raster pattern and collecting range-to-target samples at the repetition rate of the laser. If we suppose that we have a 1 kHz laser rate and each range-to-target report is a single 23-bit word, the raw data rate generated by the sensor would not exceed 23 kbps. Clearly, because of limitations in the current technology state of the art, active optical sensors currently are not capable of generating very high rates of raw data. But these are imaging sensors after all. And the kind of direct geometric information they can supply about candidate target objects permits promising implementations of efficient ATR algorithms—both template- and feature extraction-based. So it seems to be only a matter of time before the requisite technology is developed and the generation rate of raw data pushes to levels that also challenge the communication requirements.

Finally, nonimaging electro-optical sensors, such as laser range finders, are typically operated with one beam (i.e., pixel) at a time, at rates that can vary from "on-demand" to brief bursts at 1 to 20 Hz. Each report might include such information as range-to-target and perhaps a few other pieces of information, such as GPS location of the transmitter, azimuth from transmitter to target, reflected signal strength, and so on—in all, no more than a few tens of 32-bit words.

Except in highly dynamic weapons endgame situations, where the sensor-weapon coupling must be close and the data need not be communicated to other users in the sensor grid, it is difficult to believe that laser range measurements need be distributed any more often than at intervals of seconds. The resulting communication data rate requirements are minuscule—hundreds of bits per second or less!

B.2.1.8 Spectral Issues

The optical portion of the electromagnetic spectrum, stretching as it does from the ultraviolet, through the visible to the near, mid, and far infrared, encompasses a broad range of physical phenomena that can help alleviate some environmental obstacles in certain circumstances and that can be exploited to enhanced target detection and classification (e.g., ATR).

For example, although visible sensors respond only to reflected ambient light and are consequently signal starved at night, infrared systems detect the thermal radiation emitted by all the scene objects as well as reflected ambient and can produce visual-quality images at night as easily as in the daytime. Similarly, short wavelength light is scattered much more strongly than long wavelengths (e.g., the blue sky) so IR systems can often penetrate such obscurants as dust and smokes when visible cannot.

IR thermal images often indicate more about the scene than visible imagery; e.g., operating vehicles can be distinguished from nonoperating vehicles by the observable effects of the engine heat produced, and often objects that are well camouflaged in the visible show measurable contrast with the local scene background in the infrared. In general, the spectral signatures of target and background objects vary strongly as a function of wavelength throughout the observable spectrum with unique material-dependent characteristics, so a more detailed measurement of the image at several different wavelengths can provide even more useful target-background discriminants to enhance ATR performance.

Such multiband sensors have been implemented with encouraging results. For example, dual-band systems with simultaneous robust missile plume detection and excellent sun-glint rejection properties have been demonstrated. Real time, pixel-aligned dual-band, simultaneous mid- and long-wave FPAs have been developed by a number of organizations. Similar pixel-aligned, dual-polarization FPAs have also been fabricated for the exploitation of the differences between manmade and natural objects in polarization-sensitive reflection and emission. Existing space-based ground-imaging systems (e.g., the land remote-sensing satellite (LANDSAT) family) frequently collect data on as many as 5 to 20 spectral bands for discrimination purposes.

Efforts to further improve optical detection and classification of difficult targets such as land mines or nonoperating camouflaged vehicles (e.g., parked mobile missile launchers, and so on) have led to development of so-called "hyperspectral" imagers. These ambitious systems collect simultaneous data from a very large number—even hundreds—of narrow spectral bands covering much of the available optical spectrum with the hope of finding unique, distinguishable signatures that can be exploited. In a sense, hyperspectral imaging is overkill in that it is difficult to believe that hundreds of measurements are needed for every pixel in order to identify interesting objects in the scene. Much of the interest in hyperspectral lies in the "hope" that subtle characteristic spectral differences will be found when such data are collected for challenging scenarios. Perhaps, in the end, only a handful of strategically placed bands (i.e., multiband; more than one—fewer than hundreds) will prove necessary for effective detection and classification of difficult targets. Another practical reason for ultimately reducing from a hyperspectral to a multiband approach lies in the volume of simultaneous data collected. Even though the progress of computational resources will eventually accommodate hyperspectral imaging, today the data from such sensors cannot be processed in real time.

B.2.1.9 Environmental Limitations

As indicated in the discussion above, environmental factors seriously limit the usability of electro-optical sensors. Simple visible TV-like sensors are useless at night, although electronically amplified low-light sensors can be useful at night down to starlight levels of illumination—if the night is clear. Although the atmosphere is quite transparent (i.e., low absorption loss) in the visible, some portions of the infrared spectrum are strongly absorbed by the gases in the atmosphere. As a consequence, IR sensors are generally designed to avoid these regions and operate only in traditional low-absorption, atmospheric "windows"—e.g., the near IR from the visible to about 2.5 μm, the mid IR from 3 to 5 μm, and the long-wave IR from 8 to 14 μm or even to about 20 μm.

All optical wavelengths have difficulty with weather, as rain, clouds, and fog all absorb and strongly scatter the light. Due to the λ^{-4} variation of Rayleigh scattering, visible light is scattered orders of magnitude more strongly than the infrared, giving IR much better penetration through dust, fog, and other scattering media, but even IR can be limited in range under these conditions. Optical sensors are definitely not "all-weather" performers. But when they work, they produce beautiful, high-resolution images. As a result they are highly valued on the battlefield—but almost always complemented by a microwave system that is "all-weather."

B.2.1.10 Susceptibility to Countermeasures

Electro-optical sensors are quite susceptible to a variety of countermeasures. Camouflage can be very effective, as recent experience in Kosovo has demonstrated. It is often easy, by simple techniques, to make something "look" like something else—a target can be made to appear to be a portion of the background or an artifact to be a genuine target. Since optical sensors observe only the exterior aspects of objects, appearances are everything.

Jamming, in the form of a bright flare or a directed laser beam in the FOV of the sensor, can be a serious threat because optical sensors are frequently operated "wide open" in an effort to optimize sensitivity by maximizing the number of photons collected. The collected optical flux from a directed laser beam (e.g., from a tactical high-energy laser weapon, such as that currently under development jointly with Israel), operating within the IFOV of an imaging sensor, will be focused by the sensor's collection optics more or less onto a single detector in the focal plane. Under these circumstances, even modest high-energy laser power levels can physically destroy detector elements. Moreover, both the flare and the laser weapon beam, even if actual destruction does not result, can, by diffraction, cause large numbers of the detector elements around its image in the focal plane to saturate, thereby temporarily blinding large portions of the IFOV.

B.2.2 Technology Trends and Future Growth in Electro-optics

B.2.2.1 Uncooled IR Focal Plane Arrays

One of the most exciting advances in electro-optics in recent years has been the migration of microelectromechanical systems (MEMS) technology into IR focal planes. Arrays of tiny thermally sensitive structures (bolometers, as it were) can be fabricated on a silicon wafer using slightly modified integrated circuit manufacturing techniques, and along with each, an integrated on-wafer electrical measurement circuit to determine the instantaneous temperature of the microbolometric element. When an IR image is projected onto this wafer array by an optical system, the element-by-element temperature pattern that results from the local heating caused by the light is read out of the wafer as electrical signals and the device acts as an IR FPA, but with one enormous advantage over a traditional semiconductor FPA—it does not need to be cooled. Such MEMS-based FPAs operate at room temperature, with almost the same sensitivities (i.e., minimum detectable temperature differences measured in tens of millikelvins) as the liquid nitrogen-cooled semiconductors FPAs. Figure B.2 shows an inexpensive, compact, uncooled IR camera with a closeup of its MEMS focal plane and an image demonstrating a temperature sensitivity of 27 mK, comparable to the performance of a cooled IR FPA.

Add to the temperature advantage the facts that, as a close relative of a silicon integrated circuit, these uncooled FPAs are inexpensive to manufacture, are physically compact, and require very little power. For these very good reasons, uncooled IR cameras are going to find wide usage on the battlefield—as surveillance and terminal guidance sensors. In particular, they are perfect for all classes of unmanned air vehicles, including mini- or micro-, one-use, throw-away UAVs.

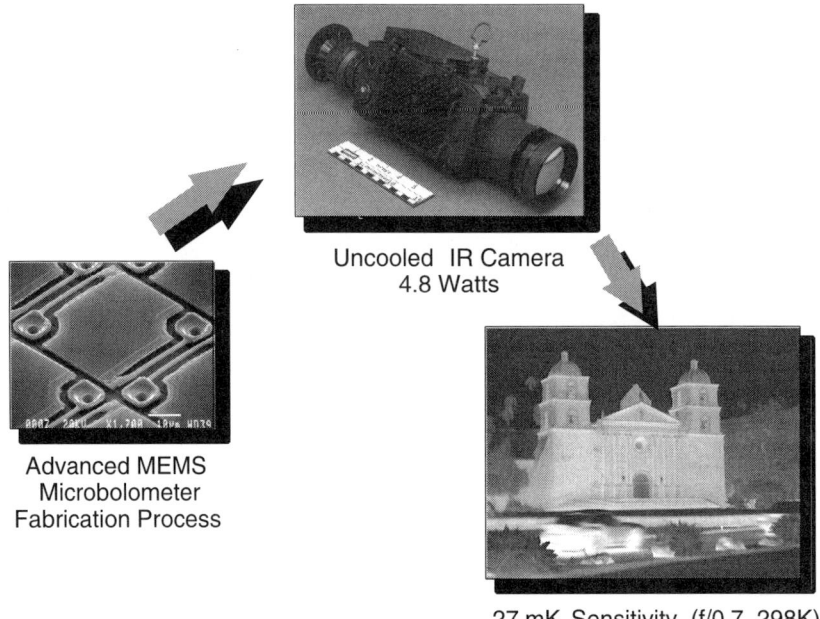

FIGURE B.2 Microelectromechanical systems-based uncooled infrared camera and temperature-sensitive image.

Their one apparent disadvantage is that the time response of an uncooled FPA is limited by thermal inertia to update rates of 30 Hz or below, whereas traditional semiconductor FPAs can operate at update rates as high as 400 Hz. For surveillance, this slow update rate is not a problem, but for terminal guidance, because of the high closing velocities and possibly rapid scene dynamics, this slow rate could be restrictive.

B.2.2.2 Advanced Focal Plane Arrays

Traditional FPAs continue to grow larger with 512 × 512 pixel HgCdTe arrays already deployed and 10^6 pixel arrays in sight. In addition to having more pixels, the FPAs are getting "smarter" with increasing amounts of on-chip sampled analog preprocessing being added.

B.2.2.3 Special-Purpose Focal Plane Arrays

Over the past decade, a number of interesting special-purpose multispectral FPAs have been developed. Both hybrid and monolithic techniques have been used to create stacked, dual-band, focal planes that produce simultaneous images in two different portions of the spectrum, say a 3 to 5 mm band and an 8 to 14 mm band, and both are perfectly pixel-aligned. This provides an ideal input to a sophisticated multifrequency ATR algorithm for target detection and classification.

Recently this pixel-aligned concept has been successfully applied to image polarization. That is, the FPA generates simultaneous, pixel-aligned images of the scene in each of two orthogonal polarizations. In all other respects, it acts like any other traditional FPA with respect to sensitivity, update rate, and so on. Since manmade objects tend to retain polarization and natural background objects generally depolarize, this special sensor offers interesting potential for target discrimination. All these possibilities are currently being explored.

B.2.2.4 Ladar Three-dimensional Imaging

The ability of laser imaging systems to obtain high-resolution, range-to-target measurements offers enormous advantages for target recognition. The sensor directly measures the geometric features of the object of interest and is not confused by scene illumination effects and the unknown distances, which trouble classical passive IR or visible-image ATR algorithms. The height of the tank or truck, or its precise orientation toward the sensor, can be directly measured without guesswork.

These advantages have been understood for decades; however, the lasers available have generally been bulky and expensive and no such system has yet been deployed, although the expected performance advantages have been demonstrated in the field with brassboard prototypes. Recently, a new generation of compact diode-pumped solid-state laser sources (e.g., Lincoln Laboratory's "microchip" YAG lasers) has evolved and interest in this promising technology has reawakened. It seems to offer unique potential for terminal guidance with automatic target selection and aimpoint determination based on geometric information about the target. Given the potential sensitivity of long-range, land attack, precision weapons to GPS jamming, such capable terminal sensors ought to be of great interest.

B.2.2.5 Hyperspectral Imaging

Multispectral imaging, with only two or three selected bands, has proven effective in enhancing the ability to detect and classify some targets. However, external appearances can often easily be altered and controlled by simple techniques (e.g., camouflage). It seems obvious that if detailed spectral information could be collected about every scene pixel, it might be possible to detect the difference between target and background pixels, particularly if it were known just which portions of the spectrum contained these crucial differences. For example, the green dyes used in the World Wars to camouflage soldiers hidden among the trees, while matching the green of the trees, had a completely different spectral response in the red—the red of autumn leaves is always present, just masked by the overwhelming chlorophyll green most of the year—and were easily detected by looking through a red filter.

Following this train of thought, so-called hyperspectral imaging sensors have been built with information collected at each image pixel, over tens to hundreds of individual spectral bands—some only a few nanometers wide. The inevitable result is a classic case of data overload with no chance at all for real-time response—at least, that is, until computer technology catches up. Given the pluses and minuses, it is not clear how valuable hyperspectral imaging will prove in the battlefield. Application to high-dynamic sensor/shooter/weapons scenarios seem unlikely, but longer-latency situational awareness might be considerably enhanced if the needed algorithms can be developed.

B.2.2.6 Optical Phased Arrays

Considering the performance advantages phased-array electronic beam steering has given to radar such that it completely dominates modern high-performance radar today, it is not surprising that considerable effort has been expended seeking ways to extend electronic beam agility to electro-optics. Given the minuscule dimensions of the wavelengths of light and the requirement to separate adjacent phase shifters by half-wavelength intervals in order to avoid grating lobes, it is easy to see that the challenge is formidable. Clearly an optical beam steering array cannot be implemented by assembling discrete elements as with a brick architecture radar phased array. Monolithic techniques, resembling those of the tile architectures, are called for.

Based on liquid crystal phase-shifting materials, which are optically transparent with electric field variable indices of refraction, combined with photolithographically deposited transparent electrode patterns, optical phased arrays have been developed and demonstrated in the past few years. Limited by current technology to small deflection angles of only a few degrees and to switching rates below 1 kHz, optical phased array technology is nevertheless impressive and promising. DARPA is now in the process of establishing a well-funded program to extend the angular capabilities of optical phased arrays to angles as large as $\pm 90°$.

B.3 SONAR

Acoustics—the propagation of sound waves through air, water, or solid ground—provides another remote sensing capability of crucial importance to the Navy because acoustic sonar permits us to "see" long distances underwater where many threats hide, but where optical sensors work poorly and radars, not at all.

As a wave phenomenon, sonar shares with radar and EO many of the same equations determining system performance, e.g., the relationship between aperture size and wavelength to beam width and angular resolution or between signal bandwidth and measurement accuracy.

Thus, from a system point of view, sonar is a familiar relative of radar and EO and can perform all the same functions—detection, classification, and localization of underwater and sea-surface targets, with the generation of situation awareness "images" of portions of the sea.

On the other hand, the medium that sonar systems have to deal with is absolutely terrible. Underwater sound propagation is almost never in a straight line because of strong medium nonuniformities associated with time and spatial variations in temperature and salinity. Reflections from the sea surface and bottom are common. And the sea is never free of acoustic noise of all kinds—from waves, from manmade objects like ships, and from biological sources like whales, porpoises, and fish.

The velocity of sound in water is very low—roughly 1 mile/s—which seriously increases the time needed to collect information for active sonar detection, classification, and localization of distant targets. Finally, the absorption of sound in water is a strong, increasing function of the acoustic frequency—low frequencies (e.g., ≤ 3 kHz) are needed for long range but cannot achieve high angular resolution because of the very large antenna sizes required and the unpredictable spatial variations in sound propagation. And although high angular resolution is possible at high frequencies (e.g., 35 to 350 kHz), it can be achieved only at fairly short ranges of several hundred meters or less. The result of these media-induced obstacles is that sonar performance in general is very slow, with image resolution and target location capabilities that degrade rapidly with range.

Because of their importance, and in spite of their many limitations as sensors, naval sonars are nevertheless ubiquitous throughout the battle group, since the underwater threats are many and real. Every ship or submarine has several—hull mounted or towed; active or passive; high frequency, medium frequency, or low frequency; and so on. And the battlespace usually contains a number of unmanned sonars—some permanently moored, e.g., strategic arrays; others temporarily drifting, e.g., sonobuoys; and others self-propelled, e.g., unmanned underwater vehicles (UUVs) or remotely operated vehicles (ROVs).

It is thus natural to consider increasing the fleet's antisubmarine warfare (ASW) capabilities by shifting from a platform-centric to a network-centric point of view. With the right communications, a widely dispersed set of sonar sensors can be made to emulate CEC and act as a single sonar system that, like the CEC radar, would be thought of as organic to a fleet of ships rather than to any individual platform. Since sonar, in contrast to radar, makes extensive use of passive detection, it is also natural to expect a mixture of passive and active modes throughout the network producing an even larger synergistic effect than would be obtained by operating all the sensors in the same mode. Clearly CEC

> **Box B.3 Hot Topics in Sonar**
>
> - Synthetic aperture sonar
> - Proliferation of cooperating unattended underwater vehicle platforms
> - Autonomous distributed systems—moored and drifting arrays
> - Ultra-broadband sonar—biologically inspired

should consider extending its operation to a mix of active and passive (i.e., bistatic) operations because of the antistealth and counter-countermeasure (CCM) advantages.

Current thrusts in sonar technology are described in Box B.3.

B.3.1 Sonar Performance

B.3.1.1 Resolution

Direct measurements of range can only be obtained from an active sonar, and the resolution is determined fundamentally, as is the case for active RF systems, by the time-bandwidth product of the transmitted signal and the signal-to-noise ratio. Because of the low operating frequencies characteristic of acoustics, time-bandwidth products (and hence pulse compression ratios) of sonar signals are typically on the order of 100 or so, whereas for radar, products of thousands to tens of thousands are common.

Although a wide range of range resolution performance is available by varying the amounts of pulse compression applied, active sonar resolution typically is controlled to match the size of the target sought in order to maximize the signal-to-noise ratio and hence the detectability of the return pulse. Too high a resolution causes the sonar to "see" only pieces of the target at a time and reduces the return signal maximum amplitude.

Angular resolution achievable by a sonar is determined largely by the diffraction properties of the antenna and again the signal-to-noise ratio. But the beam width alone is not the limit as fractional beam width accuracy is certainly possible through interferometric techniques or what is called in radar "monopulse." Because of the relatively large wavelengths associated with acoustic radiation, the phased arrays used for sonar typically do not have more than a few tens of elements (e.g., 10 to 40) along any direction, whatever the frequency range employed. Hull-mounted sonar, for example, such as the BQQ-5 submarine sonar, has a 15 ft diameter spherical array, whereas the SQS-53 sonar for large surface ships has a 16 ft cylindrical array and the SQS-56 surface ship

APPENDIX B

sonar, only a 4 ft diameter. At a frequency of 3.5 kHz, these dimensions lead to beam widths of 3° to 10°.

For high-frequency (500 kHz) imaging sonar, the antenna dimension need be only 64 cm long to obtain a 3° beam width, whereas a long-range search system emitting at 1 kHz would have to be 500 times longer (e.g., 320 m) to have the same beam width.

B.3.1.2 Field of View and Field of Regard

As phased arrays, all sonar antennas can be readily steered over large fields of regard. Linear arrays, hull mounted or towed, can achieve ± 60° or more, whereas some hull-mounted arrays are circular and can be steered through a full 360°. Long-range sonar exploits this capability by creating a few tens of beam positions at different angles so that the full FOR can be monitored simultaneously, each individual beam being a few degrees (e.g., 3° to 30°) in width, depending on the size of the array and the operating frequency. Side-scan imaging sonar generally is restricted to producing a narrow horizontal beam (e.g., 1° to as small as 1/5°) with a much broader beam in the vertical (e.g., 40°) and thus has a very small IFOV that is scanned forward by the motion of the platform. The IFOV for surveying can be as large as 15 knots or for classification (of mines, for example), as slow as 1 to 5 knots. Projected on the bottom, the IFOV of a side-scan sonar may be no larger than 30 m^2, e.g., a strip 1/5 m by 150 m.

B.3.1.3 Range

The effective range of a sonar is a strong function of the frequency used due to the quadratic increase in water absorption of sound with increasing frequency. Sonar utilizing low frequencies of 3 kHz or less can often detect targets up to 100,000 yd (~50 nautical miles) or more. But because of the unpredictable properties of ocean propagation, with possibilities for refraction down until a reflection off the sea bottom occurs (i.e., bottom bounce) or refraction down followed by refraction back up to the surface (i.e., convergence), it is very difficult to determine exactly how far away the acoustic source or reflection is. In the so-called convergence zones, detection can be excellent, even though the ranges are large. However, these zones can be quite narrow (~1 percent of the range); and between the zones, which can repeat at intervals of 40,000 to 80,000 yd, depending on which of the world's oceans the sonar is operating in, nothing much can be detected. The bending effects of the ocean gradients result in "blind" regions between convergence zones, within which targets cannot be detected by the sonar—that is, volumes of water that cannot be reached by acoustic beams radiated by the sonar because they are refracted away from and around these regions.

Passive sonar, of course, cannot directly determine range at all—only azimuth or bearing to the target can be estimated. Because of the slow dynamics of vehicle motion in the ocean, it is a common practice for the sonar, after receiving a contact on a certain bearing, to have the ship turn and take a run to a different location and then use the bearing estimate from the new location to triangulate with the earlier bearing to obtain a rough location of the acoustic source. Several cooperative passive sonars, viewing the same target simultaneously from sufficiently separated physical locations in a network-centric sensor grid, could, of course, provide instantaneous, more accurate localization.

B.3.1.4 Imaging for Mine Location

Side-scan sonar operating at high frequencies (e.g., 100 to 500 kHz) can produce good images of mines for detection and classification, with resolutions on the bottom 10 to 30 cm. However, because of absorption, such performance is limited to ranges from the platform of only 100 to 200 m—a distance that can lie within the lethal range of the mines. Often a compromise is chosen, searching first at a medium frequency (e.g., 35 to 100 kHz) which allows a reasonable standoff distance and good detection possibilities, followed by a slow-speed, closer-range, high-frequency imaging pass for classification or even the use of an ROV or swimmer. Anyway it is done, it takes an enormous amount of time. Exploitation of the two-dimensional acoustic shadows of the mines have been used successfully to produce effective ATR algorithms for mine classification.

B.3.1.5 Geopositioning Accuracy

Except at mine-detecting ranges (i.e., within a few hundred meters of the platform), the ability of sonar to determine a target's position in range and azimuth is extremely poor. With degrees of beam width, unknown paths of propagation, and great range uncertainty, sonar is of little use for geolocation. Unfortunately, for underwater targets, sometimes that is all that can be done.

B.3.1.6 Area Coverage Rate

Active sonar for medium or long range is limited by the long round-trip return time of the transmitted energy—for a range of 20 nautical miles, the round-trip time is about 23 sec. If the sonar explores the 20 nautical miles radius ± 60° FOR permitted by the phased array by 30 different 4° beams, the area coverage rate is only about 2 km^2/s. This should be compared with an above-the-surface surveillance radar that can cover a 200 nautical miles range by 120° segment in only 4 s for an area coverage rate of 36,000 km^2/s.

On the imaging side, route surveying for mines with a side-scan sonar in the push-broom mode with 2000 m wide swath and moving at 15 knots has an area coverage rate of only 1/65 km²/s. And a mine-classification high-frequency sonar, covering a swath of 150 m at a speed of 5 knots, has an even smaller area coverage rate of only about 400 m²/s, which is a minuscule 1/2500 km²/s or only 1.44 km²/h.

Finding something in or on the bottom of the ocean can take a very long time, and because of physics, our accelerating technology has not yet been able to overcome the obstacles.

B.3.1.7 Communication Data Rate Requirements

In spite of the large amounts of computational resources and time that have to be expended in order to extract meaningful information from active or passive sonar signals, the resulting data are so sparse (i.e., not many target-like objects within range at any given time) and are collected so slowly and with such low resolution that the resulting data rates to transfer one sensor's data to another location put no strain at all on an RF communication link. For example, a medium- or long-range surveillance system might need to transfer a video screen-worth of data (say, 400,000 pixels at 8 bits) every 5 or 10 s, resulting in a data rate no larger than 650 kbps. Even a high-resolution mine-hunting side-scan sonar, producing a 20 cm pixel over a 150 m swath on the bottom and moving forward at 5 knots (i.e., ~2.6 m/s) results in only about 10,000 pixels/s (say, 10 bits each) for a total rate of only 100 kbps.

On the other hand, if we wished to employ an acoustic communication link, say, between cooperating UUV platforms, the necessary data rates could stress the system, and further local processing with ATR-like algorithms and perhaps the application of data compression techniques would be called for.

B.3.1.8 Environmental Issues

The effects of the low acoustic propagation velocity, the media variability and inhomogeneity, and the frequency-dependent absorption that seriously limit sonar performance in the open ocean have been discussed. Operation in the littoral, because of the shallow depths and coastal waves, greatly aggravates an already difficult sensor problem because of the more frequent reflections from the top and bottom surfaces of the water, the absorption in and irregularity of the bottom, and the high levels of wave and surface noise characteristic of this environment. Sonar designed for the open ocean does not work well, if at all, in the littorals. New designs that properly account for the physical characteristics of the littoral need to be developed, although it is far from clear that major breakthroughs in single-sensor performance are possible. More likely the problem requires a distributed solution with large numbers of short-range sensors acting cooperatively in a network-centric mode.

B.3.2 Technology Trends and Future Growth in Sonar

B.3.2.1 Synthetic Aperture Sonar

The technology of conventional side-scan sonar for the imaging of mines has saturated, as it were, in that the range capabilities and the area coverage rate of several nautical square miles per hour are strongly limited by the acoustic velocity and attenuation properties of sea water—not by technology per se. The 10 to 20 cm resolution obtainable from practical antennas is considered more or less adequate for mine detection and classification. It is in this context that interest in synthetic aperture sonar (SAS) has been reawakened in the past several years, with the hope that the imaging resolution and range (and hence area coverage rate) might be increased through the use of low-frequency acoustic signals and very large virtual antennas, without giving up the 10 to 20 cm resolution.

Based on the very same principles as microwave SAR, the possibility for SAS was suggested early, 30 to 40 years ago. Nevertheless, very little real progress has been achieved in the intervening decades, for the same physical obstacles (e.g., media inhomogeneity, extreme temporal variability, and the slow 1 mile/sec acoustic propagation velocity) that afflict all sonar applications present even larger challenges for the coherently processed SAS. For example, the precise position of the sonar transmit-and-receive components must be known over the whole length of the virtual antenna, to fractions of the acoustic wavelength—e.g., to millimeters if a relatively low mine-hunting frequency of 100 kHz is used where $\lambda = 1.6$ cm. Because of the difficulties in achieving this positional knowledge, autofocus algorithms have been evoked which work to some extent but are so computationally expensive that they cannot yet be carried out in real time. Nor is it clear, because of media temporal fluctuations, that signal coherence can be maintained over the time it takes to traverse the full virtual antenna.

Nevertheless, interest in SAS is high at the moment, but it is fair to say that it is still "in its infancy" today. Because of the difficulties in the physics, not very much has yet been demonstrated except under closely controlled conditions.

B.3.2.2 Cooperating Unmanned Underwater Vehicles

In the spirit of network-centric operations, concepts are being actively explored for mine hunting involving multiple UUVs working in parallel. Equipped with very-high-frequency, 1 MHz or higher, side-scan imaging sensors and, perhaps in the future, high-resolution laser optical imaging systems, with ranges of only a few tens of meters at best, these mobile underwater sensors are perfect candidates for cooperative networking. Although these sensors operate independently at the moment, future cooperation through surface RF or underwater acoustic communication links, leading to a distributed and adaptive metasensor concept, should produce the synergistic multiplication of capabilities and effectiveness expected from networked concepts.

B.3.2.3 Autonomous Distributed Systems

Distributed arrays of acoustic sensors with elements less capable than the mobile UUVs suggested above are also of great interest for littoral or shallow water (100 to 500 m) area surveillance. Deployed like sonobuoys and drifting or moored in place, each node is envisioned to possess considerable on-board signal and data-processing resources and to be capable of passive automatic detection and classification as well as active signal processing. These individual capabilities, complemented by the networked acoustic/RF communications, would permit them to operate as a single large and very capable coherent distributed sonar. A typical multistatic configuration might consist of a few active sources and as many as 10 to 100 "smart" sonobuoy-like nodes distributed broadly over the area under surveillance.

In addition to the loosely coupled multistatic systems of active sources and smart sonobuoys, various physically connected drifting or moored arrays are under consideration. The autonomous drifting line array would consist of a very-low-frequency passive array of up to 100 hydrophone elements, drifting freely in the ocean currents. Data from the linear, but almost certainly not straight, aperture would be processed on board the array through battery-powered computer resources and would report detection/classification information back to the decision makers via an RF link as necessary.

Other concepts envision similar autonomous arrays moored in shallower water. Key to the success of all these concepts is the availability of significant local on-board sensor processing supported by long-life batteries.

B.3.2.4 Ultra-Wideband Sonar

In spite of our technology, animal sonar, as utilized by bats and porpoises, far exceeds our capabilities for precise location and target classification. Operating in complicated environments, in the presence of perhaps dozens of competing individuals, these animals emit complex, very broadband sonar pulses that they change rapidly as they move from detection to classification and final capture of prey, apparently adapting their internal signal-matched filter on a pulse-to-pulse basis, easily sorting out their own signals from those of other bats or porpoises.

Since the early 1990s, building on pioneering university studies of bat sonar, the Navy has sponsored efforts to develop a biologically inspired, ultra-wideband sonar, using multioctave signals and multichannel nonlinear processing with coherent recombination. Computer studies applying such processing to existing narrowband field data have already demonstrated an encouraging reduction in false alarms. The development of transducers capable of emitting the desired ultra-wideband signal is under way, and it is hoped that soon we will be able to duplicate in the littorals at least some of the extraordinary capabilities common in nature.

C

System Requirements to Hit Moving Targets

The committee presents here an example of the recommended system engineering that focuses on solving the warfighter's problems and thereby derives the characteristics of the component systems instead of starting with these characteristics as a "requirement." An acute problem at present is that of hitting moving targets on Earth's surface. Surveys show that moving targets normally constitute a high percentage of the targets in theater; tanks, armored personnel carriers, and patrol boats are examples. An important specific case is a high-value target such as a missile transporter-erector-launcher that is usually hidden when stationary and therefore vulnerable to attack only when on the move. The committee conducted an example analysis to accomplish the following:

• Quantify requirements on various concepts for end-to-end systems to hit moving surface targets, considering a range of realistic environments and target behavior;
• Explore trade-offs in how to balance the burden of performance among system elements; and
• Examine how networking concepts can be employed to achieve system requirements.

The committee is aware that the Defense Advanced Research Projects Agency (DARPA) has established the Affordable Moving Surface Target Engagement program with similar objectives. The DARPA program has just begun; when results become available, they can be used as a more concrete basis for the

APPENDIX C   385

employment of networking concepts. Until then, this analysis provides a preliminary basis.

The specific problem to be solved is that of hitting a moving surface target among randomly distributed false contacts (real physical objects that can be confused with the intended target). The intended target deliberately maneuvers to avoid engagement.

The committee considered three weapon system concepts:

1. The weapon launch platform (e.g., a manned aircraft) carries a complex sensor that can acquire the target at long range and can usually distinguish target from false contact,
2. The weapon (e.g., a future cruise missile) carries a simple seeker that can acquire the target at short range and cannot distinguish target from false contact, and
3. The weapon (e.g., a Global Positioning System (GPS)-guided bomb with command data link) is delivered without reacquisition of the target.

Since moving targets are often numerous and individually of low value, inexpensive weapons are desirable. In the first concept, the weapon could be inexpensive, but the launch platform cost and the risk to pilots are also factors. To contain aircraft cost, the Joint Strike Fighter program office is conducting trade-off studies on how much targeting capability is needed on board versus how much can be obtained from off-board sources. The weapon for concept 3 can be less expensive than the weapon for concept 2. However, targeting system cost must increase to meet the demands of the simpler weapon. This is one of the key trade-offs to be examined.

C.1 ANALYSIS OF SYSTEM REQUIREMENTS

The mathematical model employed in this analysis is explained at the end of this appendix. The model builds on one used for a previous Naval Studies Board report[1] that showed that the targeting system should provide a steady stream of reports to the weapon, as opposed to a single report. The targeting system must be able to (1) classify a target and (2) associate multiple reports with a single track. With these capabilities, a targeting system can then provide a steady stream of reports that enable a tracking filter to estimate speed and heading.

Central to the analysis are the models for target tracking and target reacquisition by the weapon or weapon launch platform. Figures C.1 and C.2 show the methodology for system requirements to kill moving targets. Figure C.1 depicts

---

[1]Naval Studies Board, National Research Council. 1993. *Space Support to Naval Tactical Operations (U)*, 93-NSB-494. National Academy Press, Washington, D.C. (classified).

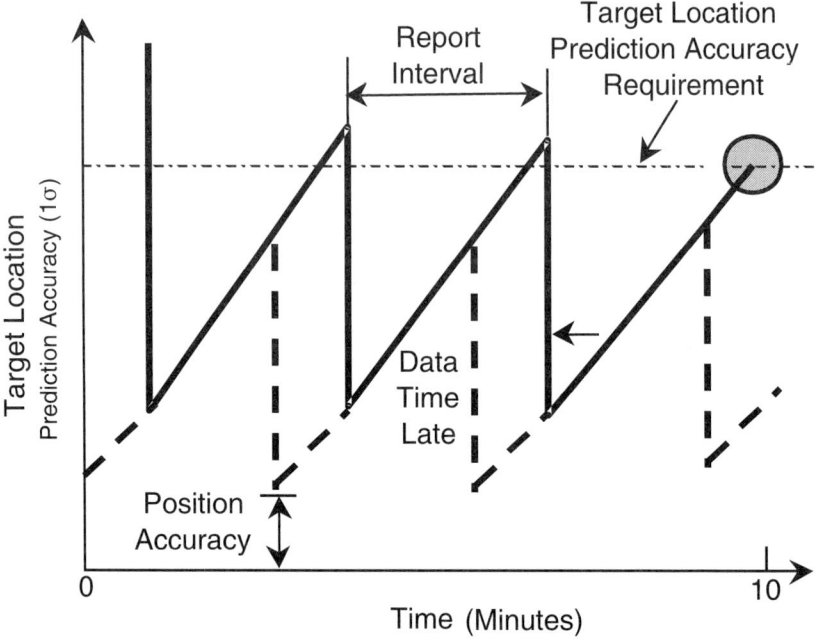

FIGURE C.1 Tracking the target.

the tracking model. The targeting system is characterized by three parameters: the position accuracy (error in measurement of target position at each update); report interval (time period between updates); and data time late (elapsed time from measurement to receipt of update by weapon or launch platform). Figure C.1 plots the target location prediction accuracy as a function of time. The committee assumes as a worst case that the weapon or launch platform arrives at the target just before an update. Figure C.2 illustrates the target reacquisition process. The assumption is that the weapon or launch platform (for weapon concepts 1 and 2, listed above) uses a search pattern that begins at the predicted target location and expands outward. Acquisition of the target requires that (a) the target is inside the sensor or seeker area of regard and (b) the search finds the intended target before a false contact is misclassified as the target. The probability of satisfying these two conditions depends critically on the target location prediction accuracy. In another report[2] is a discussion on other search patterns

---

[2]Kalbaugh, D.V. 1992. "Optimal Search Among False Contacts," *SIAM Journal of Applied Mathematics*, 52(6): 1722-1750.

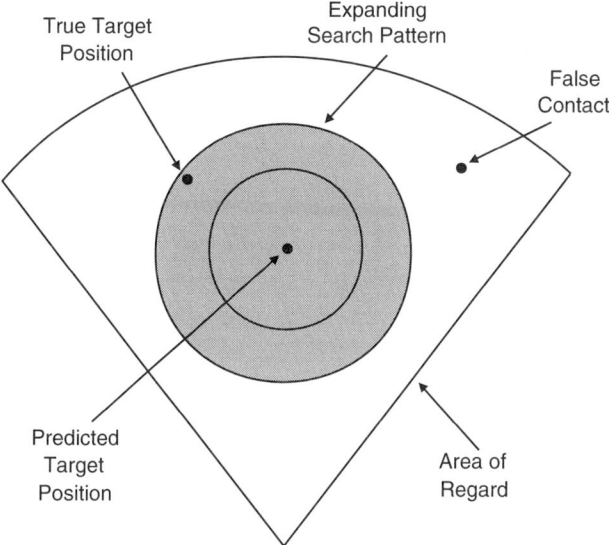

FIGURE C.2 Reacquiring the target.

and tactics one can employ to find a target in the midst of false contacts, depending on the kinds of targeting data available (e.g., the accuracy of information on location of false contacts) and mission objectives (e.g., limitations on collateral damage).

C.2 COMPLEX SENSOR FOR REACQUISITION

The most straightforward concept involving a complex sensor for target reacquisition is a manned aircraft. This is the only feasible method of hitting a moving target today. The Air Force's F-15E may currently be the most capable U.S. platform for this mission. It carries an active radar with synthetic aperture (SAR) and ground moving-target indicator (GMTI) modes and can carry an electro-optical low-altitude navigation and targeting infrared for night (LANTIRN) pod. The Navy has no tactical aircraft with SAR capability. F-18 aircraft are about to undergo an upgrade to provide a SAR capability, but the SAR output will go only to a tactical reconnaissance pod, not the cockpit. A future concept in this category is an uninhabited combat air vehicle (UCAV) with high-resolution sensors and video data link to a human controller. Each of these concepts can employ the human eye and mind in the very difficult task of target recognition.

Figures C.3 and C.4 present system requirements for the conceptual weapon or launch platform that carries a complex sensor for target reacquisition. Figure

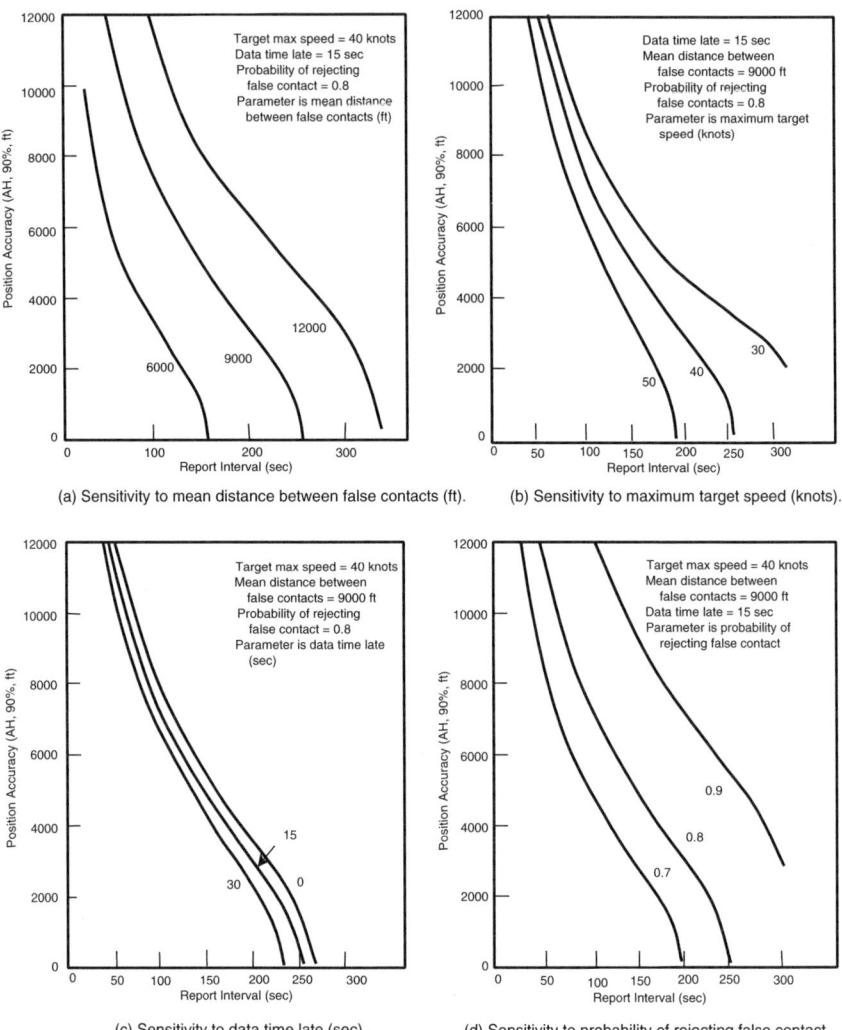

FIGURE C.3 System requirements to hit moving targets, given a complex sensor for reacquisition and a light background.

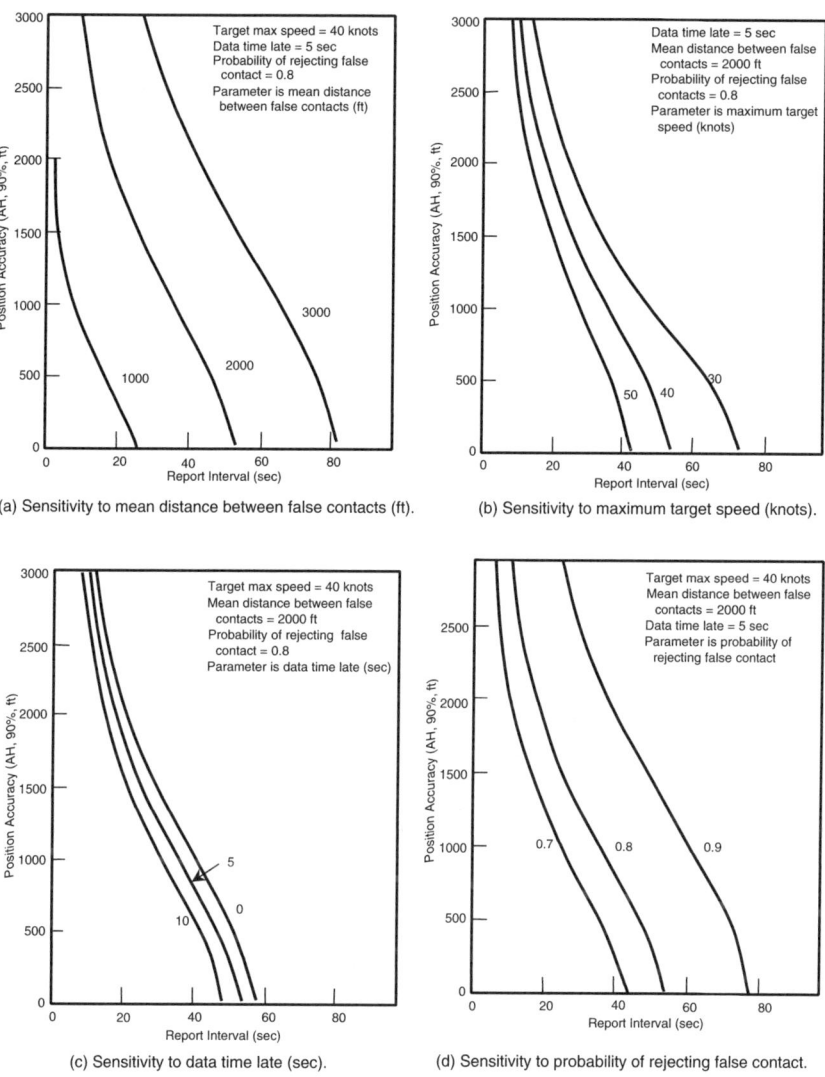

FIGURE C.4 System requirements to hit moving targets, given a complex sensor for reacquisition and a dense background.

C.3 applies for a light density of false contacts; Figure C.4, for a dense background. Each figure contains four graphs, each with a different sensitivity analysis. Graph (a) shows the sensitivity to mean distance between false contacts; graph (b), the sensitivity to target maximum speed; graph (c), the sensitivity to data time late; and graph (d), the sensitivity to the probability of rejecting a false contact. Each graph plots the combination of position accuracy and report interval yielding a target location prediction error sufficiently small that the intended target is reacquired 90 percent of the time.

It is assumed that the complex sensor can acquire the target at a range on the order of 10 nautical miles. With this detection range, the sensor's area of regard is so large that the probability of target reacquisition is determined entirely by the probability of finding the intended target before a false contact is misclassified as the target. Hence the driving parameters are the density of false contacts and the complex sensor's target recognition capability. Target recognition capability is characterized by two numbers. The question is, What is the probability that the sensor (or human-sensor combination) correctly recognizes an object under two situations: (1) given that the sensor is looking at the intended target, and (2) given that the sensor is looking at a false contact? For operational situations where the need to hit the target dominates the risk of collateral damage, the probability for the first situation must be high; the probability for the second situation can vary.

A key conclusion drawn from Figures C.3 and C.4 is that system requirements are driven by the environment. The density of false contacts is primary, and the target motion characteristics also have an influence. A major trade-off exists between (1) the targeting system's accuracy and frequency of reporting and (2) the weapon launch platform's capability to distinguish the intended target from a false contact. This is especially true in a very dense environment.

For later comparison, consider the requirements to hit a 40 knot target among false contacts 3,000 ft apart. If the launch platform with complex sensor is capable of rejecting false contacts 80 percent of the time, it requires a report every 75 s if accurate to 500 ft and less than 5 s late.

C.3 SIMPLE SEEKER FOR REACQUISITION

A possible concept involving a simple seeker for target reacquisition is a future cruise missile with targeting data link and multiple submunition packages so that the cost of a cruise missile is offset by multiple target kills. A second concept is a joint standoff weapon (JSOW) with targeting link and simple seeker. The weapon could have a video link to launch platform to help with target acquisition.

It is assumed that the simple seeker can detect the intended target at a range of 1.5 nautical miles and nominally has no capability to distinguish target from false contact. Presumably, it does not have the same high resolution as the complex sensor. Figures C.5 and C.6 present system requirements for a weapon carrying a simple seeker. The format is the same as described above for Figures C.3 and C.4.

Figure C.4 shows that, with the simple seeker's short detection range, the area coverage requirement does come into play. However, for the most part, the density of false contacts continues to be the driving parameter. Graphs (c) and (d) in Figures C.5 and C.6 show that a little time delay does not hurt and a little target recognition capability does not help. (The probability of rejecting a false contact must be on the order of 0.6 or 0.7 to relax targeting requirements significantly.) Again, system requirements are driven by the environment.

For later comparison, again consider the requirements to hit a 40 knot target among false contacts 3,000 ft apart. The weapon with simple seeker requires a report every 30 s if accurate to 500 ft and less than 2.5 s late.

C.4 NO REACQUISITION

A weapon that can be command guided into a target without benefit of a seeker is an appealing concept for cost reasons. Air-launched examples might include the joint direct attack munition (JDAM) with its unitary warhead, or a JSOW with its submunitions, each modified to include a data link. Ship-launched examples might include the extended-range guided missile (ERGM), also modified to include a link. What are the targeting requirements?

Figure C.7 shows that they are severe, especially for a precision weapon that has a small lethal radius to minimize collateral damage. The weapon needs a fresh report, accurate to tens of feet, every few seconds. Every second of delay in delivering the data hurts.

Once again for later comparison consider the requirements to hit a 40 knot target. For a weapon that does not reacquire the target, the density of false contacts is not a factor. A weapon with a 200 ft kill radius requires a report every 5 s if accurate to 70 ft and less than 1 s late.

C.5 TARGETING SYSTEM DESIGN CONSIDERATIONS

The technology currently most capable of detecting and tracking surface targets is active radar with synthetic aperture and ground-moving target-indicator modes. As mentioned, the F-15E has these capabilities for targeting its own weapons. Among wide area surveillance platforms, the Air Force's Joint Surveillance and Target Attack Radar System (JSTARS) aircraft employs these techniques today, and the high-altitude, long-endurance unmanned air vehicle Global Hawk is planned to incorporate them in the future. DARPA's Discoverer II Program has the objective of fielding two satellites to demonstrate the feasibility of an affordable constellation of satellites with SAR/GMTI capability. Difficulty in classifying targets sometimes requires use of additional data such as electro-

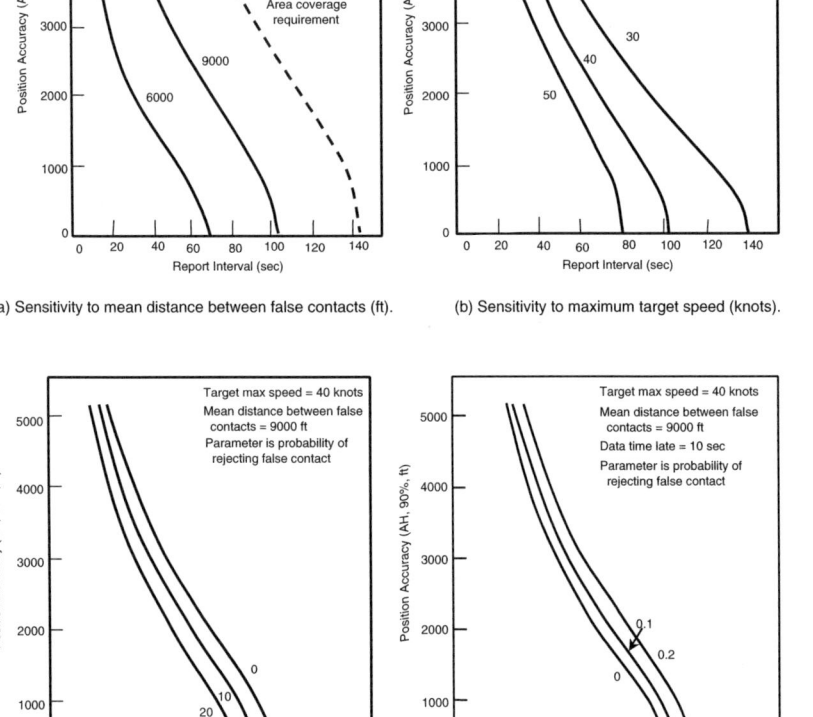

FIGURE C.5 System requirements to hit moving targets, given a simple seeker for reacquisition and a light background.

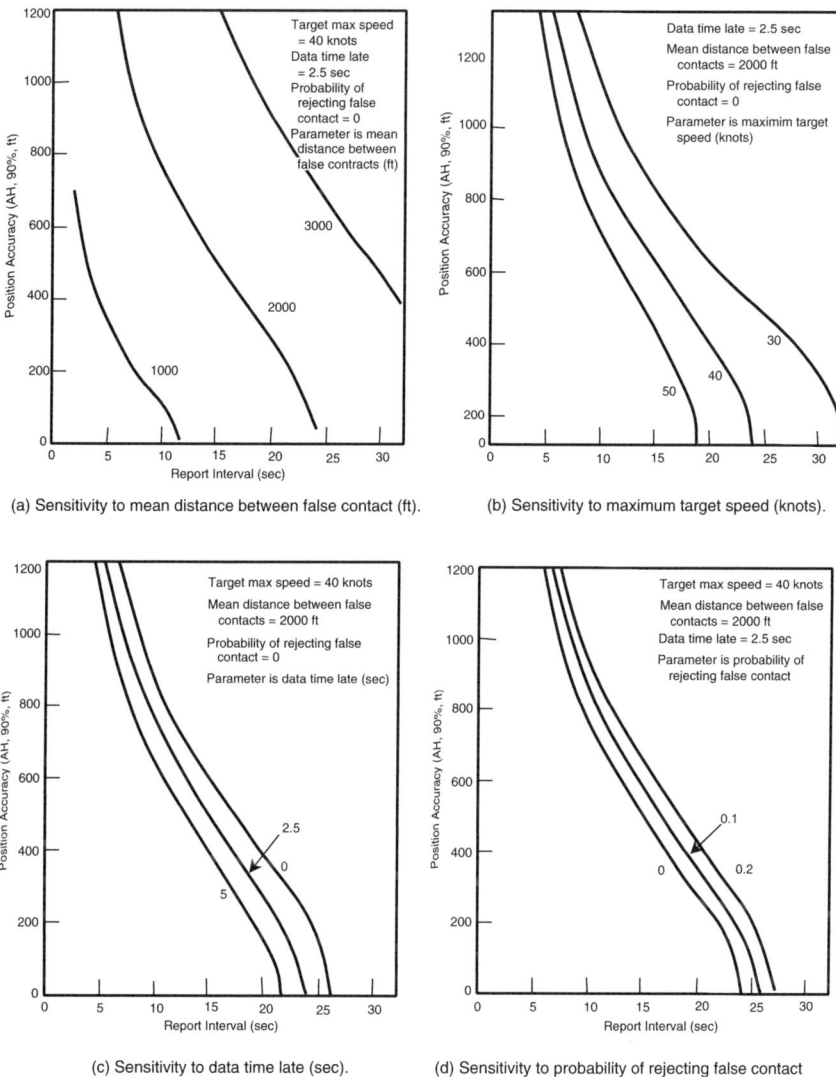

FIGURE C.6 System requirements to hit moving targets, given a simple seeker for reacquisition and a dense background.

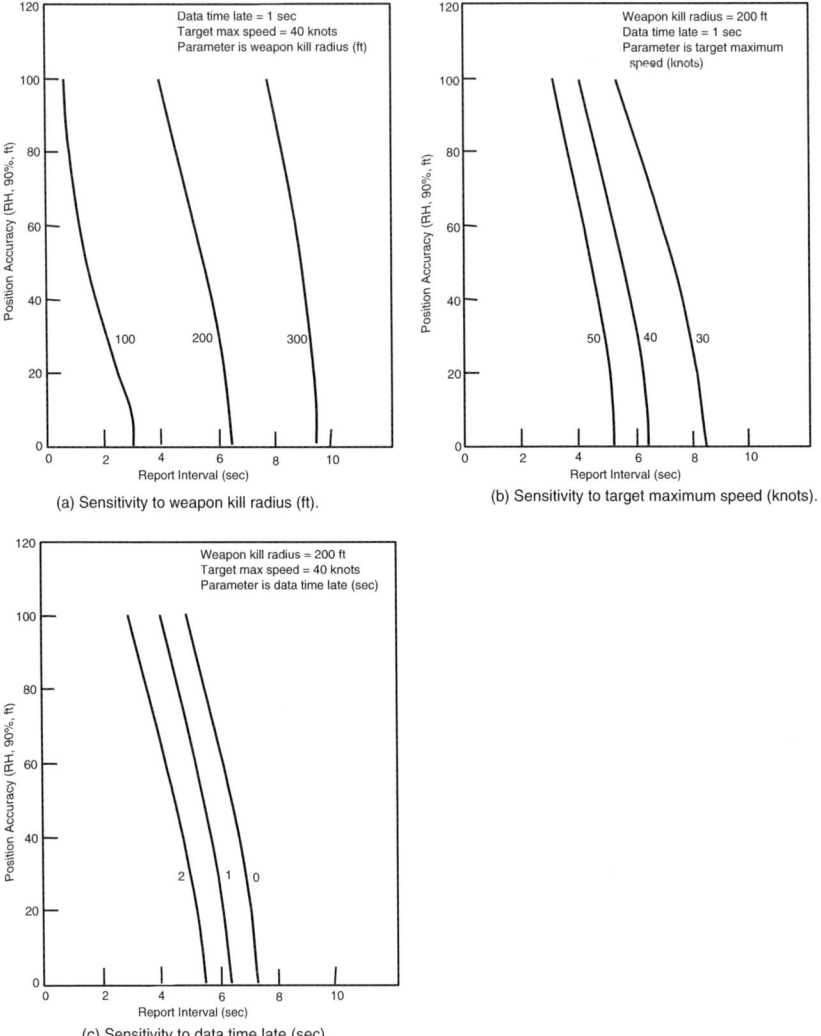

FIGURE C.7 System requirements to hit moving targets; no reacquisition.

optical imagery. SAR/GMTI sensors, especially on airborne platforms, can be subject to terrain masking in rougher terrain or urban areas. Frequency of reporting must be traded off with the size of the area surveilled; one JSTARS aircraft can provide reports every 30 s or so over thousands of square miles. Target location measurements from the aircraft are more accurate in the range dimension

APPENDIX C   395

than the azimuth; one JSTARS aircraft at long range can easily meet an accuracy of 500 ft (90 percent) in azimuth.

The committee notes in passing that the Navy has no SAR/GMTI capabilities and no formal plans to develop them. However, the E-2 Advanced Development Office has an interest in a program for a new radar that would incorporate SAR/GMTI capability in that aircraft.

C.6 EXPLORATION OF TRADE-OFFS

Given the capabilities of the JSTARS aircraft, and assuming that the capabilities of other future airborne and spaceborne systems will be similar, one can compare the three weapon concepts and draw some conclusions about them.

First, the requirements to target the weapon with simple seeker are not onerous compared with those to target a launch platform with complex sensor. In the specific case cited, 500 ft position accuracy was chosen as a baseline. Then the complex seeker reacquires the target if given a report every 75 s and no more than 5 s late. The simple seeker reacquires the target if given a report every 30 s and no more than 2.5 s late. Targeting requirements should not be difficult to meet for either of the weapon concepts that reacquire the target when distance between false contacts averages more than 3,000 ft. Note that in some important situations, e.g., a military convoy containing a high-value target, false contacts will be much closer than this (e.g., 150 ft). These situations require either much more accurate targeting or an effective automatic target recognition capability on weapon or launch platform.

The trade-off between concepts 1 and 2 then devolves to comparing the advantages and disadvantages of directing a manned launch platform with an inexpensive weapon into the target area versus having a launch platform standoff and delivering a somewhat more expensive weapon. There may be a place for both concepts, depending on the operational situation, driven by either the air defense threat in the target area or the density of false contacts. One basis for such a belief is the Navy and Air Force commitment to both JDAM and JSOW, which, against fixed targets, offer the commander a choice between direct attack and standoff from point defenses.

For the concept of a weapon that does not reacquire the target the targeting requirements depend critically on the weapon's kill radius. A weapon with 200 ft kill radius requires a report every 5 s if accurate to 70 ft and less than 1 s late. These are difficult targeting requirements to achieve.

C.7 EMPLOYMENT OF NETWORKING CONCEPTS

For the two weapon concepts described above that reacquire the target, system requirements are driven by the environment, principally the density of false contacts. How can one design a system for all likely environments? Design for very dense environments (say, false contacts less than 1,500 ft apart) would be overdesign by large margins for less stressing cases and appears to be prohibitively expensive given the nature of the targeting platforms and sensors. The answer may be to design an end-to-end system that accomplishes the following:

1. Assists a strike commander in making quick decisions on which targets should be the highest priority to hit;
2. Incorporates an online performance prediction to enable the commander to judge the likelihood of success in prosecuting an attack against a specific target at a specific time, given the current deployment of targeting system and weapon system assets; and
3. Enables the commander to move and focus targeting system and weapon system assets in near-real-time to accomplish the high-priority goals.

In other words, the answer may be to provide the commander with the tools to control assets flexibly in order to tighten the targeting-system-to-weapon-system loop when necessary.

For the concept of the weapon that does not reacquire the target, the committee found that targeting requirements were severe. Can networking enable the requirements to be met? The committee believes there are several networking concepts that may help, at least for weapons with larger lethal radii (e.g., 200 ft). Fusion of data from multiple sensors at different geometries can greatly improve the accuracy of the target position measurement, taking advantage of the radars' precise range estimates. The targeting data can be put into a common navigational coordinate system by communicating among all targeting and weapon system platforms to control the specific GPS satellites they all track.

Will an effective system for hitting moving ground targets be like the cooperative engagement capability (CEC)? Detailed studies and experiments are necessary to answer this question adequately, but one can make some observations. Figure C.4 compares CEC and a notional future architecture for a system to hit moving ground targets. There are some major differences and some strong similarities. The differences come about for two reasons. First, CEC links ships that have sensing, processing, and weapon capabilities, whereas the system to hit moving ground targets will likely be composed of *different kinds* of platforms for sensing, processing, and weapon delivery. Second, ground targets and their environment differ from air targets and their environment. The similarities come about because some of the techniques CEC uses to create tracks from measurement data provided by distributed sensors appear applicable. Prominent CEC attributes include (1) one communication system linking all participants, (2) decentralized architecture with all participants receiving all measurements and processing them all the same way to achieve the same tactical picture, (3) composite tracking (i.e., development of a target track from measurements of distributed sensors), (4) gridlock registration (i.e., comparing pictures among participants to correct for relative navigation errors and misalignments), (5) all data of one kind, (6) high communication data rates, and (7) in the future, accuracy refinement.

C.7.1 One Communication System Links All Participants

Will a system that is effective for hitting moving ground targets be like CEC in utilizing only one communication system to link participants? Answering this question requires asking who is a participant. In CEC the answer is clear; CEC links *ships* that have the needed sensor, processing, and weapon capabilities. One communication link, the Data Distribution System, suffices. A future system to hit moving ground targets will probably use one link for communicating sensor measurement data to a processing facility and another for communicating targeting data from processing facility to weapons. SAR/GMTI sensor platforms will likely use a common data link (CDL) to transmit measurements to a processing facility. For most of the cases, the Joint Tactical Information Distribution System (with its 12-s refresh rate) appears to be adequate for line-of-sight communication from processing facility to the weapon launch platform, which may use other links to the weapon to update it in flight. Over-the-horizon communication from processing facility to aircraft or cruise missiles will require satellite communications, e.g., the ultrahigh-frequency satellite communications link planned for Tomahawk Block IV, which can support report intervals at least as short as 9 s. In summary, the future system to hit moving targets will likely utilize more than one communication link.

C.7.2 Decentralized Architecture with All Participants Receiving All Measurements and Processing Them All the Same Way to Achieve the Same Tactical Picture

One of the principal advantages of a decentralized architecture is robustness; single point failures are eliminated. Decentralization is more natural when, as in CEC, all participants are similar. Future platforms to detect moving ground targets are likely to be quite diverse: large, highly capable manned aircraft; large, high-endurance unmanned aerial vehicles (UAVs); smaller medium-endurance UAVs; and perhaps a constellation of (low-cost) satellites. Expecting the same performance of each appears unreasonable. Furthermore, the ground target is more difficult to identify than the air threat, whose motion alone may betray its identity; therefore, human observation, interpretation, and decision making will be key. For these reasons, collection and processing of data are likely to occur in manned facilities, ground-based or airborne. That is, manned facilities will create and identify the tracks. Therefore it will be unnecessary for an unmanned sensor to receive measurement data from other sensors. Today's JSTARS operations include processing on the aircraft and in multiple, distributed, decentralized, ground-processing stations focused on specific geographic subareas. Geographically focused processing units may continue to be useful, wherever they are located. Each of these geographically focused units should be *capable* of receiving all measurement data from all sensors for the robustness that redundancy provides. However, it will typically *not* be processing data to achieve the same tactical picture as other units, but rather the tactical picture for one geographic locale. In summary, the future system to hit moving targets will probably not have a sensor net like CEC. It may have decentralized processing facilities that are capable of receiving all measurement data and typically process all available data for a specific locale.

C.7.3 Composite Tracking

A system designed for tracking moving ground targets will probably benefit considerably from composite tracking if several geographically dispersed SAR/GMTI platforms are covering the same area. Composite tracking will help with identification maintenance, terrain obscuration, and minimum Doppler velocity dropouts. It comes naturally when measurement data from multiple platforms are processed at one facility.

C.7.4 Gridlock Registration

Use of GPS could also solve the relative navigation issue inherent in fusing data from various platforms, but CEC's gridlock registration methodology reduces dependence on GPS and simultaneously solves misalignment problems.

C.7.5 All Data of One Kind

The future system for hitting moving ground targets will not sense and process just one kind of data. In addition to GMTI measurement data, SAR images and other attribute data will be sensed and processed, and interpretation of these data will be a key function.

C.7.6 High Communication Data Rates

As discussed above, the future system will use several data links, some of which need not have the same capability as the CEC's Data Distribution System.

C.7.7 Accuracy Refinement

As discussed above, accuracy refinement would appear to be a requirement if the weapon or its launch platform does not reacquire the target. Otherwise the feature may help with tracking in dense traffic. Additional studies are needed to determine this.

C.8 THE AUTOMATIC TARGET RECOGNITION CHALLENGES

Targeting requirements can be met rather easily provided that the weapon or launch platform has some capability to recognize the target (and density of false contacts is not too great). Target recognition is not an easy task, but imperfect capabilities can suffice. Some degree of automatic target recognition is key.

For a launch platform with a complex sensor, the committee determined targeting requirements as a function of the probability that the combination of sensor and human observer will reject false contacts, i.e., distinguish false contacts from the intended target. Although the human eye and mind are unexcelled at such a task, it is likely that the aircraft will have only a single seat, that the pilot's attentions will be divided, and that the pilot would not be, under the best of circumstances, as expert as an operator on a surveillance aircraft. The complex sensor and its processing system should give the pilot every available aid to find the target and reject false contacts.

For a weapon with a simple seeker, the key capability is the ability to recognize the target against the earth (or water, or urban street, and so on) background. The committee found that a little capability to distinguish false contact from intended target is essentially of no help. It is possible that the weapon could have a video link to a manned platform, which could enable a human eye and mind, again, to help find the target. In any case, it is imperative that the seeker have an autonomous capability.

C.9 ANALYSIS SUMMARY

For weapon concepts in which targeting system data are used to cue a sensor that reacquires the target, targeting requirements are often driven by the density of false contacts, objects that can be confused with the intended target. For environments in which false contacts are separated by 3,000 ft or more on the average, targeting requirements can be relatively easily met (e.g., with one JSTARS aircraft), even if reacquisition is accomplished by a weapon with short-range seeker and no capability to distinguish false contacts from intended target. Denser false contact environments require either a launch aircraft with complex sensor and good target recognition capability or more frequent, more accurate, and more timely targeting reports.

How can the Department of Defense design a system for all likely environments? The answer may be to design an end-to-end system that enables the commander to predict performance on line and control assets flexibly in near real time in order to tighten the sensor-to-shooter loop when necessary.

For the concept of the weapon that does not reacquire the target, the committee found that targeting requirements were severe but that several networking concepts might enable them to be met, at least for weapons with larger lethal radii (e.g., 200 ft). Fusion of data from multiple sensors at different geometries can greatly improve the accuracy of the target position measurement, taking advantage of the radars' precise range estimates. The targeting data can be put into a common navigational coordinate system by communicating among all targeting and weapon system platforms to control the specific GPS satellites they all track.

C.10 TARGET TRACKING MODEL

The committee presents here the model that was used to derive the results displayed in Figures C.1 through C.7. It applies to targeting data and target motion in the case where an on-station observer provides a sequence of position reports.

The targeting system must be able to classify a target and associate multiple reports with a single track. In general, it must be able to do this in a multitarget environment. With these capabilities, a targeting system can provide a steady stream of reports that enable a tracking filter to estimate speed and heading.

If a target is aware or suspects it is being tracked and is aware of the technical characteristics of the tracking system (update rate and position accuracy), it can maneuver in order to maximize the tracking system's error. In doing so, it must trade off its ability to advance rapidly. This analysis assumes that the target velocity consists of a steady component and maneuver component. The steady component is a fixed-speed and uniformly distributed heading held constant throughout the search. The maneuver velocity is a random process with root mean square value set so that the target neither loses ground on its intended track nor moves faster than its maximum speed (most of the time). The target's frequency of maneuvering is then set to maximize the tracking system's error.

In what follows, attention is confined initially to a single axis of motion, and then, with appropriate choice of parameters, the results are extended to two dimensions.

Define

x = target position,
S = steady component of velocity, and
v = maneuver component of velocity.

Assume equations of motion as follows:
$\dot{x} = S + V$,
$\dot{S} = 0$, and,
$\dot{v} = a - \omega v$,

APPENDIX C

where
  a = white noise acceleration, and
  ω = maneuver frequency.

Now
$$E\{a(t)a(t + \tau)\} = \sigma_a^2 \delta(\tau),$$

where $\delta(\tau)$ is the Kronecker delta function, and

$$\sigma_a^2 = 2\omega\sigma_v^2,$$

where σ_v^2 is the steady-state variance of the maneuver velocity.

If a state vector $\underline{x}$ is defined such that

$$\underline{x} = \begin{pmatrix} x \\ s \\ v \end{pmatrix}$$

and covariance matrix $\underline{P}$ such that

$$E\left\{(\underline{x} - \underline{\bar{x}})(\underline{x} - \underline{\bar{x}})^T\right\} = \underline{P},$$

then $\underline{P}$ satisfies the equation

$$\underline{P} = \underline{AP} + \underline{PA}^T + \underline{GBG}^T$$

between observations, where

$$\underline{A} = \begin{pmatrix} 0 & 1 & 1 \\ 0 & 0 & 0 \\ 0 & 0 & -\omega \end{pmatrix},$$

$$\underline{G} = \begin{pmatrix} 0 \\ 0 \\ 1 \end{pmatrix}, \text{ and}$$

$$\underline{B} = 2\omega\sigma_v^2,$$

and the equation

$$\underline{P}^+ = \underline{P}^- - \underline{K}\,\underline{M}\,\underline{P}^-$$

at observations where

$\underline{P}^-$ = covariance of $\underline{x}$ just prior to an observation,
$\underline{P}^+$ = covariance of $\underline{x}$ just after an observation,
$\underline{M} = (1\ 0\ 0)$,
$\underline{K} = \underline{P}^-\underline{M}^T(\underline{M}\,\underline{P}^-\,\underline{M}^T + \underline{R})^{-1}$, and
$\underline{R} = \sigma_N^2$ = variance in measurement of target position.

The differential equation for $\underline{P}$ can be solved and the step equations simplified. Define

$$q_1 = P(1,1) = E\{(x-\bar{x})^2\},$$

$$q_2 = P(1,2) = E\{(x-\bar{x})(s-\bar{s})\},$$

$$q_3 = P(1,3) = E\{(x-\bar{x})(v-\bar{v})\},$$

$$q_4 = P(2,2) = E\{(s-\bar{s})^2\},$$

$$q_5 = P(2,3) = E\{(s-\bar{s})(v-\bar{v})\},\ \text{and}$$

$$q_6 = P(3,3) = E\{(v-\bar{v})^2\}.$$

We initialize the problem as follows:

$q_1 = 10^4$ (or other very large number),
$q_2 = q_3 = q_5 = 0$,
$q_4 = (.33\ V_{MAX})^2$, and
$q_6 = (.29\ V_{MAX})^2 = \sigma_v^2$.

This choice of parameters allows the target neither to lose ground on its intended track nor to attempt to move faster than its maximum speed (most of the time).

We then apply the following recursive algorithm:

$$k_N = \sigma_N^2 / (\sigma_N^2 + q_1^-),$$

$$q_1^+ = k_N q_1^-,$$

$$q_2^+ = k_N q_2^-,$$

APPENDIX C

$$q_3^+ = k_N q_3^-,$$
$$q_4^+ = q_4^- - q_2^- q_2^- / (\sigma_N^2 + q_1^-),$$
$$q_5^+ = q_5^- - q_2^- q_3^- / (\sigma_N^2 + q_1^-),$$
$$q_6^+ = q_6^- - q_3^- q_3^- / (\sigma_N^2 + q_1^-), \text{ and}$$
$$\underline{q}^- = \underline{C}\underline{q}^+ + \underline{d},$$

where

$C(1, 1) = 1,$
$C(1, 2) = 2T,$
$C(1, 3) = 2(1-\rho)/\omega,$
$C(1, 4) = T^2,$
$C(1, 5) = (2T/\omega) \cdot (1-\rho),$
$C(1, 6) = (1-\rho)^2/\omega^2,$
$C(2, 2) = 1,$
$C(2, 4) = T,$
$C(2, 5) = (1-\rho)/\omega,$
$C(3, 3) = \rho,$
$C(3, 5) = T\rho,$
$C(3, 6) = \rho(1-\rho)/\omega,$
$C(4, 4) = 1,$
$C(5, 5) = \rho,$
$C(6, 6) = \rho^2,$ and
$\rho = \exp(-\omega T),$

and

T = time between updates

and

$d_2 = d_4 = d_5 = 0,$
$d_1 = (\sigma_v^2/\omega^2) \cdot (2\omega T - 3 + 4\rho - \rho^2),$
$d_3 = (\sigma_v^2/\omega) \cdot (1 - \rho)^2,$ and
$d_6 = \sigma_v^2 \cdot (1 - \rho^2).$

On the last iteration of the algorithm the time between updates T is replaced by

$$T = T + t_D,$$

where t_D is the data time late, i.e., the elapsed time from measurement to receipt of update by weapon or launch platform.

D

Weapons

Because this study is focused on networking, the committee is interested in the range of weapons (that is, the extent of the area over which they can exert influence), in their information requirements for guidance (that is, what information sensors must see and what information the infrastructure must deliver promptly to the shooter or to the weapon), and in their command support requirements (that is, what needs derive from planning and coordination functions). However, the committee first categorizes weapons according to their missions and required information support and then discusses *guidance, acceptable time late, volume of data required, and allowable target location error* as a function of their target sets.

D.1 MISSIONS, WEAPONS, AND REQUIRED COMMAND AND INFORMATION SUPPORT

D.1.1 Power Projection

Naval doctrine publications define power projection as "application of offensive military force against an enemy at a chosen time and place." In consonance with the *Joint Vision 2010* definition of precision engagement,[1] the top-level requirements for power projection may be said to include the following:

---

[1]Shalikashvili, GEN John M., USA. 1997. *Joint Vision 2010*. Joint Chiefs of Staff, The Pentagon, Washington, D.C.

APPENDIX D   405

- Capability to
  — Strike anytime, anywhere,
  — Incapacitate many targets of many different kinds,
  — Predict and assess results, and
  — Sustain operations in a high tempo,
- While limiting
  — Losses to own force,
  — Collateral damage, and
  — Force size and cost.

Current and programmed naval weapons for power projection are launched from manned attack aircraft, surface combatants, and attack submarines.

D.1.1.1 Attack Aircraft and Air-Launched Strike Weapons

The Navy's principal attack aircraft today are the carrier-launched F/A-18 C/D and F-14, and the vertical-take-off-and-land (VTOL) AV-8, used principally for close air support.

The F/A-18 E/F is scheduled for initial operational capability in the next few years, and the Joint Strike Fighter program is in the advanced development stage. As an example of weapons payload capability, the F/A-18 E/F will deliver several thousand pounds more than 500 miles from the carrier without refueling.

Naval tactical aircraft will rely principally on the Joint Tactical Information Distribution System (JTIDS) as the link for updating target information for situational awareness and strike coordination. The Global Broadcast System may also provide key data for targeting and situational awareness. The naval aviation community recognizes the value of connecting aircraft to external sources. For example, a major component of the Joint Strike Fighter program is a trade-off study to determine how much command, control, communications, computing, intelligence, surveillance, and reconnaissance (C4ISR) equipment the aircraft should carry on board and how much it can rely on off-board sources to provide. Reducing aircraft cost is a driver in the study.

Air-launched weapons are often categorized by their flight range into direct attack ($R < 15$ nautical miles), standoff from point-defense ($15 < R < 60$), and standoff from area defense ($R > 60$). The joint direct attack munition (JDAM) has become the weapon of choice for many direct attack missions. A JDAM is built by attaching a kit with Global Positioning System/Inertial Navigation System (GPS/INS) guidance and fin controls to an existing 1,000 or 2,000 lb free-fall bomb. (A kit for 500 lb bombs is planned for later introduction.) Other older man-in-the-loop precision-guided munitions (e.g., Maverick with electro-optical (EO) or infrared (IR) guidance) will remain in the inventory but production will cease. In the near future, the joint standoff weapon (JSOW), a gliding weapon with a GPS receiver and an INS, will become the weapon of choice for standoff from point defense missions. The rocket-propelled high-speed antiradiation missile (HARM) will also be in this range category. It is being upgraded to add GPS to its INS. For standoff from area defense the Navy prefers the soon-to-be-fielded standoff land attack missile-expanded response (SLAM-ER). The Air Force favors the joint air-to-surface standoff missile (JASSM) that is also being developed and tested for compatibility with carrier aircraft and the carrier environment. These are jet-powered cruise missiles guided with GPS/INS and imaging IR terminal seekers. The JASSM will have automatic target recognition (ATR) capability. The SLAM-ER will initially have a data link for man-in-the-loop operation; ATR is planned for later introduction.

Guided ("smart") submunitions represent a category of future possibilities. The Army and Air Force have under development various systems that the Navy could employ. The Brilliant Anti-Armor Technology submunition uses a combination of acoustic and infrared sensors to search a several-miles wide area and home on motorized ground targets. The bomb live unit (BLU)-108 submunition, used in the Air Force's sensor-fused weapon and the Army's search-and-destroy armor munition (SADARM) 155 mm artillery munition, dispenses spinning disks that find targets with IR sensors and then fire projectiles into them.

Another possible future concept is an uninhabited combat air vehicle (UCAV), which can be thought of as in between (in size and capability) a recoverable cruise missile and a pilotless attack aircraft. Some UCAV concepts assume development of a new class of 250 lb munitions. The Defense Advanced Research Projects Agency (DARPA) has a program under way to develop a prototype, and the Naval Air Systems Command's Advanced Development Office has initiated an exploratory effort.

Another possibility is a Mach 5 to 7 hypersonic weapon (ramjet, possibly with supersonic combustion) with perhaps 300 nautical mile range to kill time-critical and buried targets. The Office of Naval Research (ONR) and DARPA are cooperating on an exploratory development effort.

D.1.1.2 Combatant Ships and Ship-launched Weapons

The Navy's Aegis cruisers and destroyers and DD-963 class destroyers can carry Tomahawk cruise missiles. Most of these ships have the vertical launch system (VLS), which stores missiles in 90 to 120 cells. Tomahawk and all versions of standard missiles are stored one per VLS cell.

Since its first use against Iraq in January 1991, Tomahawk has become a weapon of choice for deep strikes in limited military actions and for defense suppression prior to major military actions utilizing manned aircraft. Tomahawk is a 1,000 nautical mile range jet-powered cruise missile with GPS/INS guidance, terrain-contour matching, and an optical map-matcher for terminal update. A new variant, Block IV, is under development for initial operational capability around 2003. Block IV will enable in-flight retargeting through ultrahigh-frequency satellite communications and will reduce weapon cost, among other things. Block IV will enhance considerably U.S. tactical responsiveness and flexibility in unmanned deep strike operations against fixed and relocatable targets.

The surface Navy currently has almost no capability to provide fire support to defend U.S. Marine Corps light maneuver units. The 5-inch guns on Aegis ships have an effective range of about 14 nautical miles. However, the surface Navy is developing the extended-range guided missile (ERGM) by adding rocket power and GPS guidance to a submunition-dispensing artillery shell. The ERGM will enable accurate fires to a range of 63 nautical miles. In a remanufacturing program, the Navy is adding GPS to convert existing, obsolete standard missiles (built originally for air defense) into the land-attack standard missile (LASM). The LASM will enable accurate fire to over 100 nautical miles. ERGM and LASM will be retrofitted to Aegis ships.

The Navy has a major program under way to develop a new destroyer class, the DD-21, which will have strike as its principal mission. The DD-21 may carry as many as 700 ERGM rounds in its magazine. Like the Joint Strike Fighter program, the DD-21 program has significant efforts under way to trade onboard and offboard C4ISR capabilities in the interest of reducing platform cost.

The DD-21 appears to be the Navy's opportunity to address its naval surface fire support problems in a fundamental way. An advanced gun system (AGS) is in the early stages of concept exploration as part of the DD-21 program. The Navy also appears committed to developing an advanced land attack missile (ALAM) as an associated program within the DD-21 Program Executive Office. The Army Tactical Cruise Missile System (ATACMS), a GPS-guided, submunition-dispersing tactical ballistic missile, is a candidate for this role. Another possibility is a smaller rocket that the DD-21 could carry in greater numbers.

D.1.1.3 Attack Submarines and Submarine-launched Weapons

U.S. Navy attack submarines are capable of launching Tomahawks. Those of the 688-I class have a dozen vertical tubes in the bow for carrying Tomahawks, and the missiles can be fired from their torpedo tubes as well. Attack submarines have participated in many of the operations this past decade in which Tomahawks have been launched.

In recent years, the submarine community has shown an interest in expanding the submarine's role in strike operations. Conversion of ATACMS for submarine launch has been considered in order to give the attack submarine a naval fire support capability. Others have questioned whether the submarine's few weapons could make a difference in a naval fire support role.

D.1.1.4 Primary Command and Information Support Requirements of Strike Weapons

A key requirement for effective, precise operations with many of the above-mentioned weapons is a rapid, accurate means of obtaining GPS coordinates for large numbers of targets. A number of development efforts are under way that address this issue. The Navy's ability to kill emergent or time-critical targets is limited primarily by targeting and command and control (C2) time lines, but if these time lines were improved, then missile fly-out times would limit effectiveness.

The Navy's growing reliance (and that of the other Services) on GPS is a source of concern because of the inevitability of a significant GPS jamming threat. A number of complementary ways to deal with the GPS threat are as follows:

- Increase the resistance of GPS receivers to jamming by use of advanced signal recovery algorithms and tighter coupling with the inertial navigation system;
- Develop low-cost, low-drift-rate microelectromechanical inertial navigation systems;
- Use controlled radiation pattern arrays to maximize antenna sensitivity in the direction of the GPS satellites, while steering an antenna null in the direction of jammers;
- Use alternate means of navigation update, e.g., precision terrain map-matchers; and
- Attack the jammers, at least the powerful ones.

The Navy's means of killing moving targets are effectively limited to man-in-the-loop aircraft operations at close range. Killing moving targets at long range will require new weapons or adaptations to existing weapons as well as new targeting and command and control capabilities. Sections D.1.1.4.4 to D.1.1.4.7 address this issue in more detail.

Mission planning and mission rehearsal systems for tactical aircraft are becoming increasingly sophisticated and increasingly important to mission success. They rely on databases for photography, threat location and capabilities, terrain elevation, and weather forecasts.

Tomahawk mission planning, now performed strictly at just two facilities ashore (Cruise Missile Support Activities), is moving to carrier and then to cruiser and destroyer operations. Will control of Tomahawk, ERGM, and LASM be integrated on Aegis ships? Will control of Tomahawk, AGS, and ALAM be integrated on the DD-21? The Navy is grappling with these issues at present.

Systems for dynamic battle management do not exist. Such systems are needed for coordination of tactical aircraft strikes, Block IV Tomahawk strikes,

APPENDIX D   409

and naval surface fire support. They are especially needed because the Navy is, and perhaps always will be, limited in firepower. Efficiency is essential. Tactical aircraft can carry only a few weapons into the fray; surface ships are developing new capabilities to project more power at greater ranges, but they too will be limited in any extended operation. Another significant requirement is deconfliction, ensuring that U.S. forces do not fall victim to friendly fire.

D.1.1.4.1 Target-Weapon Pairing

Targets influence the choice of weapons and consequently the data and network support needed to allow their launch and successful execution of mission.

D.1.1.4.2 Fixed Targets

Fixed targets are either structures (buildings, bridges, dams, tunnel entrances, large antenna structures, and so on) or distributed facilities (railroad switch yards, fuel storage areas, warehouses, entrenched troops, and so on).

Weapons that are used to attack fixed structures usually have relatively large unitary warheads, which for greatest effectiveness, need to be delivered to their aimpoint with an accuracy of a few meters. Examples are bridge abutments, transformer banks that feed a large factory complex, or command bunkers used to direct an adversary's operations. The weapons set used for such targets include the Tomahawk, JDAM, SLAM-ER, laser-guided bombs, and, where precision is not a prerequisite, Mark-80 series bombs. The guidance for weapons in this class is based on the use of one or more of the following: GPS/INS, image correlation for terminal guidance, semiactive laser guidance, or in the case of Maverick, man-in-the-loop.

Attacks on fixed targets are generally not time critical. The information needed to support an attack or re-attack with such weapons must be supplied in times that are compatible with the generation of the daily air tasking order (ATO). Although the data rates needed to support the employment of such weapons may be modest, the total amount of information needed may be large. For image-guided weapons such as the Tomahawk, the number of images needed to plan an individual strike is about 10. Depending on the size of the field of view and contrast and resolution needed, each image may contain between 10^7 and 10^8 bits. If 100 weapons are launched in a day, the amount of data needed to launch such an attack will be between 10^{10} and 10^{11} bits.

If the targeting network does not incorporate distributed, locally available databases, the transfer of that many bits of data from a central data repository to a forward-deployed strike planning cell in a fraction of a day will require the availability of data links that can support data transfer rates of at least 1 to 10 Mbps. Afloat planning cells for Tomahawk missions normally contain imagery libraries that contain much of the imagery needed to support strike mission planning. Operational experience indicates that the supplementary imagery needed to support mission planning can be transmitted in a timely fashion through the use of existing long haul data links.

One may imagine future Tomahawk strikes that involve the release of 1,000 rather than 100 weapons within a day. Such strikes would require either data transfer rates that are 10 times current rates or distributed databases that are significantly larger than the imagery libraries currently available on forward-deployed platforms.

Increasing the size of deployed imagery libraries would appear to be perfectly compatible with both existing and projected data storage technology.

D.1.1.4.3 Fixed-area Targets

Targets in this category are usually attacked with weapons carrying submunitions dispensing warheads. Weapons used for attacks upon distributed targets include ATACMS, JSOW, ERGM, and, if procured by the Air Force, JASSM. These weapons transport different size payloads of submunitions. Thus their lethal footprint varies from a few hundred meters for ATACMS to a few tens of meters for ERGM.

The information support needed for weapons in this class is relatively modest. The weapons generally have GPS/INS guidance, and target location errors of a few tens of meters can be tolerated. Thus, only modest amounts of data need to be provided prior to release of weapons in this class. The situation may change if GPS jamming forces a change in the guidance system to terrain-aided navigation. In such circumstances, the data transfer problem would be comparable to the data transfer problem needed to support the planning of a large Tomahawk strike.

D.1.1.4.4 Nonfixed Targets

Nonfixed targets are grouped by the committee into three subclasses designated as relocatable, ephemeral, and moving. This distinction is made because these three classes of targets tend to tolerate rather different levels of latency in the information networks that support their launch.

D.1.1.4.5 Relocatable Targets

That class of targets that may be moved at the adversary's discretion is designated as relocatable targets. Targets in this category include, for example, tanks, trucks, armored vehicles, and missile batteries. Although targets in this category often remain in a single location for extended periods of time, they are difficult to attack on the same basis as is used in an ATO cycle devoted to attacks on stationary targets. Reports from both Operation Desert Fox and the Kosovo

APPENDIX D   411

campaign indicate that each adversary, knowing the ATO cycle time and the time of passage of imaging satellites or of unmanned aerial vehicles (UAVs), would relocate his assets by several hundred meters before the expected arrival of NATO strike weapons.

Unless a weapon delivery capability is co-located with, or is under the tactical command of, a sensor platform that is capable of redetecting relocated targets, the rate of success encountered when targets of this class are attacked with GPS/INS-guided weapons will be low. A network that can cope with target relocation must be rather specialized. The ideal sensor platform for the detection of target relocation is the Joint Surveillance and Target Attack Radar System (JSTARS) whose moving-target indicator (MTI) radar detects moving targets provided that they are not masked by terrain obscuration or dense foliage. When the targets stop moving, they are no longer detectable by MTI radar. However, the location of the point where the radar lost contact with the moving target is the current target location. That point can be attacked if a weapon-equipped platform is within weapon range and has data links that allow it to be cued by JSTARS.

Weapons available to attack relocatable targets are well matched to their target set. They include JDAM, JSTARS, sensor-fused weapons, ERGM, HARM, and so on. These weapons should perform well provided that an appropriate tactical network is established that permits attacks on targets immediately after they have been located by a sensor system.

D.1.1.4.6 *Ephemeral Targets*

Ephemeral targets are normally hidden from detection by conventional sensors (radar, EO/IR, and signal intelligence) and are exposed to possible detection and attack for periods of a few minutes while they execute their mission. Examples might be the Iraqi "shoot-and-scoot" Scud missiles; North Korean artillery that emerges from a cave, fires a few rounds, and then retreats into concealment; or a guerrilla band that undertakes an attack and then disbands and becomes indistinguishable from the general population.

Currently, the U.S. Navy has no weapon sensor combination that can effectively engage an ephemeral target from any significant standoff range. Admittedly, if an ephemeral target emerges when an armed and ready F/A-18 is within weapon range, the target will have a high probability of being destroyed.

A component of the Navy (ONR) is working on a sensor weapon system based on the scenario of using a surveillance UAV that patrols an area where an ephemeral target is thought likely to emerge. Upon detection of a valid target, the UAV alerts a firing platform at sea and passes on the target's coordinates. With no C2 delays, the remote platform is presumed to fire a supersonic weapon capable of traveling at an average Mach number of about 11.25. The weapon would arrive at the target (300 miles distant from the launch platform) within 2.5 min of initial target detection.

Obviously significant technological advances would need to be realized before such a scenario could be reduced to practice. In the interim, the Navy will probably opt for pragmatic solutions. Sensor platforms accompanied by well-armed weapon carriers will patrol areas that have high probabilities of hiding ephemeral targets. The weapon of choice will probably be something as modest as JDAM and the data link will be JTIDS.

D.1.1.4.7 Moving Targets

Targets in motion at the time of their detection by a remote sensor present a particular problem to a launch platform that attempts to attack them from beyond line of sight or long standoff distances. A subsonic weapon that is launched at a target from a 30 km distance might have a time of flight of about 1.5 min. During the weapon's time of flight, a ground target moving at 40 km/h will have moved about 1 km. Under such circumstances, the probability of target kill will be low unless the target's motion was compensated for by choice of aim point prior to weapon release or unless the weapon had a sensor and data link that allowed continuous update of its aim point.

D.1.1.4.8 Suppression of Enemy Air Defense

The weapons employed for the suppression of enemy air defense (SEAD) include the AGM-88, the high-speed anti-radiation missile, and a number of other air-to-surface missiles including the Hellfire missile used on the Apache Helicopter, the AGM-65 Maverick, the JSOW, and the JDAM.

Employment of these weapons for SEAD operations requires network support. HARM homes on the radar signals used to guide enemy air defense missiles, has a relatively short kinematic range, and for best results must be provided with target location uncertainty of less than about 2 km. Radiating targets may be localized by EA-6B aircraft, theater electronic intelligence (ELINT) aircraft, or national sensors. If the HARM is launched by the same EA-6B that provided the localization of the radiating target, then the supporting network and weapon release command authority is contained within a single air frame.

A true information network is required when the aircraft that launches the HARM depends on off-board sensors mounted on remote aircraft or national assets. Hostile radars do not radiate continuously. Thus, if such a radar is geolocated by a sensor system that is remote from the HARM launch platform, the network latency must be extremely low so that the geolocation information can be passed to the weapon launch platform before the target radar has ceased to radiate.

Upgrades are being considered to increase the HARM weapons system's propulsion range, improve its capability against radars that cease to radiate, and increase its warhead lethality. Although these projected improvements will provide better overall weapon performance, they should not result in reduced requirements for network support. HARM will continue to need relatively accurate target geolocation information, and with increased kinematic range, it will be even less tolerant of network latency than it currently is.

Air-to-ground SEAD weapons that are used to attack nonradiating air defense weapons (antiair or small shoulder-fired infrared (IR) guided missiles) are completely dependent on information networks for knowledge of target location. Aircraft that launch air-to-ground missile (AGM)-type weapons generally do not have sensors that provide detection and localization of nonradiating camouflaged weapons with a high area sweep rate. To the extent that fixed, nonradiating, camouflaged targets can be detected by EO/IR or synthetic aperture radar (SAR) imaging sensors, aircraft or national sensors specifically configured for the purpose are usually tasked to perform the function. The information derived from such sensors is passed to a SEAD battle manager who assigns aircraft and weapons to specific targets. The committee notes in passing that no operational sensor can detect shoulder-fired missiles on a consistent basis. In Kosovo, keeping aircraft at sanctuary altitudes and standoff ranges that could not be reached by either antiair or shoulder-fired missiles solved the problem.

Evolutionary improvements are scheduled for all AGM-type missiles. As in the case of the HARM, these improvements will result in significant improvements in weapon performance, but they will not result in a reduction of network support requirements.

SEAD also includes the requirement to eliminate defensive enemy fighters through air-to-air combat or through their destruction while on the ground. Since air-to-air combat is a component of fleet air defense, it is discussed in the next section.

D.1.2 Theater Missile and Air Defense

Littoral warfighting produces many challenges for the theater missile and air defense (TMAD) component of naval warfare. Mission effectiveness and efficiency in use of TMAD weaponry require the following:

- Common awareness of the operational situation,
- Ability to effectively coordinate defensive measures, and
- Capacity for defense in depth.

Defense in depth is accomplished by two primary methods. The first method is through deployment of missile defense units into a geographic arrangement that assures multiple opportunities to engage and re-engage threats. The second method is through deployment of weaponry with varying fly-out ranges from a central location. Capable planning tools are needed to support development of doctrine and tactics and force laydown. Additionally a suitable array of sensor capabilities must be available to support timely detection, control, and engagement for both techniques of providing defense in depth.

This discussion begins with the innermost layer of defense in depth, ship self-defense, and works outward through the progressively longer-range layers.

D.1.2.1 Ship Self-Defense

The primary threat for shipboard self-defense systems is the antiship cruise missile. Challenging aspects of the evolving threat include lower target signatures in the radio frequency (RF) and infrared spectra, higher speeds, programmed and responsive maneuvers, onboard countermeasures, ability to conduct time-of-arrival control during raid attacks, and difficult-to-destroy payloads. These threats may be launched from aircraft, surface or submerged vessels, and land-based locations within the littoral environment.

The weapons for ship self-defense include the Close-In Weapons System (a closed-loop gun system) and short-range surface-to-air guided missiles such as the rolling airframe missile (RAM) (rocket-powered and guided by IR and passive RF). Both weapons are designed for short-range operations and require targeting to come from the defending ship. Off-board information will be used to support development of a composite target track suitable for cueing of local detection and tracking systems.

D.1.2.2 Area Air Defense

The primary threat for area air defense systems are the antiship cruise missile and the supporting systems such as launch platforms and countermeasures (e.g., standoff jamming). Challenging aspects of the evolving missile threat are the same as for ship self-defense. The added challenge is to avoid degradation in performance due to standoff countermeasures and raids.

The weapons for area defense include the family of standard missiles (rocket-powered and guided in terminal homing by semiactive RF and, on the IVA version, IR) and the evolved sea sparrow missile (ESSM) (rocket-powered and guided by active RF). These weapons are designed for longer-range fly-outs and could be supported with targeting from off board the firing ship. Off-board information will be used to support development of a composite target track suitable for cueing of local detection and tracking systems, management of sensor resources, and scheduling of engagements.

D.1.2.3 Air-to-Air Combat

Combat air patrols are established for the purpose of destroying enemy air-

APPENDIX D   415

craft before they can come close enough to the U.S. surface force to release their weapons.

Success is dependent on many factors including pilot training and tactics, aircraft agility and signature suppression, airborne sensors, support by early warning sensors on surveillance aircraft, the effectiveness of electronic countermeasures, and lastly the effectiveness of air-to-air weapons.

Modern concepts of air superiority are based on an integrated air defense network that provides information derived from sensors on airborne early-warning aircraft such as the E-2C or Airborne Warning and Control System (AWACS), to a battle manager who assigns targets to patrolling fighter aircraft. (Note that the battle manager is generally located on either the AWACS or the E2-C aircraft.) Cued fighter aircraft use their on-board sensors (radar and forward-looking infrared) to acquire hostile aircraft and launch their weapons. For air-to-air engagements, the concepts and doctrines of network-centric warfare are well established. Although the performance of existing network links has some limitations, air-to-air weapons are well supported by information networks.

Although the AIM-54 Phoenix weapon with a kinematic range of 150 km has been available for many years, because of restrictive rules of engagement, its use in combat has been rare. In recent years, the basic weapons for air-to-air engagements have been derivatives of the AIM-9 family of IR-guided missiles and the AIM-120 active radar-guided missiles. AIM-120 missiles currently have kinematic ranges on the order of 20 nautical miles (about 39 km). AIM-9 class missiles are reported to have kinematic ranges of about 10 km.

The overall thrust of the Navy and Air Force plan is to remain with existing AIM-9X and AIM-120 class missiles and to concentrate on pre-planned product improvements (P3I) in the areas of propulsion (kinematic range), off-boresight capability, hard-kill countermeasures, and integration into a network-centric model. The stated long-range goal is to neck down to a single, dual-range, air-to-air weapon that might be introduced about 2015.

In the continuum of air warfare, U.S. capabilities that include network support capabilities, AWACS and E2-C sensors, ELINT, National systems, aircraft performance, electronic warfare, and pilot training have given the United States an edge that has resulted in an enviable record in recent air combat.

In the area of seeker performance acquisition range and off-boresight performance are the critical performance parameters. U.S. air-to-air weapons hold advantages in ordnance lethality over those of other nations, and the P3I programs in this area are exciting, particularly with respect to accuracy and payload size. Missile size is important, particularly in stealth platforms with internal carriage requirements. If the missile cannot be accommodated internally in a stealth aircraft, much of the advantage of the stealth treatment may be lost.

The air-to-air weapons program is inherently designed to be evolutionary in nature. Performance improvements have been incremental but steady. The technology being used in the program is at the forefront of propulsion and warhead technology. New and exciting leaps in technology are being realized. Current weapon performance is significantly better than it was 10 or 15 years ago. Although the goals of current weapon improvement programs such as a 25 percent increase in weapon range (for the same weapon volume) and a 15 percent increase in weapon velocity may seem relatively modest, they may well mean the difference between success and failure in air-to-air combat.

Air-to-air weapons are relatively mature in the sense that current weapon capabilities may be well up on the curve of realizable performance. As long as incremental performance improvements can be achieved at reasonable cost, support for such work is likely to continue. Although one may applaud the results achieved by a high-quality incremental program, one cannot escape wondering whether new approaches to air-to-air combat based on improved network capabilities should be explored.

As an example, success in a short-range air-to-air encounter depends among other things on how far off boresight an IR-guided weapon can be fired. This issue is being addressed with significant success. Nevertheless, if one asks what an incremental improvement in an off-boresight capability translates into in the time domain, the answer is generally about a few tenths of a second.

The dependence of future air superiority on such marginal incremental capabilities is not very reassuring. Clearly, until better concepts based on network-centric operations are developed, such incremental efforts should be continued. However, one would hope that the development of improved information networks, improved sensor resolution, and weapons will allow (even under restrictive rules of engagement) air targets to be engaged at much longer standoff ranges than current capability permits. Ultimately, with the potential advantages of network-centric operations, close-range air-to-air engagements should not be allowed to occur. The survival of U.S. aircraft should not be allowed to depend on a marginal enhancement in current off-boresight capability.

D.1.2.4 Land-attack Cruise Missile Defense

The primary threats for land-attack cruise missile defense (LACMD) systems are surface-target-strike cruise missiles. Challenging aspects of the evolving missile threat are the same as for ship self-defense.

The weapons for LACMD include the standard missile as well as aircraft-launched air-to-air missiles such as Sidewinder (rocket-powered and guided by IR) and the advanced medium-range air-to-air missile (AMRAAM) (rocket-powered and guided by active RF).

D.1.2.5 Tactical Ballistic Missile Defense

The primary threat for tactical ballistic missile defense (TBMD) is medium- to long-range tactical ballistic missiles, including conventional weapons as well

APPENDIX D   417

as weapons of mass destruction. Challenging aspects of the evolving threat include lower target signatures in the radio-frequency and infrared spectrums, higher speeds, programmed and responsive maneuvers, unintentional and intentional countermeasures, ability to conduct time-of-arrival control during raid attacks, and difficult-to-destroy payloads.

Shipboard weapons for TBMD include the Standard Missile-2 Block IVA (area tactical ballistic missile defense) and the Standard Missile-3 (theater-wide tactical ballistic missile defense). The SM-3 features a hit-to-kill exo-atmospheric interceptor with electro-optical guidance.

D.1.2.5.1 Primary Command and Information Support Requirements of Air Defense Weapons

As discussed above, mission effectiveness and efficiency in the use of theater missile and air defense weaponry require the following:

- Common awareness of the operational situation,
- Ability to effectively coordinate defensive measures, and
- Capacity for defense in depth.

The committee discussed how to achieve defense in depth. Operational situational awareness is facilitated by gathering battlefield information from distributed sensors (in space, in the air, on land, or at sea). The content, accuracy, and latency of the information may vary widely. Additionally, information about portions of the theater or the threat may be quite sparse. The challenge is to develop a robust network of capabilities for gathering, processing, and disseminating the best possible information needed by the warfighters. Since warfighter needs do vary with time (e.g., planning versus real-time combat) and type of unit, the network must provide for flexibility in information access and display.

Coordinating defensive measures in a fully developed theater is needed to avoid ineffective and inefficient use of the limited number of guided missiles available, as well as to avoid incidents of friendly fire. A classic technique, in the absence of a distributed capability for fire control, is to divide the battlespace into regions or sectors. A combatant is assigned a portion of the battlespace; any threat entering the battlespace is to be engaged. This approach is generally slow to adapt to changes in availability of defensive units, provides little lead time to shooters, and retains significant inefficiencies.

The need exists for a dynamic system capable of the following:

1. Assessing sensor and weapon resource availability,
2. Determining probability of kill for each shooter,
3. Providing sufficient lead time and fire-control quality data to the preferred and next-in-line shooter,

4. Ensuring a means to command or determine that a shot has been taken against the designated threat missile, and
5. Accurately confirming threat missile kills.

The area air defense commander (AADC) system is being developed to meet these needs.

A single integrated air picture is a requirement for theater missile and air defense in order to accomplish the following:

- Support planning and doctrine development, and
- Permit effective real-time coordinated engagements (e.g., make weapon assignments, confirm engagements, issue re-engage orders).

Weapons in the theater-missile and air-defense mission require accurate fire control information in order to do the following:

- "Gridlock" sensor-to-shooter coordinate and time-reference frames,
- Discriminate threat from associated countermeasures and environments using multispectral and spatially separated sensors (including space-based sensors), and
- Make real-time kill assessments and distribute that information for use in coordinated defense.

TMAD weapons require timely fire control information in order to do the following:

- Maximize battlespace and defense in depth by providing fire-control quality data before the shooting ship is capable of detecting and engaging on a local target track;
- With lower-quality data, permit cueing of local (or other) detection and tracking systems; and
- Distribute target and engaging missile track data to permit effective coordinated engagements (up to and including forward-pass concepts).

The previously mentioned cooperative engagement capability is being developed to meet these needs.

D.1.3 Undersea Warfare

The weapons of undersea warfare are primarily torpedoes and mines. Navy emphasis in this warfare area at the moment is on defense, particularly to detect and counter sea mines. In the future, however, the submarine threat is expected to re-emerge as a high priority.

D.1.3.1 Mine Warfare

Sea mines are a class of weapon that has significant deterrent effects on submarine and surface-ship operation in areas where they have been deployed. In effect, they are weapons that have been programmed to detonate on recognition of the signature of specific targets. They respond to signatures detected by their organic sensors that have been subjected to the thresholds set by internal logic gates.

The U.S. Navy does not have an aggressive program for the introduction of new classes of sea mines. Improvements are programmed for sensor performance, signal processing, acoustic signature reduction, and resistance to sweeping and countermeasures. Although there are no plans to incorporate sea mines into a network-centric warfare concept, there are R&D programs working toward this objective.

D.1.3.2 Countermine Warfare

Expeditionary forces prefer to deal with mines by detecting and avoiding them. If covert detection is not required, airborne sensors can usually detect some classes of mines more rapidly than they can be neutralized. Sweeps, both mechanical and influence, detect as they neutralize, but their coverage rate is small.

Marine Corps doctrine calls for standoff forces that hold large stretches of coast at risk and then make rapid attacks before the enemy can reinforce the intended landing area. This doctrine requires that mine clearing be performed in stride and not alert the enemy of the intended landing area.

Among the weapons that are being developed to support in-stride clearing and breaching in very shallow water are large, rocket-deployed nets festooned with explosive charges. The charges clear a landing lane by detonating or displacing the mines.

D.1.3.3 Antisubmarine Warfare

Current undersea weapons include variants of the Mark-46 and Mark-48 torpedoes and sea mines. Although U.S. torpedoes can be used in an antisurface ship mode, they are primarily configured as antisubmarine warfare (ASW) weapons. Sea mines can be employed in either an ASW or in an antisurface ship mode.

Torpedoes are large and expensive weapons. Only a relatively small number are carried in ASW patrol aircraft or in submarines (where they compete with Tomahawk missiles for the available launch tubes). Thus they are not released unless a significant probability exists that a target submarine has been correctly classified and has been localized to within the acquisition capabilities of the torpedo's sensors and kinematic range.

As the radiated signatures of submarines decreases, and as stealth technology reduces their active sonar cross section, opportunities to use torpedoes will tend to decrease. Information networks that combine the output from multiple spatially separated sensors will become increasingly necessary to position a firing platform close enough to its target that a torpedo can be released and subsequently be guided to the victim submarine.

ASW has always been fought as a form of network-centric warfare. First, a database must be established that identifies the training, deployment, and maintenance cycles of an adversary's submarines. Overhead imagery, communications interceptions, and human observers provide information concerning the submarine's predeployment status. A submarine's departure from port can be monitored by similar means. Its objective and likely area of deployment often can be inferred from the current political situation or from historical patrol patterns.

If an undersea surveillance system can provide occasional detections, and partial track information, then mobile reacquisition platforms (submarines and ASW aircraft) can be vectored to a projected point on the enemy submarine's assumed track. Once in an area where there is a high probability of encountering an enemy submarine, local networks of passive and or active acoustic sensors supplemented by nonacoustic sensors, may be established to allow close enough localization of the target to allow release of a wire-guided torpedo. Some modern torpedoes contain high-frequency sonars that define the target structure with sufficient fidelity to allow aimpoint selection. Other torpedo sensors are based on wake homing. When a submarine launches a torpedo, information and guidance commands between the weapon and launch platform are transferred over a fiberoptic link. In the case of an air-launched torpedo, communications between the weapon and launch platform requires a fiber-optic umbilical between the torpedo and a surface buoy that is in radio contact with the airborne launch aircraft.

Although difficulties still exist, and not all platforms have full or continuous connectivity, the data links that are necessary to support a network-centric concept of ASW and torpedo usage exist. The most difficult links to operate are those to a deeply submerged submarine, to an air-launched torpedo, and to a networked tactical sensor field. The data rates for these links are typically no more than a few kilohertz. In the case of a submarine that cannot, for operational reasons, deploy a surface-piercing antenna, communications must be on a scheduled broadcast basis. Network broadcasts of this type have of course been employed almost since the beginning of submarine operations.

The U.S. Navy has no current funded program to introduce an entirely new torpedo. Current weapons will be subject to incremental improvements that will provide improved engines (better speed and range) improved on-board signal processing (enhanced resistance to countermeasures, improved target resolution and aimpoint selection), and increased stealth to avoid alerting the target before the torpedo has closed to within a range that precludes escape by target maneuver. None of these foreseen or programmed improvements should stress the bandwidth or latency requirements of existing ASW networks.

The technology exists to develop an antitorpedo torpedo. Such devices are capable of traveling at speeds approaching 200 knots (about 370 km/h). If launched in a timely fashion, they could be used to save a ship or submarine from an incoming torpedo. The concept of operations would involve a closed network that employs an acoustic sensor to detect an incoming torpedo and a processing node that correctly classifies the torpedo and alerts a weapon-release control authority. Since minimum latency could be tolerated between detection and release of the antitorpedo torpedo, the system would need to operate in a largely autonomous mode. To date, no R&D program in support of this concept has been funded.

Although ASW sensors and tactics may be unique, ASW is an important warfare element of the tactical situation and is one of the battle group's concurrent tasks. In contrast to the other warfare elements, ASW operations concentrate less on weapons delivery and more on detection and classification-situational awareness. Since ASW situational awareness will be derived from networked distributed sensors, ASW could again become an important beneficiary of network-centric technology.

The dramatic progress in threat quieting and the shift from blue water to the complex transitional littorals is driving the development of ASW initiatives toward active acoustics, nonacoustics, and network-centric concepts. Littoral ASW operations tend to be asset intensive, and battle group assets are required to operate in concurrent multiwarfare situations. The fundamental problem, both in blue water and littorals, is the loss of long-range continuous target tracking using platform-centric, legacy-sensor, and traditional operations. Today, operators are presented with short-range intermittent detection opportunities in a high-clutter environment, making classification difficult and dynamic. The response to this problem is to emphasize off-board sensors, distributed field processing, and network-centric information processing. A network-centric approach to ASW offers the potential for improved detection, classification, and asset allocation through sensor data fusion, collaborative analysis, and joint planning.

The battle group is required simultaneously to maintain the subsurface, surface, air, and land battle scenes while allocating assets in dynamic and shifting circumstances. The battle group must conduct ASW while conserving platforms and staffing for concurrent missions. Significant progress against the quiet littoral threat may result from recent advances in computing and communications technologies that support a network-centric approach to ASW. Generating a common tactical picture would make it possible for a target to be detected, classified, tracked, and engaged faster than currently is achievable with today's platform-centric approach. Battle group elements would become part of a grid of sensors and processing stations. Their positions, search tactics, and sensor setup could be optimized for sensor performance and environmental conditions. Clutter and false alarms could be resolved more readily through the use of a composite information base derived from all platform sensors over the course of operations. Target signature features across the whole search area could be evaluated to generate potential target tracks and a clutter map. Environmental drivers such as detailed bathymetry and sound propagation characteristics could be collected and analyzed to optimize the battle force ASW disposition. For all this to happen, it becomes necessary to provide an architecture that facilitates vertical and horizontal transfer of sensor information, a coherent tactical picture, hypothesis of intent, and assessment of potential options. The architecture would enable collaborative planning, multiple tactical decision aids, data fusion, advanced displays, and vulnerability assessments.

D.1.3.4 Primary Command and Information Support Requirements for Undersea Weapons

Information support for mining operations is more in the realm of intelligence and environmental support than tactical support. Situational awareness is needed to protect the mine-laying platform.

The most important information in support of countermine activity is the information that can be obtained about mine locations without revealing U.S. interest in the area. The use of covert mine reconnaissance from undersea platforms, aircraft, and space is generally intelligence preparation for littoral operations. Networked mine countermeasures will be important for situational awareness during amphibious operations.

Success in ASW is increasingly dependent on the ability to fuse seemingly disparate information and to reject false alarms from high-clutter littoral environments. Stealthy targets in these environments defy long-range continuous detection and tracking by legacy sensors. "Sniffs and whiffs" may accrue from different sensors on different platforms at different times, and revealing their common origin depends on network-centric fusion processing. In addition, contact fusion must extend beyond kinematic information alone. The summary of the target features derived from individual sensors can be shared, providing a more complete understanding of the target and enhancing the target classification process. The operational force must exploit environmental conditions, historic patterns, operational intelligence, event relationship, classification clues, and subjective evaluations. The operational success depends on being able to collect and share appropriate information across the force.

Common tactical decision aids and means for collaborative planning in this warfare area are required to provide the force commander with timely tactical interpretations, force planning, and tactical option reduction. Tactical decision aids should incorporate previous search results from all platforms and should help the force ASW commander resolve potential target contacts, given the limitations on force assets, sensor performance, and requirements for continued search. The common tactical picture provides the scarce resource; real-time cueing by the stimulating platform will be needed.

D.2 SUMMARY OF NAVAL WEAPON SUPPORT NEEDS

This section summarizes information support requirements for weapons across all warfare areas, characterizing the needs in terms of accuracy of target location, data timeliness, and volume of data.

Several attributes characterize weapons and determine which weapons are appropriate for targets of interest. Weapon guidance type strongly influences the complexity and volume of information required for weapon employment. Weapon range and average speed establish the critical time span over which targeting information must remain current, whether it is provided at launch or updated in flight. Finally, either the weapon's seeker field of regard or its warhead lethal radius determine the accuracy with which location of the target must be provided to the weapon. Other attributes of the target such as its hardness, size and shape, dwell time, and signature are important factors in making the appropriate weapon/target pairing. In addition, the target environment may have a strong impact on weapon selection. In urban warfare, for example, high priority may be placed on controlling collateral damage. The following paragraphs examine naval weapon and target attributes in appropriate combinations.

D.2.1 Weapon Guidance

There are four broad categories of weapon guidance: open loop, geodetic, closed loop, and ATR. Figure D.1 categorizes naval weapons expected to be in inventory circa 2010.

D.2.1.1 Open Loop

In open-loop guidance, a trajectory is imparted to the weapon that is calculated to cause it to hit its intended aim point. The weapon makes no in-flight corrections of any kind. Naval weapons in this category include ballistic artillery and gun rounds, rockets, bombs, and aircraft cannon. Engagement times (time of flight) of these weapons range from seconds to tens of seconds.

D.2.1.2 Geodetic

Geodetic weapons are programmed to be guided to a two- or three-dimensional coordinate in space and to dispense submunitions, detonate by contact fusing, or detonate by internally generated command. The weapon may maintain its geospatial reference through continual GPS updates. This category includes

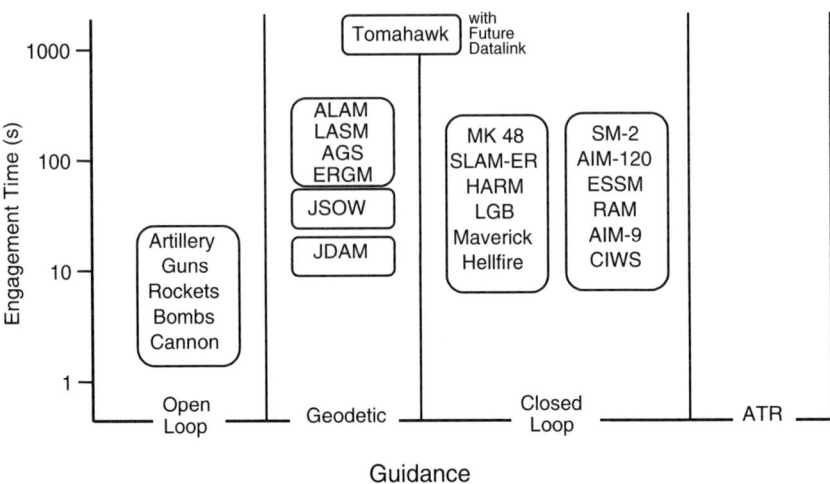

FIGURE D.1 Categories of weapon guidance and the naval weapons expected to be in inventory about 2010. Acronyms are defined in Appendix H.

many new weapons, e.g., JDAM, JSOW, ERGM, LASM, Tomahawk Block III (in certain modes), and others in the conceptual phase, e.g., the round for the advanced gun system and the advanced land-attack missile. Typical engagement times of these weapons span from tens of seconds through hundreds of seconds to about 2 h for Tomahawk.

D.2.1.3 Closed Loop

Weapons with closed-loop guidance usually have some form of seeker on board. Its implementation may either be self-contained or retain a person in the loop via a data link. Weapons in this category include the Mark-48 torpedo; many strike weapons such as SLAM-ER, HARM, laser-guided bombs, Maverick, and Hellfire; and antiair weapons such as the SM-2, ESSM, RAM, AIM-120, and AIM-9. Engagement times are normally tens of seconds to several minutes. A few weapon systems employ closed-loop guidance without a seeker. For example, the close-in weapon system (CIWS) shipboard radar measures projectile miss distance and corrects the gun's aim. The projectile's time of flight can be a fraction of a second to several seconds. As another example, Tomahawk's digital scene matching area correlation (DSMAC) is not a seeker (it never "sees" the target), and yet the navigation corrections it provides and the use of a relative coordinate system to guide to the target location qualifies it as a closed-loop system. The future addition of a data link to the Tomahawk will permit providing target updates to the weapon, in effect providing closed-loop guidance.

D.2.1.4 Automatic Target Recognition

Automatic target recognition is an extension of closed-loop guidance wherein the seeker is capable of recognizing an intended target among false contacts. That is, the seeker can automatically recognize and reject objects protected by rules of engagement. The Navy currently has no weapons in its inventory capable of ATR.

D.2.2 Acceptable Time Late

Tolerable latency of targeting data varies significantly with the dwell time or speed of the target, as shown in Figure D.2. Data supporting the targeting of fixed targets (both area and structure) may be days, months, or years old. However, current identification must be provided, as was illustrated in the recent NATO bombing of the Chinese embassy in Belgrade. Relocatable targets require more timely targeting, varying from hours to minutes, depending on the mobility of the target. For example, an attack against complex surface-to-air missile sites or tent encampments may be successful with data hours old, whereas counterbattery attacks against missile launchers may require targeting-quality data available to an appropriate weapon system in minutes. Weapons employed against airborne targets need targeting data no more than seconds or milliseconds old.

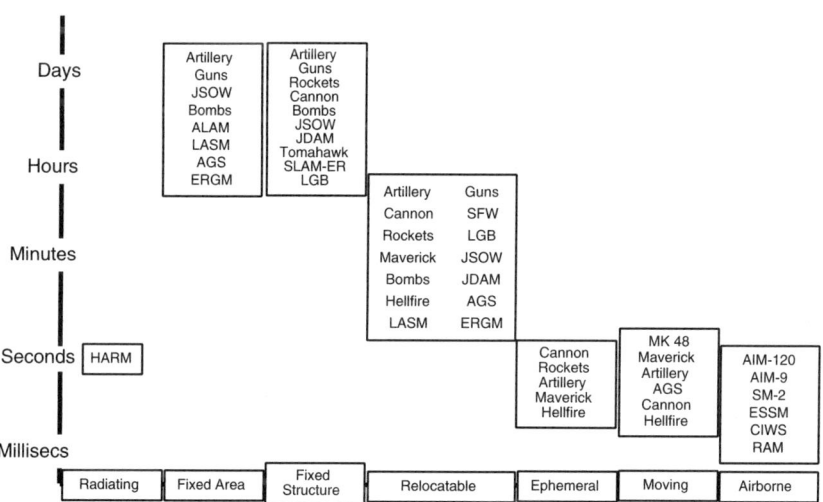

FIGURE D.2 Acceptable data time late versus target types. Acronyms are defined in Appendix H.

D.2.3 Volume of Data Required

Figure D.3 attempts to summarize the volume of data required to target individual naval weapons. Developed to measure stress on communication and data links in time of crisis, it considers the total number of bits required to transmit perishable information needed to target the weapon.

Operational drivers include the following:

- In-flight updates on target location to some air defense weapons throughout flight (SM-2, AIM-120, and ESSM);
- Imagery transmitted from surveillance sensor to image-processing station for identification of a target and precise extraction of its geographic coordinates. The committee assumes this is the principal method of targeting the JDAM, JSOW, LASM, ALAM, and Tomahawk (in GPS-only mode);
- Imagery transmitted from surveillance sensor to image-processing station for identification of a target and rough extraction of its geographic coordinates and retransmission of the image to a mission rehearsal system. The committee assumes this is the principal method of targeting SLAM-ER and one of the methods used for targeting unguided bombs;
- Seeker images transmitted to aircraft crews controlling SLAM-ER or Maverick;

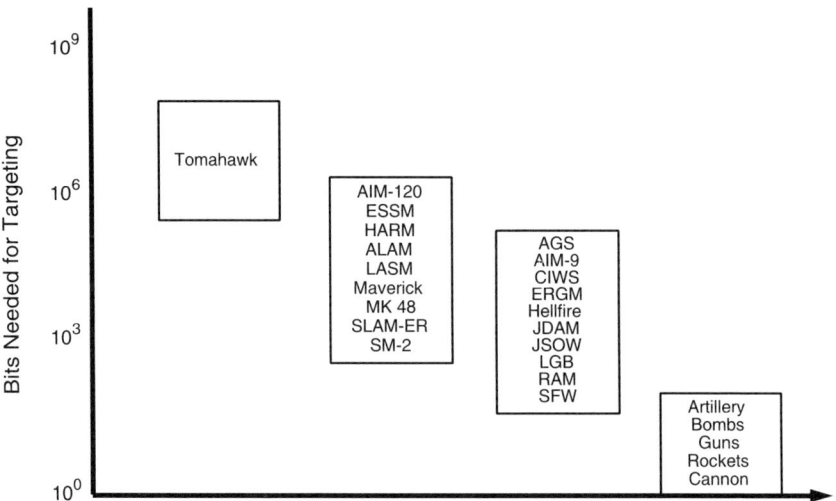

FIGURE D.3 Data load required to target weapons groups. Acronyms are defined in Appendix H.

APPENDIX D   427

  • Passing of geographic coordinates from on-scene forward observers to naval fire control systems and on to weapon-launch platforms (ERGM, AGS, artillery, guns);
  • Passing of geographic coordinates from ground controller to close support aircraft (rockets, bombs, cannon);
  • Imagery transmitted from surveillance sensor to image-processing station for identification of target, creation of image references to be loaded onto the weapon prior to launch, and precise extraction of the relative coordinates between scene centers and targets (Tomahawk in DSMAC mode); and
  • Transmission of Tomahawk missions from planning center to launch platforms.

Figure D.3 does not include data already generally available (e.g., weather data) or nonperishable data (e.g., terrain elevation data that may nevertheless be vital to the mission).

D.2.4 Acceptable Target Location Error

Guided weapons are normally provided with an aim point. For many reasons, the aim point may be imprecisely located or defined. Targets that are selected from well-registered imagery may be geolocated to within a meter or less. Targets that are located by signal intelligence (SIGINT) or radar bearings may be located within an error ellipse whose longest axis might be 1 or 2 km. Moving targets must be tracked continuously by a sensor in order to provide a weapon with a target location.

The maximum allowable target location error (TLE) will depend on the warhead of the weapon used, the guidance system, and the nature of the target.

Figure D.4 displays permissible TLE for naval weapons. As a weapon that homes on a radiating signal, HARM can operate successfully with a TLE as great as 1 km. Its probability of target destruction is much higher if the weapon is launched in a target-range-known mode than if it is launched in a target-range-unknown mode.

The TLE for fixed area targets varies with warhead size and with the number of rounds that will be fired at a given area. A weapon such as ERGM that has a relatively small warhead is most effective if the TLE is 20 m or less. A weapon with a large warhead such as ATACMS, which has a lethal diameter of 200 m, can be effective even if the TLE is 100 m.

Air-to-ground weapons with unitary warheads generally are most effective if the TLE is less than 10 m. Air-to-air and air defense weapons that operate under closed-loop guidance can be effective even if their initial TLE was between 100 and 1,000 m. On the other hand, a weapon such as CIWS will be ineffective if the TLE exceeds 1 m or so.

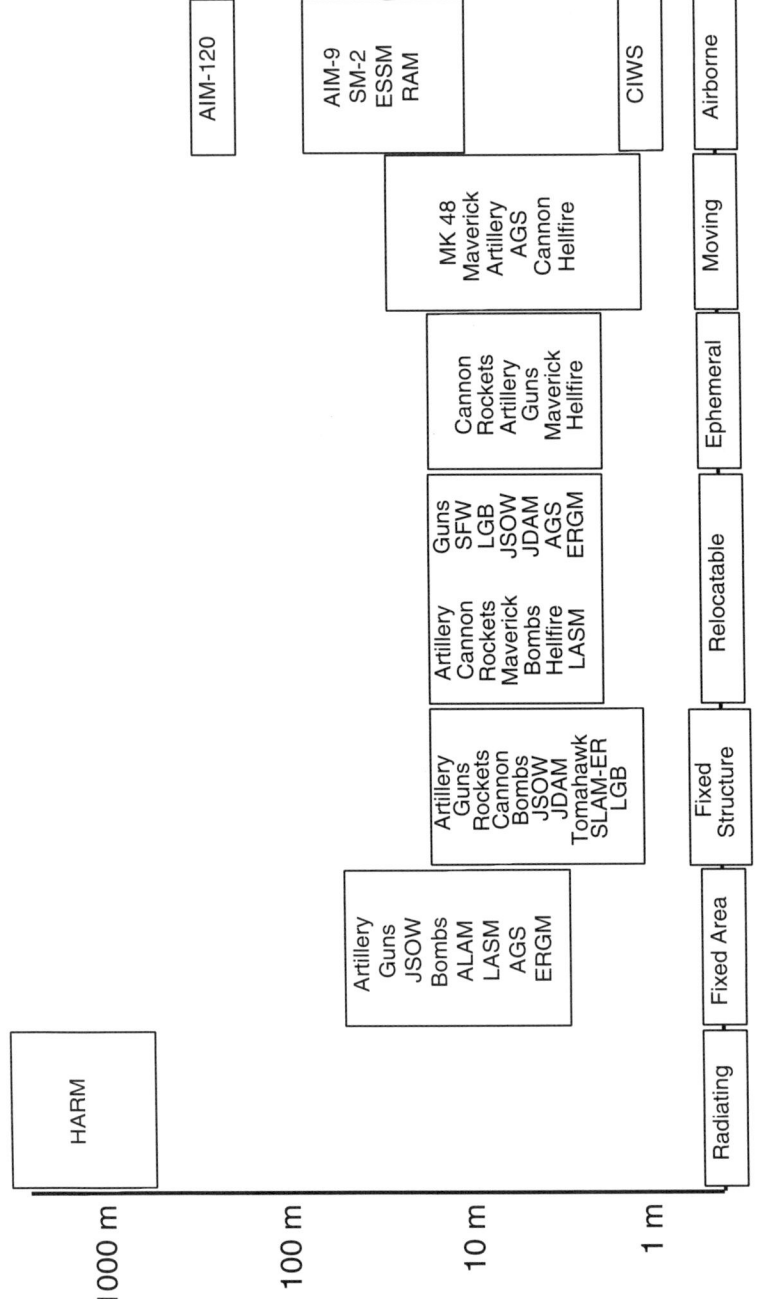

FIGURE D.4 Target location error versus target types. Acronyms are defined in Appendix H.

E

Tactical Information Networks

This appendix adds detail to the treatment of tactical information networks in the main body of the report. Considering the general aspects of network design, it shows that there are significant inadequacies in the Navy's current design for tactical information networks and the planned communications links may not be a good match for the information networking demands placed on them. It points out ways in which commercial networking technology could be used within the Navy's tactical networks and those areas in which commercial technology will probably need to be augmented with military-specific technology. The committee does not pretend to design the Navy's tactical information networks. That will require a great deal of architectural design and system engineering work. However, this appendix does show that a comprehensive, unified approach to the design of tactical information networks is missing in the Navy's planning. Such an approach is a necessary ingredient for network-centric operations.

E.1 CHARACTERISTICS OF TACTICAL TRAFFIC

It is often said that, to a network, "bits are just bits"—but that is not entirely true. Although it is not the job of a network to interpret the contents of the bits that pass through it, the network must provide the services that its traffic requires. In general, each traffic type within a network has somewhat different requirements. In the specific case of tactical networks, the different types of traffic have quite widely varying needs. Hence, different types of traffic must be marked as such and the network must handle each traffic flow according to the rules established for that type of traffic. In the most extreme cases, entirely separate networks ("stovepipes") may need to be set up for some types of traffic to ensure that their service requirements can be met.

E.1.1 Varying Types of Tactical Traffic

Earlier sections of this report describe current and planned tactical information network use in some detail. This section briefly recapitulates and then proceeds to translate these traffic considerations into more general architectural demands on the tactical network.

• *Sensor feeds.* A wide range of sensor data flows into a tactical network. These feeds range from such heavy flows as imagery and synthetic aperture radar (SAR) from airborne platforms, through lesser flows such as moving-target indicators (MTIs), and down to sporadic and light traffic from cues such as unattended ground sensors or underwater sensors. Most types of imagery do not need particularly accurate transmission; they can withstand fairly high bit error rates. But more highly processed information, such as MTI tracks, require quite reliable delivery.

• *Weapons control.* Real-time control of weaponry can range from the exceedingly time-critical "in-the-loop" applications, such as shooting down incoming missiles in the cooperative engagement capability (CEC) system, to relatively undemanding "update" applications for tracking a slowly moving target. Traffic delivery deadlines can thus range from milliseconds, in the most demanding cases, to tens of seconds in the least demanding. In either case, however, extremely high reliability is required in the delivery of control commands to an in-flight weapon.

• *Common tactical picture.* The common tactical picture is a human-visible representation of the current situation, delivered with the appropriate level of detail so that the operations personnel can understand those aspects of the situation relevant for them and make decisions accordingly. This type of application can be implemented in a number of different ways, and each will impose different types of requirements on the tactical network. As an example, the U.S. Army's Force XXI Battle Command Brigade and Below (FBCB2) program uses a hierarchical reporting mechanism in which each moving platform reports its position to a server, at a rate depending on how fast it is moving. The entire picture is then periodically distributed to everyone and filtered down to its relevant details at each receiver. This particular implementation imposes a fairly heavy "background hum" of position (and related) information on the network, with short messages and required latencies on the order of 1 s. However, high reliability is not required because messages are constantly being resent. If one is lost, the next will probably get through.

• *Tactical command and control (C2).* Many tactical commands will be given by voice. Some fraction, however, will be delivered as data messages.

APPENDIX E   431

TABLE E.1 Notional Summary of Tactical Traffic Requirements

Traffic Type	Distribution	Acceptable Delay	Reliability	Bandwidth
Imagery	Few platforms	High	Moderate	High
Cooperative engagement capability	Few platforms	Very low	Very high	High
Weapon control	Single weapon	Low	Very high	Low
Common tactical picture	All/most platforms	Moderate	Moderate	Moderate
Command and control	Few platforms	High	High	Low

These messages may be created and interpreted by people—for example, typed on a keyboard and later printed and read on paper. Alternatively, they may be created and interpreted entirely by machine. For instance, platforms may automatically report fuel consumption to logistics databases. Different types of messages can have radically different kinds of service requirements: some (such as a call for fire) are urgent and must be reliably delivered; others are more routine or could be dropped if necessary.

- *Notional summary of traffic types.* Table E.1 is not meant to be precise or encyclopedic, but marshals the requirements sketched above into a tabular format to show the degree to which tactical traffic flows differ.

E.1.2 Conclusions Drawn from the Traffic Requirements

This section draws some conclusions based on the tactical traffic mix discussed above. These conclusions help guide the critique here of the Navy's current plans for tactical networks; the committee makes its recommendations in Chapter 4.

E.1.2.1 A Mixture of Predictable and Unpredictable

The first conclusion is that a significant portion of tactical traffic is unpredictable. It can be very "bursty," surging one moment and dying away the next, in response to ever-changing and unforeseeable operational needs.

CEC radar tracks are, for instance, highly predictable. There will be a known number of CEC-enabled ships within a tactical arena, and once the network has formed, the traffic within the CEC's Data Distribution System defined below is quite predictable. The presence of incoming missiles is not easy to predict, but the CEC system is designed so that its traffic load does not change much as it transitions from an inactive state to a fully engaged state.

The flow of imagery also seems at first relatively predictable. There are only so many sources of imagery and they can only be tasked at such a rate, and so the flow of images seems predictable. This sense of predictability may, however, change as smaller unmanned aerial vehicle (UAV) sensors are introduced. One can imagine that sensors would be deployed at relatively short notice; one day there might be only one or two UAVs capturing imagery and the next day, there might be quite a few. In addition, individual imagery streams might turn on and off abruptly as sensors maneuver over target areas, detect activity, and so forth. Finally, variable efficiency compression (e.g., in a JPEG file) can change the traffic load unpredictably.

Distribution of the common tactical picture also seems, initially, to be reasonably predictable. In its simplest form, every platform could report information about itself at regular intervals, and the summary could be redistributed (if necessary) at equally regular intervals. In practice, however, platforms could well adjust the rate at which they sent updates to best match their current mobility, and the number of platforms in an area would be likely to change fairly dramatically over time; indeed the indications of enemy activity would likely change unpredictably.

In summary, a noticeable percentage of the traffic flowing through tactical networks is likely to be unpredictable and/or bursty. Such traffic surges, then dwindles, then surges again, changing from moment to moment. Since tactical networks have very finite capacities, this gives strong motivation to design a network that can take advantage of statistical multiplexing—to let one type of traffic surge fill the bandwidth that is momentarily being left unused by another traffic type.

E.1.2.2 Delay, Priorities, and Reliability Requirements

The next conclusion is that a tactical network must be able to accommodate varying delay requirements, priorities, and reliability requirements. When translated to information-networking mechanisms, these requirements generally devolve into two distinct kinds of priority—*delay priority*, which measures degrees of time-sensitivity, and *reliability*, which defines which messages should be accorded extra error correction when transmitted across radio frequency (RF) links and preferentially retained in memory during overload conditions. With such prioritization, it is inevitably the case that there must be some administrative mechanism for deciding on the relative priorities of traffic and establishing these policies.

E.1.2.2.1 Delay Priorities

Each host is responsible for marking each message, i.e., datagram, with the appropriate delay priority. Each node along the forwarding path must maintain sufficient queuing and forwarding discipline so that it forwards all packets with more stringent requirements before those with less stringent requirements. Possibly datagrams must be marked with expiration times, and routers must delete expired packets from the system; this would help during overload conditions. Possibly the network should implement flow control and not accept packets that it could not deliver in time. Stringent delay requirements will also strongly influence design of radio waveforms. In particular, they affect the way in which the channel is accessed (in multiaccess channels) and the maximum size of a frame that is transmitted in one contiguous piece. In particular, such requirements often lead to waveforms with short frames and bounded-delay channel access schemes.

E.1.2.2.2 Reliability

Each host is also responsible for marking each datagram with its relative requirements for reliability. This translates into how reliably the datagram must be delivered across an RF link, and hence can alter the error-control strategy (e.g., optional fast Ethernet channel (FEC) and/or automatic retransmission queue (ARQ)) for transmission across that link. It also affects how queues are managed within network nodes (primarily routers) when memory is running low: messages with the lowest priority tags are dropped first. Note that queues *often* fill up in routers—this is a normal, rather than exceptional, case. The simplest case is when a fast link leads into a router, but a slow link leads out; for example, a flood of traffic arrives very quickly on an Ethernet but must then be forwarded over a slow radio link. Queue overflows, and thus packet discards, are normal and inevitable in such cases. Important traffic can be protected, however, by assigning it a higher reliability than other traffic.

Stepping back from this detailed discussion, it is apparent that a rather large administrative burden is looming in the background. If indeed traffic is to be prioritized, someone must make all the administrative decisions as to those priorities. Since this involves deciding whose traffic is more important than whose, such decisions have often proved rather difficult to make and enforce. Such problems will not magically disappear as tactical networks become widespread.

E.1.2.3 Future Types of Tactical Traffic

The final conclusion that can be drawn is this: There is currently no "strong family resemblance" among the different traffic types in the network, and hence future traffic added to the network will probably look rather different from the traffic types that are well understood today. This should translate to a relatively heavy emphasis on designing a network that is flexible, open ended, and very easy to modify. Unfortunately, this conflicts with the equally admirable goal of efficiency. If the current set of traffic were going to remain unchanged for decades, it would make excellent sense to optimize every detail of the network and RF waveforms to support this traffic mix. But such is not likely to be the case. It is very difficult to predict exactly what new traffic will be flowing through tactical networks in the coming two or three decades, but very easy to predict that there will be something new—and quite a bit of it, most likely. Hence now is not the time to optimize; now is the time to be flexible.

E.2 CURRENT PLANS FOR TACTICAL COMMUNICATIONS— "HOW IT IS"

The Navy is currently deploying, or planning to deploy, several different types of data communications subsystems on its tactical platforms. This section briefly reviews the most important of these current or planned tactical data links, with an eye to those technical details that make the data links more or less suited to smooth integration into an overall tactical network architecture. It is interesting to note that the Navy's "canonical view" of the emerging architecture for network-centric warfare is not much more than a relabeling of two of its deployed systems and that a number of other tactically significant naval systems—now being planned or envisioned—simply do not fit into this canonical view.[1]

E.2.1 The Canonical View

The canonical "architecture" for network-centric operations consists of three layers. Two of these, the Joint Composite Tracking Network (JCTN) and the Joint Data Network (JDN), are meant to carry tactical traffic. The third layer, the Joint Planning Network (JPN), is presumed to be too slow to carry tactical traffic, although in some futuristic concepts it does appear as part of the tactical infrastructure, e.g., to perform reachback from forward sensors to a Global Broadcast System (GBS) or even a Global Positioning System (GPS) uplink so that the sensor information can be distributed for tactical purposes by satellite.

The Joint Composite Tracking Network is in fact synonymous with the cooperative engagement capability (CEC), and so these terms are used interchangeably here. The CEC is a tightly integrated set of distributed radar systems, each of which is mounted on a different platform. Its internal radio networking system is the data distribution system, which is perhaps best thought of as a specialized set of highly reliable, time-critical protocols that provide a distributed "radar backplane." The Joint Data Network, similarly, is another name for the Joint Tactical Information Distribution System (JTIDS). JTIDS radios are installed on a range of platforms and organized into "nets." Every platform within a given net can share data with all other platforms in the same net. Table E.2

---

[1]Mayo, RADM Richard, USN, Office of the Chief of Naval Operations, N6B, "Information Systems Development, Plans, and Programs," briefing to the committee, January 27, 1999, Washington, D.C.

APPENDIX E

TABLE E.2 Data-carrying Characteristics of the Joint Tactical Information Distribution System (JTIDS) and Cooperative Engagement Capability (CEC) Radio Systems

	Intended Uses	Type	Raw Speed	Open?
JTIDS	Force control messages, common tactical picture, and so on	Time division multiple access	Up to 115 kbps	No
CEC Data Distribution System	Internal distribution of CEC tracks and coordination	Special reliable flood	Classified	No

provides a thumbnail sketch of the data-carrying characteristics of these two radio systems.

E.2.2 Tactical Links Outside the Canonical View

As mentioned above, the "canonical architecture" for networked operations defines exactly two tactical communications subsystems for the Navy, namely CEC and JTIDS. By implication, it would seem these two subsystems are expected to carry all tactical data traffic. In practice, however, several other radio subsystems are also employed, or being seriously considered, and indeed carry rather different types of tactical traffic than those carried via CEC or JTIDS. Table E.3 summarizes each of these noncanonical links. Thumbnail descriptions of some canonical and noncanonical links are given in the following sections.

E.2.3 Joint Tactical Information Distribution System

The Joint Tactical Information Distribution System, also known by the related but not interchangeable terms Management Information and Data System (MIDS), Link 16, or TADIL-J, is the Navy's chosen radio subsystem for distributing "force control" messages within the tactical arena. These messages include surveillance tracks, weapons coordination, air control, target information, precise position location and identification, and even digitized voice networks.[2] JTIDS

---

[2]All information on JTIDS has been derived from two sources: Welch, LCDR David, USN. 1999. "Overview of Links 4, 11, 16, and 22," Office of the Chief of Naval Operations, N62G, Washington, D.C., February 17; and Program Executive Office (Air and Missile Defense), U.S. Army and Army Aviation and Missile Command, Life Cycle Software Engineering Center, and Missile Research, Development and Engineering Center, U.S. Army. 1999. *Introduction of JTIDS (U)*. Available online at <http://jtids.redstone.army.mil/JTIDS_101/sld001.htm>.

TABLE E.3 Summary of Noncanonical Tactical Links

Link	Intended Uses	Type	Raw Speed	Open?
Common data link	Imagery and synthetic aperture radar (SAR) collection	Point to point	200 kbps up, up to 274 Mbps down	Yes
Tactical common data link	Imagery and SAR collection	Point to point	200 kbps up, 10 Mbps down	Yes
Global Positioning System	Weapon target location updates	Broadcast	Unknown	No
Vehicular radio communication-99	Command and control, video conferencing for the Marines	Multihop network	Unknown	Yes

radios—or their MIDS variants—will be installed on a variety of aircraft, surface ships, and submarines over the next 7 years, as well as in Patriot and theater high-altitude area defense (THAAD) forces. Table E.4 lists JTIDS (Link 16) characteristics.

Certain technical characteristics of the JTIDS waveform have important effects on the types of networks that can be built with JTIDS radios, and so such characteristics are briefly described in the following sections.

E.2.3.1 Waveform

JTIDS operates in the L-band. It divides the spectrum into 51 channels between 969 MHz and 1,209 MHz, with a channel spacing of 3 MHz. Certain portions of this spectrum are also used for identification friend or foe (IFF), tactical air navigation (TACAN), distance measuring equipment (DME), and Mode S, which excludes two subbands and imposes some restrictions on exactly how JTIDS can be used in noncombat situations. In particular, time slot duty cycles for JTIDS must be restricted to no more than 20 percent under "normal" conditions. Exercise conditions do not have duty cycle restrictions, and full combat conditions have no restrictions.

JTIDS uses a time division multiple access (TDMA) waveform. Every 24-hour day is divided, in the JTIDS waveform, into 112.5 epochs. Each epoch lasts 12.8 minutes and is subdivided into 64 frames of 12 seconds apiece. Each frame is further subdivided into 1,536 time slots. Each time slot is thus 7.8125 ms long. Time slots within frames are organized into three distinct sets, labeled A, B, and

TABLE E.4 JTIDS (Link 16) Characteristics and Those of Other Tactical Digital Information Links (TADILs)

	TADIL A Link 11	TADIL C Link 4A	TADIL J Link 16	TADIL J Link 22
Anti-jam	No	No	Yes	No
Crypto secure	Yes	No	Yes	Yes
Data rate (kbps)	1.3 to 2.25	5.0	28.8 to 115.2	2.4
Message standard	M series	V/R series	J series	J series
Participants	20	4-8	128+	40
Critical nodes	Yes	Yes	No	No
Voice circuits	No	No	2	No
Architecture	Radio broadcast	Radio point-to-point	TDMA[a]	TDMA[a]
Frequency[b]	HF/UHF	UHF	UHF/spread	HF/UHF spread

[a]TDMA, time division multiple access.
[b]HF, high frequency; UHF, ultrahigh frequency.

C. Time slots within a frame are identified as A-0, B-0, C-0, A-1, B-1, C-1, ..., A-511, B-511, C-511. A given radio ("terminal") may have up to 64 blocks of time slots assigned to it. Each time slot block (TSB) is defined by a triplet: set (A, B, or C); index (0 to 511); recurrence rate (0 to 15). Each assignment for a given terminal is designated as transmit, receive, or relay.

A JTIDS net is a group of terminals that exchange messages among themselves. In other words, it is a group of terminals whose time slots have been defined so that when one member of a net is transmitting, every other member of the net is receiving. Obviously this requires careful planning to ensure that indeed all the other members are receiving at that time, that only a single radio is granted a "transmit" time slot at a given time, and so forth. The JTIDS architecture allows 127 different nets (numbered 0 through 126) to be active simultaneously within the same RF spectrum. Since JTIDS is a frequency-hopping radio, each net is made mutually exclusive by assigning a unique frequency-hopping pattern for transmissions.

E.2.3.2 Access Modes

As defined, JTIDS provides three distinct access modes for a terminal that needs to transmit: dedicated access, contention access, and time slot reallocation (TSR) access.

- *Dedicated access* is the mode described above. In this mode, the network planners ensure—by preparing the corresponding time slot plan for a given network—that a given JTIDS terminal has exclusive use of an assigned TSB. This has the advantage that the terminal is guaranteed access to the network at regular intervals, but it also has the corresponding disadvantage that the time slot goes wasted if the terminal has nothing to say at a given moment.

- *Contention access* is quite different. In this mode, a given net provides a pool of time slots available for any terminal's use. Any terminal that needs to transmit will randomly select a time slot from this pool and transmit in that time slot. This scheme has a number of advantages: it is easy to plan, makes it quite simple for terminals to enter or leave the net while the net is in operation, and provides some of the traffic efficiencies of statistical multiplexing for traffic that is bursty or hard to predict in advance. Its main disadvantage is that multiple terminals may transmit during the same time slot, which can result in lost messages and/or some terminals hearing one transmitter while others hear a different one.
- *Time slot reallocation access* is the most complex mode. As with contention access, all terminals share a single pool of time slots. Rather than transmit at will, however, the terminals perform a distributed algorithm to apportion the time slots. Each terminal transmits its bandwidth needs periodically, and every terminal performs identical algorithms to ensure that the pooled time slots are apportioned as per the needs. It is unknown whether this access scheme has been implemented in practice.

E.2.3.3 JTIDS Data Rates

Each JTIDS time slot has the following components. The time slot begins with a variable start jitter delay, then synchronization and time-refinement patterns, the payload (message header and data), and finally dead time to allow for RF propagation. This discussion concentrates only on the message data portion of a time slot. Each data portion can contain 3, 6, or 12 75-bit words, depending on the exact encoding of the message. Thus each time slot can carry anywhere from 225 to 900 bits of data payload, giving an aggregate data rate for a given JTIDS net of between 28,800 and 115,200 bps. Some of this "raw" capacity is used for housekeeping and so is not available for tactical traffic, but these numbers give a useful yardstick for the approximate capacity of a JTIDS net.

By comparison, current commercial phone-line modems run at roughly 53,000 bps in the downstream direction. Thus one JTIDS net has a raw capacity ranging between roughly one-half and two times that of a phone-line modem. Since JTIDS divides its available L-band spectrum up into 51 channels, the extreme upper bound on the number of bits per second that can be transmitted simultaneously from all JTIDS terminals in a tactical arena is thus $51 \times 115{,}200 = 5{,}875{,}200$ bps. (This assumes that all available spectrum is devoted to JTIDS, that all terminals use the maximum possible data rate, and that all time slots in all channels are used for transmission and ignores the overhead of housekeeping bits.) Working from the previous calculation, JTIDS achieves 5,875,200 bps in 51×3 MHz of RF spectrum, for an aggregate spectral efficiency of 0.0384 bps/Hz. Partly, of course, this is driven by the tactical need for very robust anti-jam features. To a noticeable extent, though, it is driven by the basic short-frame TDMA structure of the JTIDS waveform where rather short payloads are surrounded by the dead times of synchronization patterns and propagation allowances.

E.2.4 Common Data Link and Its Variants

The common data link (CDL) is a family of radios that provides standardized, wide-band, line-of-sight (LOS) communications between airborne reconnaissance sensors and their users.[3] The common data link has been mandated by the Assistant Secretary of Defense (Command, Control, Communications, and Intelligence) (ASN (C3I)) as the data link for all imagery and signals collection systems. The Navy is currently acquiring ship-mounted CDL terminals, under the name common high-bandwidth data link (CHBDL).

At the RF level, CDL has both X- and Ku-band options. In the X-band, the command link (up) is 9.750 to 9.950 GHz, and the return link (down) is 10.150 to 10.425 GHz. In the Ku-band, the command link is 15.15 to 15.35 GHz, and the return link is 14.40 to 14.83 GHz. The tuning increment is 5 MHz in both bands. Typical airborne antennas are 7- or 9-inch directional dishes, though some applications use omnidirectional antennas. Typical surface antennas are 1 m or 6 ft dishes.

The CDL command link runs at 200 kbps. It employs binary phase shift key modulation, Viterbi convolutional encoding (rate 1/2, constraint length 7), interleaving, and pseudo-noise spreading. The command link uses a 40-bit frame to structure a sync channel, link command and control channel, audio channel, and from 1 to 10 subchannels for Prime Mission Executive Command. The CDL return link runs at a selectable data rate: 10.71 Mbps, 137 Mbps, or 274 Mbps. It employs offset quadrature phase shift key modulation, Viterbi convolutional encoding (rate 1/2, constraint length 7), and interleaving. The return link multiplexes a number of data channels, each running at a fixed rate, onto the overall link.

TCDL is a Defense Advanced Research Projects Agency (DARPA) program that will develop multiple qualified sources for a lightweight, low-cost, CDL-

---

[3]All information on CDL and TCDL has been derived from two briefings presented to the committee's Tactical Networks and Resources Panels on April 20, 1999: Schuh, CDR Paul, "CHBDL: Common High Bandwidth Data Link," Office of the Chief of Naval Operations (N641), Washington, D.C.; and Preziotti, Gerry, "Common High Bandwidth Data Link—Surface Terminal (CHBDL-ST)," Johns Hopkins University, Applied Physics Laboratory, Laurel, Md.

interoperable data link for small UAVs. TCDL systems will be less capable than full CDL systems; in particular, they will only implement a slower return link.

At least one manufacturer provides a CDL variant with asynchronous transfer mode (ATM) interfaces. This allows transmission of arbitrary ATM data between aircraft and the surface ship. Since the Internet Protocol (IP) can be deployed over ATM links, it also allows tying the aircraft into a wider-area military internet at the appropriate security level.

E.2.5 Networks for the Land Forces (Marines, Army)

In the near future, the U.S. Navy will need to connect its networks with a number of tactical land-based networks. In particular, Marine Corps and Army forces will be important partners in tactical networks, and the Navy's networks must be designed to integrate smoothly with the networks that will be provided by these two Services. In the longer term, the U.S. Navy must also interoperate with an ally's networks; but this whole area is sufficiently ill-defined that it is not discussed here. It is only noted in passing that the armies of two important allies (Canada and the United Kingdom) have both determined that their tactical networks will employ Internet technologies, which does provide a clear path toward interworking with the Army and the Marine Corps.

The Marine Corps is experimenting heavily as it moves toward using tactical networks. Two efforts deserve particular note. The first is their acquisition of the new Tactical Data Network (TDN). The other is the recent advanced concept technology demonstration (ACTD), Extending the Littoral Battlefield (ELB). Each is discussed in turn.

The TDN is built from standard Internet technology, using ruggedized Ethernet hubs, conventional servers for e-mail, (commercial) routers, and standard network management tools. Networks are linked together using the Internet protocols running over standard line-of-sight trunk radios (AN/MRC-142) or satellite transponders.

The ELB demonstration aimed to show how the Marines would fight in the future. Dispersed forces ashore went beyond line of sight and called back (via voice and data networks) for fire. A number of different types of communications technology were employed on an experimental basis.[4] Most interestingly, the Marines created a wide-area, mobile ad hoc network using a combination of VRC-99 packet radios with commercial 802.11 WaveLAN radios. The entire network was based on Internet technologies. Radios were deployed at sea, aloft,

---

[4]All information on ELB is derived from the following report: Cole, Ray. 2000. *Office of Naval Research Demonstration Manager's Campaign Plan: Extending the Littoral Battlespace (ELB) Advanced Concept Technology Demonstration (ACTD)*. Office of Naval Research, Arlington, Va., forthcoming.

and on the ground to form the network links. The VRC-99 radios provided a self-organizing radio network that acted as a long-range "backbone" for data communications. Local wireless communication tied into this backbone via the commercial WaveLAN radios. Although the demonstration encountered a number of technical problems, it was the first real instantiation of the Marines' vision for how they will use networking.

The U.S. Army's tactical networks can be divided into a few basic categories: far-forward, highly mobile networks (the Tactical Internet); movable command centers (Tactical Operations Centers); and rear, relatively fixed networks (Warfighter's Internet—Terrestrial). The first two have clearly fixed upon the Internet architecture. The devices in these networks are standard PCs or UNIX computers connected to the network via standard serial lines or Ethernets. Routing is performed by a mixture of commercial routers and specialized military routers. The Warfighter's Internet has not yet been completely defined, but it clearly will interoperate with the other Internet-based Army networks as well as the more fixed strategic networks (SIPRNET).

E.2.6 Future Tactical Communications Systems

A number of other tactically significant communications subsystems are currently in the planning or ideation stages. These subsystems are not discussed here, but it is important to recognize that they are on the drawing boards and may within a decade become real components of the Navy's tactical networks. These systems range all the way from Discoverer II, envisioned as a low-Earth-orbiting (LEO) constellation of radar-imaging satellites that provides near-real-time imaging to forward-deployed forces, down to acoustic communications links between ships and submerged submarines. One can also imagine commercial networks as part of this overall blend, e.g., transporting tactical data via Iridium or GlobalStar satellite phones. The associated link speeds range across many orders of magnitude, and the types of traffic expected across these links do not fit well into the Navy's canonical view of its "emerging architecture."

E.3 THE VISION—HOW IT SHOULD BE

The network-centric vision depends strongly on one's vantage point. At the highest level, it encompasses such operational benefits as self-synchronization of distributed forces. At the application level, it encompasses concepts such as distributed radar systems, real-time targeting of weapons in flight, and the like. The vision as seen at the *networking* level is noticeably more prosaic but, the committee argues, equally important. This section explores the network-centric vision from the perspective of Layer 3 of the Open Systems Interconnection (OSI) model, i.e., in terms of networking architecture and technology.

E.3.1 Network Services That Are Medium-Independent

The medium is *not* the message. Any type of traffic should be able to flow across any kind of communications link, and all decisions about which types of data should flow across which radio networks should be *policy* decisions rather than cast-in-concrete engineering design decisions. If there is some reason that imagery should flow across JTIDS nets, it should require nothing more than changing a few policy rules. And one should not need to upgrade JTIDS radios in order to carry new types of messages, any more than one upgrades one's Ethernet transceiver in order to receive new types of Internet traffic.

E.3.2 Seamless Connectivity Across Joint Forces

All types of application traffic should be able to flow smoothly across the tactical, and indeed strategic, networks of the joint forces. The Navy afloat should be able to exchange imagery, tracks, and the common tactical picture with the Marines and Army on the ground, with the Air Force, and indeed with allies and coalition forces. It should not matter that the forces use different types of radios; the messaging should be independent of such lower level details. Connecting one Service to another should be a "plug-and-play" operation, with all addressing and routing taking place automatically, rather than a laborious, time-consuming process that relies on specialized gateways. At the strategic level, each Service's network connects in a simple, straightforward way. There is no "protocol translation gateway" needed at a special connection point in order for the Army to access the Navy's Web sites. Common protocols and a common addressing plan obviate this need. The same should be true for each Service's tactical network.

E.3.3 Greatly Reduced Planning and Management

To the largest extent possible, tactical networks should be plug and play. Extensive planning and finicky configuration are a bad match for the chaotic, fluid tactical environment. Commanders-in-chief (CINCs) should be able to compose forces in ways that make operational sense without being constrained by artifacts of the communications systems. The networks should gracefully adapt to the offered traffic load as it changes from minute to minute, switching automatically from handling heavy imagery flows across its RF links at one moment to handling radar tracks and location information the next.

E.3.4 Network Security

Tactical networks must be secure against intruders, disruption, and eavesdropping. Loss or compromise of network nodes must not imperil the remainder

APPENDIX E 443

of the network. Intruders should not be able to insert false data into the networked information systems, nor should they be able to deny service to network participants.

E.3.5 Open Systems That Take Advantage of Commercial Standards

Finally, insofar as it is feasible, the Navy's tactical networks should employ open systems based on commercial standards. These days, military networking technology cannot begin to compare with its commercial counterparts, and the gap is growing wider month by month. The Navy cannot afford to design and build all its own network technology; and even if it could, the military technology would be obsolete long before it was deployed.

E.4 DEFICIENCIES IN THE NAVY'S TACTICAL NETWORKING PLANS

The communications links that the Navy plans to acquire do not add up to a coherent network architecture. They remain instead a compendium of discrete, separately purchased radio systems. Although the Navy often portrays these systems as an emerging architecture, there is in fact no underlying system architecture for these links and, hence, no actual network. And since the systems fail to leverage commercial networking technology, they have already fallen far behind commercial standards in networking.

E.4.1 Stovepipes

Military stovepipe systems have been decried for years, so this discussion is brief. Each naval RF data subsystem has been bought and engineered as a stand-alone solution to some unique problem. Each is a point solution. To a large extent, a given type of service *equals* the corresponding radio equipment. The overall result is that each RF data system is rigidly structured to provide a specialized set of services. Thus the JTIDS radio provides target information, the CEC radio (DDS) provides radar track distribution, and so forth. Conversely, if a user wants target information, he or she must install a JTIDS box.

Until a few years ago, wireless services in the commercial world were similarly fragmented. One type of radio was used for paging, another for cellular telephony, various others for specialized data services, and so forth. But these distinctions have rapidly blurred as service providers move to converge on only two networking schemes that are deployed across a very wide variety of radio technologies: *telephony-compatible voice services* (across standard waveforms such as the Global System for Mobile Communications (GSMC), code division multiple access (CDMA), or proprietary waveforms), and *Internet-compatible data services* (via General Packet Radio Service (GPRS), IS-95B, or proprietary solutions). The Navy should understand these trends and draw the appropriate lessons for its own networks.

E.4.2 Closed, Proprietary Systems

A further problem is that these stovepipe systems are built as closed, and generally proprietary, solutions. In standard networking terms, this means the following:

- The networking protocols are not *open*. As opposed to most commercial networking protocols, there are relatively high barriers to entry for new companies that wish to provide compatible, interoperable technology for naval RF systems.
- The networking protocols are not *layered*. Modern networks are built of layered protocols, where each protocol is carefully isolated from the layers below and above it in the protocol stack. This layering is generally described in terms of the OSI reference model, although existing network technology does not exactly match the layers used in this descriptive model. Naval RF systems generally do not make sharp distinctions between the layers, with the result, for instance, that it requires major effort to ensure that a JTIDS radio can convey a new type of message. Contrast this with the Internet, where new forms of messaging and communications are freely invented and can be carried across the existing network structure with no modification to the network's routers.
- These systems employ *little or no commercial networking technology*. As such, they fail to leverage what is possibly the fastest-developing technology of the current era. By contemporary standards, "networks" such as JTIDS already seem archaic. Since they are missing out on the extraordinarily rapid development of networking at the current time, it is hard to imagine how out-of-date these naval systems will be when they are fully deployed. The Navy long ago stopped designing and building its own computers; now it is time to stop designing and building its own network technologies.

E.4.3 Lack of Flexibility

A further characteristic of naval RF systems is that they require a great deal of advanced planning before they can be deployed. A certain amount of planning is inevitable—frequencies must be allocated, keys must be distributed, and network names and addresses must be assigned. Systems such as JTIDS, however, introduce a level of planning complexity that makes it difficult to freely "mix and match" tactical forces in the field. JTIDS frequency and TDMA slot planning is complex enough so that they take days or weeks to complete. A task force sets under way with (say) five JTIDS plans. If a contingency arises that cannot be accommodated by one of these preset plans, improvisation is difficult.

Certain aspects of this planning problem could have relatively straightforward technical solutions. For instance, organizing the RF links into a small set of broadband links—rather than a large set of narrowband links—would help reduce the frequency planning problem. Switching from TDMA channel access to prioritized carrier sense multiple access/collision avoidance (CSMA/CA) channel access, or a hybrid TDMA plus CSMA/CA scheme, would eliminate or reduce the slot-planning problem. This is certainly quite different from the current types of radio technology being deployed by the Navy, but technically quite feasible. In fact, such schemes are widely used in the commercial world, e.g., in wireless local area networks (LANs).

Other aspects of the planning problem are much harder to solve. In fact, current commercial networking technology actually introduces new planning problems. Conventional data networks typically involve a good deal of planning (of addresses, names, and so on); and since such networks generally change rather slowly, there is little commercial emphasis on plug-and-play networks or on good tools to help with network planning.

E.4.4 No Good "Data Haulers"

Compared with the commercial world, the Navy will be very bandwidth-poor. This is partly inevitable, given the exigencies of the tactical environment. To some extent, however, the scarcity is avoidable. The Navy is simply not planning to buy many high-capacity tactical radios.

Calculating the relative bandwidths of tactical radios and commercial data radios makes this point quite apparent. Commercial wireless networking distinguishes between "access" networks such as wireless LANs, and "backbone" networks that are typically point-to-point links. Current wireless LANs run over small omnidirectional antennas at a maximum speed of about 10 Mbps, although they often fall back to much slower speeds (e.g., 1 Mbps) due to interference, multipath, and the like.[5] Current point-to-point links run over directional antennas at speeds between about 10 Mbps and 155 Mbps. The Navy's plans for the CDL and DDS radios are sufficiently specialized that it is probably unrealistic to think of these radios as generic "data haulers" for naval tactical networking. This leaves JTIDS as the only plausible radio for transporting generic data across the battlespace. A JTIDS terminal's maximum throughput of roughly 115 kbps thus provides only 11 percent of the bandwidth of a slow wireless LAN (1 Mbps), only 1 percent of the capacity of a fast wireless LAN (10 Mbps), and less than 1/10 of a percent of the capacity of a high-speed (155 Mbps) commercial RF link.

Thus one can summarize by stating that the Navy's tactical data radios will

---

[5]Wireless LANs are power constrained by Federal Communications Commission regulation; some of these problems could be solved by simply boosting the transmitter power, antenna, and so on.

provide bandwidth that is somewhere between *one and three orders of magnitude less* than those used in the commercial world. It is true that low-probability-of-interception/low-probability-of-detection (LPI/LPD) and anti-jamming (AJ) features of the waveform account for some of this difference. The bottom line remains, however, that the Navy will not—with current plans—have a great deal of bandwidth for its tactical networks.

E.4.5 No Coherent Plans for Information Network

The problems outlined in previous paragraphs are symptomatic of a much larger issue: the Navy has no overall plans for its tactical information network. Such plans would include the following:

- A unified addressing and naming plan for nodes in the network;
- An overall list of the types of tactical traffic that flow across the battlespace, the sources and destinations of this traffic, and their required service characteristics;
- A routing plan for the network;
- Designs for how the tactical network fits into the larger picture of each other Service's tactical network, the Joint Planning Network, and so forth; and
- Engineering of RF links and spectrum to meet network needs.

E.4.6 No Overall Plan for Network Security

The Navy has no overall plan for security in its tactical networks. Security is part of the *foundation* of an information network architecture—it is extremely difficult to add after the fact. Years of work have gone into adding security to the Internet after the fact, but the results are still far from satisfactory. Web sites continue to be hacked, and intruders continue to gain access to poorly protected computers. The Navy should take care that its tactical networks are at least as secure as those provided by commercial vendors.

The key point here is that networked naval warfare introduces new vulnerabilities. Enemy capture of a network node means that the enemy is now *inside* the Navy's network. If this seems a remote possibility, remember that the naval tactical networks will be closely linked into ground networks for the Marines and the Army. Overrun of a tactical node can be something as simple as capture of a single wheeled vehicle. Once an enemy has captured a functioning network node, it can take some time before anyone notices and excises that enemy from the network. During this time, an enemy can cause a great deal of damage; some of it may last far longer than the node itself. For instance, they may spoof the common tactical picture, adding fictitious elements to it. They can also engage in various types of network denial-of-service attacks.

Link encryption plus over-the-air rekeying is *not* a sufficient answer to this problem. Full-scale network protection requires a number of additional technologies, including firewalls, intruder detection, layered defenses, and so forth. Commercial technology provides limited help in this area. It can supply firewalls, an infrastructure for public-key management, key exchange protocols, authenticated routing protocols, and so forth. But at present no commercial technology can adequately deal with the "enemy capture" problem. DARPA is tackling a wide range of issues with its Information Assurance program, but even so, it is unlikely that tactical networks will be sufficiently well protected in the near to mid term. This should be of major concern to the Navy.

E.5 HOW TO ACHIEVE THE VISION

This section provides a series of high-level observations on how the Navy might proceed as it develops its overall tactical network architecture. There are no startling insights here; rather, these observations elaborate on the usual steps of systematic network design. However, the committee has seen no signs that the Navy has yet gone through these steps, and so encourages the Navy to do so before beginning to design or roll out the operational subsystems. In particular the committee makes note of the following:

- The Navy should recognize that every network needs an architecture and immediately establish an overall tactical network architecture that includes published (and enforced) architectural standards documents and layered protocol stacks. This architecture should recognize that the tactical network is a "network of networks" and explicitly plan for independent upgrade paths of all components and communications subsystems within this overall network.
- The Navy should explicitly determine which points on the stovepiped-open continuum are suitable for interconnecting its various communications subsystems, with justifications as to why those points are the best design choices.
- The Navy should explicitly determine the best blend of commercial and military networking technology for its tactical networks, with detailed justifications for why military technology is required for any given part of the design. The committee believes that the commercial Internet Protocol should be the basis for this design.
- The Navy should provide open, commercial-compatible interfaces to certain of its tactical communications systems (JTIDS, CDL) so that standard networking technologies can be used in conjunction with these subsystems.

The committee has not found any definitive, Navy-wide list of arguments against using commercial technology in tactical networking. However, a number of arguments seem reasonably widespread, such as: Tactical links require anti-jam protection and Type 1 encryption, tactical traffic is life-critical and delays must be guaranteed bounded, and tactical RF links do not have enough capacity to carry conventional networking traffic. These issues are important and deserve serious attention. However, it is by no means evident that they rule out commercial technology. Accordingly, the next section presents the committee's view on the blend of commercial and military technology. After that, two further matters pertinent to achieving the overall vision are discussed—opening radio architectures and areas of needed research.

E.5.1 Blend of Commercial and Military Technology

E.5.1.1 Lower Layers (Radios)

The Navy, which employs a wide variety of radios (OSI layers 1 and 2) in its tactical networks, needs lower layers (radios) of the following broad classes:

- Multiple-access (shared-channel);
- High-capacity, point-to-point; and
- Satellite links.

At the very least, the Navy needs anti-jam protection for most of its tactical links. In many cases, it is also desirable that its radios have LPI and LPD. None of these features is likely to be implemented to any significant degree in commercial radios, and so it is likely that the lower layers (radios) in the Navy's tactical network will be predominantly military. Some links, however, may be implemented via commercial radios, and even in military-specific radios, many of the component technologies may be commercial.

E.5.1.2 Upper Layers (Applications)

Layers 4 and above of the protocol stack, i.e., the applications and the protocols that they use, will likely be a mix of commercial and military. Some of the tactical applications are really indistinguishable from normal commercial uses, for example, transfer of large imagery files. There is very little justification for creating specialized military applications or protocols for such purposes. Other tactical applications have no commercial analog. A good example is distribution of the common tactical picture. This will require a highly specialized application program and indeed most likely a specialized protocol to distribute the necessary data; so this application and protocol will likely be military specific. Many other applications will involve a blend of commercial and military networking technologies.

E.5.1.3 Middle Layers—The Missing Piece

The lower layers will be predominantly military; the upper layers will be commercial mixed with military. What about the "networking" layer, i.e., Layer 3 of the OSI stack? Chapter 4 recommends using the IP for all links where feasible and specialized military protocols for the rest. Hence, here both its strengths and its weaknesses are discussed briefly in this context.

E.5.1.3.1 Strengths of the Internet Architecture

The main strengths of IP as a standard "bearer" protocol are that it is ubiquitous, cheap, and very rapidly developing in the commercial world. Since virtually everyone is now aware of the astonishing growth of the Internet, this point is not belabored here. But it is a very real advantage.

Furthermore, an IP-based tactical network architecture will connect seamlessly with the Navy's IT-21 network, the Marine's TDN, and all of the Army's forward area networks, as well as the networks being deployed by the UK and Canadian armies. These networks are all Internet-based and so, given the appropriate policies and attachment points, traffic should be able to be freely exchanged across all these networks (or restricted or even forbidden, if the policy so dictates).

The Internet architecture also offers excellent network security tools. Its IPSec protocols provide a highly extensible framework for performing end-to-end security with a variety of authentication and privacy features. Internet firewalls are commodity items. New security techniques, including developments in DARPA's Information Assurance program, are geared toward the Internet. No other networking technology has anything near this range or depth of security apparatus.

Finally, training would be simplified if the Internet architecture were adopted for the Navy's tactical networks. Since Internet technology is a commodity item, books, videotapes, and training courses abound. The Navy would not need to develop special training materials to teach its staff how to configure, manage, or troubleshoot its tactical networks.

E.5.1.3.2 Weaknesses of the Internet Architecture

The Internet architecture does have weaknesses, however, ones that are highlighted in the tactical networking environment. The key weaknesses of IP for tactical environments are as follows:

- *IP header overhead.* A "bare" IP datagram requires 20 bytes of header information; to this, the User Datagram Protocol upper layer adds 8 bytes, or the Transmission Control Protocol (TCP) adds 20 bytes. If the IP packet's payload is very short, this header overhead can be significant. However, this problem can be greatly reduced by employing some variant of the standard IP header compression protocols. On average, such protocols can reduce the IP header to a few bytes.

- *Poor transmission control protocol performance on tactical RF links.* The TCP is sensitive to dropped packets and large degrees of jitter in packet arrival times; in particular, it treats such events as if they were caused by network congestion and responds by sharply reducing its throughput. This gives rise to poor TCP performance over many types of low-speed (e.g., 10 kbps or lower) radio links, including commercial cellular phone systems and tactical radios. Although some engineering can help to lessen these problems, the wisest course might be to add sufficient forward error correction to higher bandwidth links to make packet loss fairly rare, and to forgo using TCP across low-bandwidth tactical links.

- *Problems with mobility.* The Internet Protocol suite implicitly assumes that a computer does not move very often from one network to another. When this assumption is violated, a number of protocol problems occur. For instance, operating systems may need to be power cycled to obtain new IP address information, existing data sessions may need to be torn down and restarted, name to address mappings may become obsolete, and so forth. This problem is too complex to discuss here, and there are some efforts to revise the Internet protocols to lessen or eliminate this problem (IP mobility). However, issues still exist and it is unclear whether IP mobility will in fact become widely adopted. This area should be noted as a concern for those tactical platforms that move from one network to another.

- *Configuration.* Typical Internets require a great deal of "behind the scenes" configuration to make them work. A wide variety of servers and routers need separate, but consistent configuration. Network addresses and masks must be consistently assigned, dynamic host configuration protocol (DHCP) and domain name service (DNS) servers must be properly set up, management terminals must be correctly identified for all the manageable elements of the system, a wide variety of options must be set on commercial routers, and so forth. Misconfigured computers and routers are relatively common in the commercial world and would likely be even more common in tactical environments. This area would need serious, sustained effort to ensure that the administrative burden is not excessive. (On the other hand, most types of networking introduce a comparable set of configuration and administration issues.)

E.5.2 Potential for "Opening" Existing/Planned Radios

Previous sections point out that the Navy's current radio links are in essence stovepipe systems and that at present these links tightly couple the RF technology with the types of messages that can be transported across the links. These sections then go on to note the advantages of using a modern, layered network architecture, and the desirability (where possible) of employing a single, universal bearer protocol such as IP across all the RF links.

What would it take to "open up" the Navy's existing and planned data radios so that they could carry any type of message? The internal details of these radios are not known, and so the committee cannot give a detailed answer to this question. However, the committee has had experience with a similar task for various other kinds of radios, and in general, it has been technically feasible and not extensively costly. It is a question that the Navy should be posing to its contractors.

Table E.5 summarizes the committee's views about opening the Navy's three major tactical data radios.

TABLE E.5 Opening the Navy's Major Tactical Data Radio Systems

	Joint Tactical Information Distribution System	Common Data Link, Tactical Common Data Link, etc.	Data Distribution System
Current use	Track distribution	Intel sensors (imagery, synthetic aperture radar, . . .)	Cooperative engagement capability "backplane"
What is required to make radio open	New Internet Protocol message format, standard interface	Contractor already has assisted transfer mode, Ethernet interfaces	New IP message format, standard interface, very fast/reliable cutover to radar
Good points	Most widespread data radio (sea, air)	High-bandwidth surface/air link; good for relaying communications	High-bandwidth link among some ships and planes; excellent anti-jamming
Bad points	Spectrum issues; Time division multiple access channel access	No reliable anti-jamming? Lacks encryption?	Development costs? High nonrecurring expense test costs; spectrum issues
Should it be opened?[a]	Yes	Yes	Unclear

[a]Response summarizes committee's views.

E.5.3 Research Opportunities

Most of the transition to a tactical network suitable for network-centric operations can be accomplished with a solid network architecture and sound systems engineering, followed by intensive development work on the various communications subsystems. Such work is hard, but it is not rocket science. Some new tactical networking technology is desirable, though, and will require an R&D effort as it will most likely not be developed in the commercial sphere.

E.5.3.1 Better Channel Access for JTIDS

The JTIDS waveform defines several channel access mechanisms. None is very close to the current commercial practice, e.g., as embodied in the Institute of Electrical and Electronics Engineers (IEEE) 802.11 wireless LAN standard. The 802.11 waveform conducts a transmission via four discrete steps: (1) the transmitter sends a request to send (RTS); (2) the receiver sends a clear to send (CTS); (3) the transmitter sends a variable-length data frame; (4) the receiver sends an acknowledgment. This waveform, together with its associated state transition diagrams, gives a reasonable CSMA/CA channel access protocol that guards against the classic hidden terminal problem. Research and experimentation with similar waveforms could bring significant benefits to the JTIDS waveforms in terms of allowing JTIDS to carry higher levels of bursty traffic than they can now successfully transport.

E.5.3.2 Potential Bandwidth Increases

Bandwidth is always a scarce commodity in the tactical world, but it is likely that the Navy will encounter quite severe bandwidth shortfalls in the future. There are a number of technical approaches that could help, and research in this area could prove very useful. Current naval plans for tactical networking expect that the data radios will be running at quite high power to ensure that the receiver is within earshot of the transmitter. This is an effective technique but leads to inefficient use of the RF spectrum. In particular, it does not encourage local spatial reuse, i.e., the reuse of the same RF spectrum at geographically separated regions. The Navy's plans are perhaps most clearly highlighted by contrast with commercial cellular systems, which impose tight power controls on their transmissions in order to reuse the same frequencies multiple times within a metropolitan region. If the cellular systems worked like Navy radios, they would support only a few dozen simultaneous phone calls over an entire city. In addition, JTIDS' overall capacity as a data hauler is quite low; a different waveform designed for wideband, packetized communications could likely do better and still have the requisite anti-jam properties.

E.5.3.3 Ad Hoc Networking

Ideally, Navy tactical networks would form automatically to include all nodes in a given area. In addition, nodes would automatically begin to act as relays when needed to forward traffic on to further nodes that are outside radio range of the original transmitters. Such networks are termed ad hoc networks or sometimes mobile ad hoc networks (MANETs). There has been little commercial interest in such networks, as commercial service providers can usually arrange to carefully plan their RF base stations, bring wireline connectivity to the base stations, and so forth. This obviates the commercial need for such ad hoc networking technology. DARPA and other military agencies, on the other hand, have been funding research into ad hoc networks for some years, and the first or second generation of workable ad hoc networks is now up and running. These networks include near-term digital radio (NTDR), VRC-99, Global Mobile Information Systems, and others. These networks appear to work reasonably well; indeed, the NTDR has even been adopted for use in the Army's First Digitized Division. The technology is far from mature, however, and R&D would likely have very useful payoffs.

ANNEXES TO APPENDIX E

Annex 1 Imagery and SIGINT Dissemination Characteristics

This annex presents reference information on direct dissemination from imagery and SIGINT platforms (Table E.A.1) and indirect imagery dissemination systems (Table E.A.2).

TABLE E.A.1 Imagery/SIGINT Platform Characteristics (Direct)

Platform	Sensor	Data Link	Link Data Rate	Mode
U-2	SAR	CDL	274 Mbps	LOS
	EO/IR	CDL	274 Mbps	LOS
	Other	CDL	274 Mbps	LOS
	SAR	ETP	274 Mbps	Sat
Global Hawk (HAE UAV)	SAR, EO/IR, MTI	CDL	274 Mbps	LOS
	SAR, EO/IR, MTI	Ku	50 Mbps	Sat
F-14	EO (TARPS)	CDL	137 Mbps	LOS
F/A-18	SAR (APG-73)	CDL	137 Mbps	LOS
	EO/IR (ATARS-USMC)	CDL	137 Mbps	LOS
	EO/IR (SHARP-USN)	CDL	274 Mbps	LOS
Predator (MAE UAV)	Video (EO/IR)	Legacy	Analog, ~1.5 Mbps	LOS
	SAR	Legacy	3.0 Mbps	LOS
	Video (EO/IR)	Legacy	3.0 Mbps	Sat
	SAR	Legacy	3.0 Mbps	Sat
Pioneer (TUAV)—USN	EO/IR/Video	Legacy	Analog, ~512 kbps	LOS
Pioneer (TUAV)—USMC	Video	Legacy	Analog; currently tape only	LOS

APPENDIX E

Surface Terminal	Data Link Receiver	Processor	Remarks
JSIPS-N/TEG	CDL-Navy-ST/TIGDL	CIP	
JSIPS-N/TEG	CDL-Navy-ST/TIGDL	CIP	
BGPHES-ST	CDL-Navy-ST/TIGDL	BGPHES	
CONUS	Unique		Indirect only via NIS/CA
MCE/JSIPS-N/ TEG	CDL-Navy-ST/TIGDL	CIP	GH now limited to 50 Mbps rate
Satellite receiver (ashore)	TIGDL	CIP	Satellite relay—actual data rate function of satellite transponder power
NAVIS/JSIPS-N	CDL-Navy-ST/TIGDL	CIP	Some TARPS sensors still film based; not all TARPS have data link; TARPS sensor to be replaced by SHARP
JSIPS-N/TEG	CDL-Navy-ST/TIGDL	CIP	
JSIPS-N/TEG	CDL-Navy-ST/TIGDL	CIP	Includes medium-altitude electro-optical and low-altitude electro-optical and infrared line scanner sensors
JSIPS-N/TEG	CDL-Navy-ST/TIGDL	CIP	CIP upgrade for SHARP planned
Legacy	Legacy	N/A	Data link upgrade planned (TCDL)
Legacy	Legacy	OBP	Data link upgrade planned (TCDL)
Legacy	Legacy	N/A	Data link upgrade planned (TCDL)
Legacy	Legacy	OBP	Data link upgrade planned (TCDL)
Legacy	Legacy	N/A	Pioneer, a legacy system, to be phased out when VTUAV becomes operational
USMC G/S (control link)	Legacy	N/A	Analog 512 kbps data link currently not used; Pioneer, a legacy system, to be phased out when VTUAV becomes operational

Table continued on next page

TABLE E.A.1 Continued

Platform	Sensor	Data Link	Link Data Rate	Mode
VTUAV	Video	TCDL	1.5 to 10.71 Mbps	LOS
P-3/EP-3	Video	Legacy	Analog, ~256 kbps	LOS
	SIGINT	Legacy		LOS
Joint STARS	MTI	SCDL	41 to 56 kbps	LOS
	SAR	SCDL		LOS
Rivet Joint	SIGINT	TADIL A and J	256 kbps	LOS

NOTE: Approximately 28 ships (12 carriers, 12 large-deck amphibious assault ships, and 4 command vessels) will be outfitted with JSIPS-N, CDL-Navy, Challenge Athena, and Global Broadcast System (GBS). This is the core of USN imagery afloat, and it supports timely direct and indirect imagery tasks. Additionally, GBS will be used to expand imagery dissemination to the larger fleet with a lower level of imagery functionality and smaller image products. The Annual UAV Report and the Manned Airborne Reconnaissance Plan previously published by Defense Airborne Reconnaissance Office (DARO) are useful sources of unclassified data regarding aircraft and related sensor data.

DEFINITIONS:
  AWACS, Aircraft Warning and Control System
  BGPHES, Battle Group Passive Horizon Extension System
  CDL, common data link-Navy; formerly called common high bandwidth data link
  CDL-Navy-ST, common data link-Navy surface terminal; ~29 terminals funded
  CGSM, Common Ground Station Module, receives JSTARS data
  CIP, common imagery processor
  CTT, commander's tactical terminal
  ETP, Extended Tether program
  HAE UAV, high-altitude endurance unmanned aerial vehicle (UAV)
  JSIPS, Joint Services Imagery Processing System; USMC and USAF ground station;
    1 funded for USMC
  JSIPS-N, Joint Services Imagery Processing System-Navy; USN surface station;
    ~29 afloat and 4 ashore funded
  JTT, joint tactical terminal; replacement for CTT and tactical receive equipment (TRE)
  Legacy, program-unique systems

Surface Terminal	Data Link Receiver	Processor	Remarks
TCS/JSIPS-N/TEG	TCDL	N/A	Pioneer replacement; under development
Legacy	Legacy	N/A	Data link upgrade planned (TCDL)
Legacy	Legacy	N/A	
CGSM/afloat	SCDL	N/A	USN planning for JSTARS transitioning SCDL to TCDL
CGSM/afloat	SCDL	N/A	
CTT	TADIL	N/A	CTT being replaced by JIT

LOS, line of sight
MAE UAV, medium-altitude endurance UAV, Predator
MCE, mission control element; HAE UAV ground station for advanced concept technology demonstration (ACTD) activities
NIS, National input segment
OBP, on-board processor
SHARP, Shared Reconnaissance Pod; previously Super Hornet Airborne Reconnaissance Pod
TARPS, Tactical Airborne Reconnaissance Pod System
TCDL, tactical common data link
TCS, Tactical Control System; tactical UAV ground station; may merge into both JSIPS-N and TEG in future
TEG, Tactical Exploitation Group; USMC ground station; 3 funded
TGIF, Tactical Ground Intercept Facility
TIGDL, tactical interoperable ground data link
TIS, tactical input segment; part of JSIPS-N; TIS = CDL-Navy-ST and CIP and screener workstation and support equipment
TUAV, tactical UAV
VTOL, vertical takeoff and landing

SOURCE: Compiled from data courtesy of National Imagery and Mapping Agency, Bethesda, Md., 1999.

TABLE E.A.2 Imagery Dissemination Systems (Indirect)

Dissemination Architecture	Communication System	Communications Transport	Receiver Type
DDS			
Shore sites	DISN	Fiber/SATCOM	DE
GBS	GBS	SATCOM	GBS
MTACS			
Ship	Fleet SATCOM	UHF SATCOM	Fleet SATCOM
Shore	DISN	Fiber/SATCOM	POS/DSCS/ Tri-band
Shore	Trojan Spirit II	SATCOM	TS II
JCA			
Shore	DISN	Fiber/SATCOM	POS/DSCS/Tri-band
Ship	CA	SATCOM	CA
Ship	GBS	SATCOM	GBS
DCS			
Shore	DSCS	SATCOM	DSCS
Ship	DSCS	SATCOM	DSCS

NOTE: All indirectly disseminated imagery is independent of collection source. Approximately 29 carriers and light amphibious assault ships will be outfitted with JSIPS-N, CDL-Navy-ST, Challenge Athena III, and GBS. This is the core of USN imagery afloat, and it supports timely direct and indirect imagery tasks. Additionally, GBS will be used to expand imagery dissemination to the larger fleet.

DEFINITIONS:
    CA, Challenge Athena; high-bandwidth (for ships) communications system that will support DDS; CA used for more than just imagery
    DCS, Defense Communications System; part of the DISN; provides defense satellite connectivity to ships and ashore facilities
    DDS, Defense Dissemination System, National Imagery and Mapping Agency (NIMA) Program Office. NIMA inserts imagery data into specific communication systems to reach specific customer's receive equipment (DE).
    DE, dissemination element; receive capability hardware and software within the DDS
    DISN, Defense Information Systems Network: DISA networked communications infrastructure; includes DATMS, SIPRNET, Intelink, JWICS, and other systems
    DSCS, Defense Satellite Communications System; one of the long-haul components that make up DISN; may also be used by DDS and Trojan Spirit II on a location-by-location, scenario, and deployment-dependent basis
    GBS, Global Broadcast System

Number of Receivers: Current	Total Number of Receivers in POM	Capability	Imagery Use	Remarks
~9	~10	1.5-45 Mbps	1.5-45 Mbps	DDS to ships via JCA
~50	~300	1.5-24 Mbps	0.768-6 Mbps	
All ships	All ships	64 kbps	Small	
All fixed and garrison sites	All fixed and garrison sites	0.128-45 Mbps	0.128-45 Mbps	
3	3	1.5 Mbps	128+ kbps	
All fixed and garrison sites	All fixed and garrison sites	0.128-45 Mbps	0.128-45 Mbps	
16	20	1.5 Mbps	768 kbps	Data rate can vary from 384 to 1544 kbps; normally 768 kbps
~50	~300	1.5-24 Mbps	0.768-6 Mbps	
STEP sites	STEP sites	768 kbps	128 kbps	
13	44	1.544 Mbps	Varies	CV/CVN, LHA/LHD, CG, LSD-41, LPD-17

JCA, Joint Services Imagery Processing System-Navy (JSIPS-N) Concentrator Architecture; USN future imagery dissemination "system"; will utilize both GBS and CA II communications system and will replace DDS for the USN. NIMA will send all imagery to USN JCA in Suitland, Maryland; JCA will then disseminate it to individual ships and user sites.

MTACS, Maritime Tactical Communications System; Service-maintained, networked communication infrastructure supporting the USMC and interfaces with the USN; provides a user connection to the DISN and a communications link between command component afloat and ground component ashore

POM, program objective memorandum

POS, point of service; connection between local networks and DISN, could be SATCOM ground terminals like T-MET, Tri-band, or STEP, or fiber-optic land line

STEP, standard tactical entry point; provides tactical communications entry into the DISN via the DSCS; to be upgraded to teleport concept, which also will include expanded capacities, protocol interfaces, and commercial connectivity

TS II, Trojan Spirit II; U.S. Army deployable SATCOM terminal system; USMC also owns and operates several terminals

SOURCE: Compiled from data courtesy of National Imagery and Mapping Agency, Bethesda, Md., 1999.

Annex 2 JPN, JDN, and JCTN Network Components

This annex provides a schematic diagram (Figure E.A.1) of the networks composing the Joint Planning Network, the Joint Data Network, and the Joint Composite Tracking Network.

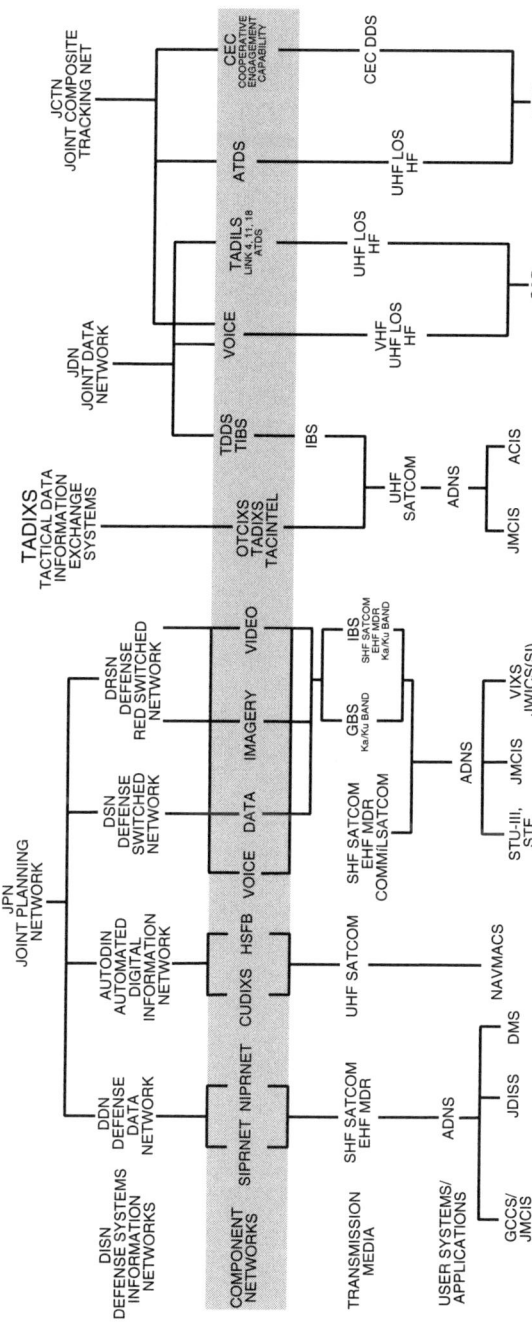

FIGURE E.A.1 Joint Planning Network, Joint Data Network, and Joint Composite Tracking Network network components. SOURCE: Modified from "Figure 3-1 TAD Communication Networks" in Bragg, Rick, PMW176-2, 1997, *AADC Afloat Communications Requirements Analysis*, Program Executive Office Theater Air Defense, Space and Naval Warfare Systems Command, San Diego, Calif., May 30, p. 3-2.

F
The Organizational View of the Recommended Operations Information and Space Command

A major benefit of the recommended new functional type commander, Operations Information and Space (see Chapter 7, Section 7.7.2), is to make permanent the process of the naval forces' change to network-centric operations. In the proposed reorganization (Figure F.1), the functional type commander, Operations Information and Space, would report to the fleet commanders and would be the focal point for all network-centric operational support matters involving fleet operations. This would ensure a single point of contact for fleet support operations and provide a focus for the real-time allocation of assets for information operations. Reconfiguration of networks ashore or afloat could be managed in real time worldwide, and computer network defense, or offense, would be enhanced. The command should have a single operations center providing visibility of all ashore and afloat networks used by the Navy.

The commander should also serve as the community manager for the entire cadre of information operations specialists, both officer and enlisted, including personnel now in each of the following commands: Commander Naval Security Group Command, Commander Naval Space Command, Commander Space and Naval Warfare Systems Command, and so on as shown in Figure F.1.

The proposed reorganization would remove the above commanders from the direct control of the Chief of Naval Operations and shift that control to the fleet commanders. These considerable resources would then be directly available to the fleets, and information operations personnel would focus more directly on fleet operations.

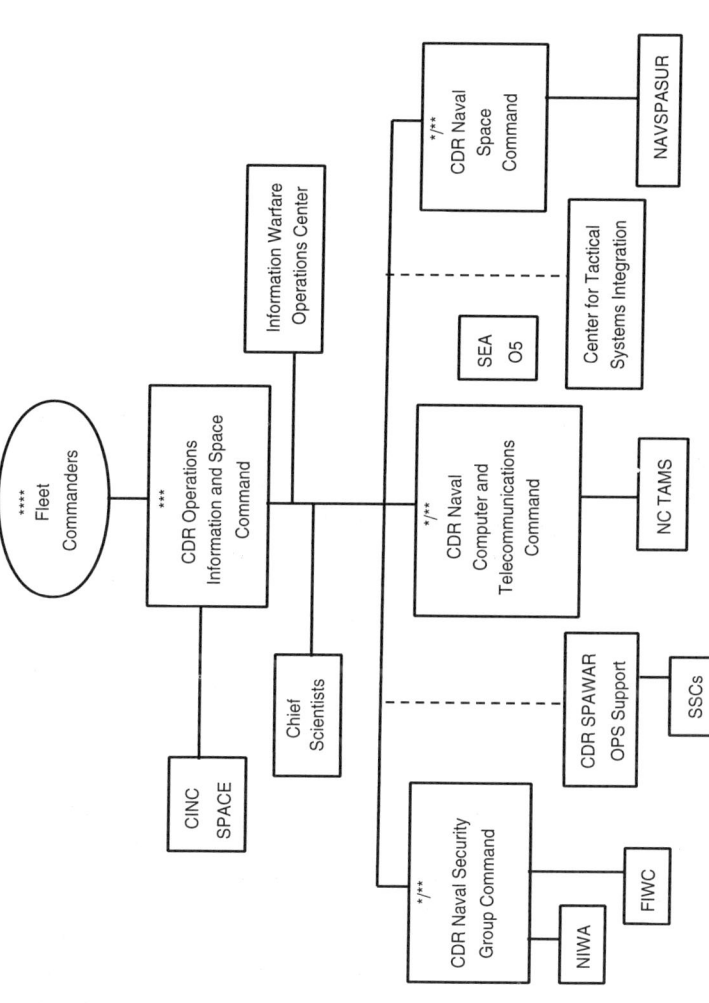

FIGURE F.1 Proposed reorganization with addition of a new functional type commander, Operations Information and Space. Acronyms are defined in Appendix H. The asterisks indicate the recommended number of stars the admiral who holds that position should wear: Admiral, O-10, an officer of four-star rank; Vice Admiral, O-9, an officer of three-star rank; Rear Admiral (upper half), O-8, an officer of two-star rank; Rear Admiral (lower half), an officer of one-star rank.

G

Committee Biographies

Mr. Vincent Vitto is president and chief executive officer (CEO) at C.S. Draper Laboratory. Mr. Vitto's expertise is in defense technology and its application to naval forces, particularly in space communications and radar technologies. Mr. Vitto's previous positions were with the Lincoln Laboratory of the Massachusetts Institute of Technology (MIT), where he rose from head of the Communications Division to assistant director of the Laboratory. He was responsible for technology and system concept development programs including surface surveillance and communications technology. Mr. Vitto has participated extensively in activities of the Air Force Science Advisory Board and currently is a member of the Defense Science Board. Mr. Vitto is chair of the National Research Council's (NRC's) Naval Studies Board.

Dr. Alan Berman, a private consultant, currently consults for the Applied Research Laboratory of Pennsylvania State University (ARL/PSU), where he provides general management support and program appraisal. He also consults for the Center for Naval Analyses (CNA), where he assists with analyses of Navy R&D investment programs, space operation capabilities, information operations, and command, control, communications, computers, intelligence, surveillance, and reconnaissance (C4ISR) programs. Dr. Berman has an extensive background in defense research. His previous positions include dean of the Rosenstiel School of Marine and Atmospheric Sciences at the University of Miami and director of research at the Naval Research Laboratory (NRL). He is a member of the NRC's Naval Studies Board.

APPENDIX G 465

Mr. Gregory R. Blackburn is assistant vice president of the Technology Research Group at Science Applications International Corporation (SAIC). Mr. Blackburn's background is in Department of Defense (DOD) information operations and space programs. Prior to joining SAIC, Mr. Blackburn served as the director of information operations for the Office of Secretary of Defense (OSD), where he directed efforts in policy for DOD information operations and space control programs. Mr. Blackburn's experience includes over 29 years as a naval cryptologic officer doing threat analysis and assessing vulnerability of state-of-the-art C4ISR, intelligence, information operations, and information programs.

Dr. Norval L. Broome is a senior Navy C4ISR systems analyst for the Naval and Technology Division at MITRE Corporation. Dr. Broome's expertise is in telecommunication systems engineering and naval C4I systems with emphasis in tactical and strategic submarine communications. He formerly held positions as director of MITRE's San Diego operations and department head for Naval Communications Systems, where he managed MITRE's work program with SPAWAR. Prior to joining MITRE, Dr. Broom was a senior engineer with Honeywell, Inc., Marine Systems Center where he provided detailed design and performance predictions for sonar subsystems used in Navy torpedoes. Dr. Broome's professional career includes 20 years of Navy service as an engineering duty officer. Dr. Broome is a member of the Naval Studies Board.

Dr. John D. Christie is a senior fellow at the Logistics Management Institute. Dr. Christie has an extensive background in DOD acquisition policy and program analysis. From 1989 to 1993, he was director of Acquisition Policy and Program Integration for the Office of the Under Secretary of Defense (Acquisition). In his role as director, Dr. Christie directed the preparation of a comprehensive revision to all defense acquisition policies and procedures resulting in the cancellation and consolidation of 500 prior separate issuances. He has been an active participant in NRC studies, and, most recently served as co-chair of the Committee on Shore Installation Readiness and Management.

General John A. Corder retired as a Major General from the U.S. Air Force (USAF) in 1992. General Corder has extensive experience in U.S. Air Force operational and joint issues. Since retiring, General Corder has been employed as an independent technical advisor. He has served as a member of the Air Force Scientific Advisory Board on the subjects of theater battle management, new world vistas, and tactical ballistic missile defense. General Corder's military career also includes assignments as a fighter wing commander and commander of the USAF Warfare Center. He was also deputy commander for Air Combat Operations for the Central Command Air Forces in the 1991 Persian Gulf War. General Corder was responsible for the planning and execution of 3,000 combat sorties per day—an effort that involved the coordination of Air Force, Navy, Marine Corps, and allied aircraft from nine other nations.

Dr. John R. Davis, a private consultant, most recently worked as director of engineering, TRW Logistics, Support and Test Evaluation Division. Dr. Davis's background is Navy technology and requirements development, particularly in regard to command and control and information systems. He retired from government service as technical director and chief scientist for the Navy's Space, Command and Control, and Information Warfare Division. He was responsible for the acquisition of an advanced technology program for a sensor-to-shooter system prototype to provide real-time intelligence, surveillance, and reconnaissance planning and in-cockpit execution, support for maritime power projection. Dr. Davis managed the redirection of Navy telecommunication programs into the evolution of multimedia, multiservice networks for space and terrestrial military and commercial communications. Dr. Davis has been employed at the Naval Research Laboratory in a number of positions, including superintendent of the Information Technology Division.

Dr. Paul K. Davis is a senior scientist at RAND, where he works primarily on defense planning. Dr. Davis's background is in strategic planning and program analysis and evaluation. Dr. Davis is also on the faculty of the RAND Graduate School and is chair of its international security and defense planning program. Earlier he served a 5-year tour as corporate research manager for defense and technology planning and as program manager for strategic planning and assessment. Prior to joining RAND, Dr. Davis was a senior executive in the Office of the Secretary of Defense Office of Program Analysis and Evaluation. Dr. Davis is a member of the Naval Studies Board.

Mr. Seymour J. Deitchman is a private consultant on national security, research and development management, and systems evaluation. A mechanical and aeronautical engineer by training, Mr. Deitchman served at the Institute for Defense Analyses (IDA) as vice president for programs. As a senior research associate at IDA, he was responsible for projects on tactical aviation and man-in-space and man-in-the loop simulation. He served as assistant vice president for research and vice president for planning and evaluation. Mr. Deitchman once served as director of overseas defense research at the Advanced Research Projects Agency (ARPA), where he was responsible for planning and executing ARPA's R&D program on counterinsurgency and related technical matters in support of U.S. military operations in Southeast Asia and some operations in the Middle East. He has been a member of various government advisory panels on national security matters.

Dr. John F. Egan is a private consultant who retired in 1998 as vice president for corporate development at Lockheed Martin Corporation. Dr. Egan has been responsible for providing support to three successive chief executives in defining and implementing strategic plans to consolidate the defense industry. Dr. Egan has a broad understanding of Navy programs, business and strategic planning, and acquisition and policy. An electrical engineer by training, Dr. Egan is a former chief scientist for the Chief of Naval Operations and has extensive experience with electronic and information warfare. Dr. Egan is a current member of the Chief of Naval Operations Executive Panel. Dr. Egan has participated in numerous NRC studies. He recently served as co-chair of the Committee on Shore Installation Readiness and Management.

Mr. Brig "Chip" Elliott is a lead scientist with Bolt, Beranek, and Newman (BBN) responsible for designing, developing, and integrating tactical and commercial communications systems for government and commercial organizations. His expertise includes wireless network technologies and tactical and commercial communication systems. As the technical lead scientist at BBN, he uses Internet technology to build networks for international corporations and U.S. government agencies. Mr. Elliott was the chief architect who designed the networking component of the U.S. Army's Near-Term Digital Radio (NTDR) program. This program is a fully mobile, secure, and highly survivable telecommunications system that will form the backbone of the Army's tactical Internet. Mr. Elliott has served on an Army Science Board study on tactical Internet design.

Dr. Edward A. Feigenbaum is Kumagai Professor of Computer Science and Co-Scientific Director of the Knowledge Systems Laboratory at Stanford University. Dr. Feigenbaum, a member of the National Academy of Engineering (NAE), has an extensive background in artificial intelligence, knowledge-based systems research and applications, and defense research and development. Dr. Feigenbaum served as chief scientist of the U.S. Air Force from 1994 to 1997. He is a former chairman of the Computer Science Department and director of the Computer Center at Stanford University. Dr. Feigenbaum has served on numerous government and scientific advisory boards, including the NSF's Computer Science Advisory Board and an ARPA Committee for Information Science and Technology.

Admiral David E. Frost, a retired Vice Admiral, U.S. Navy, is president of Frost and Associates, a consulting firm serving both commercial and government industries. His background is in U.S. Navy space planning, missions, and operations. Before retiring in 1996 from the Navy, Admiral Frost served as Deputy Commander-in-Chief of the U.S. Space Command, as well as Vice Commander-in-Chief of North American Aerospace Command (NORAD). NORAD is a command involving the United States and Canada that provides warnings of missile and air attack against both of its member nations, as well as safeguards the air sovereignty of North America, and provides air defense forces for defense against an air attack. Other assignments include commander, Naval Space Command and Commanding Officer of the USS *Seattle* and USS *Saratoga*.

Admiral Robert H. Gormley, a retired Rear Admiral, U.S. Navy, is president of the Oceanus Company, a business development and technology advisory firm serving U.S. and foreign clients in fields of aviation, defense equipment, and electronics. Admiral Gormley has an extensive background in naval operational issues, particularly in regard to aviation systems. His interests are primarily in airborne reconnaissance systems, unmanned aerial vehicles, vertical/short-take-off landing aircraft, weapon system combat survivability, military requirements formulation, and test and evaluation planning. Earlier, Admiral Gormley commanded the aircraft carrier USS *John F. Kennedy* as well as an air wing and fighter squadron during the Vietnam War.

Dr. Frank A. Horrigan recently retired as a member of the Technical Development Staff for Sensors and Electronic Systems at Raytheon Systems Company. He is an expert in radar and sensor technologies. Dr. Horrigan, a theoretical physicist, has more than 35 years' experience in advanced electronics, electro-optics, and computer systems. He has a wide knowledge of all technologies relevant to military systems and has experience in planning and managing information R&D investments and in projecting future technology growth directions. He is a member of the NRC's Naval Studies Board.

Dr. Richard J. Ivanetich is director of the Computer and Software Engineering Division at the Institute for Defense Analyses, where he has spent most of his professional career. Prior to that he was assistant director of the System Evaluation Division. Through his work at IDA, Dr. Ivanetich has a broad understanding of defense systems technology and operations analyses, particularly regarding computer and information systems; command, control, and communications (C3); modeling and simulation; crisis management; and strategic and theater nuclear forces. Dr. Ivanetich began his professional career as an assistant professor of physics at Harvard University. He is a member of the DARPA Information and Science Technology (ISAT) study group and a member of the NRC's Naval Studies Board.

Admiral Wesley E. Jordan, Jr., a retired Rear Admiral, U.S. Navy, is vice president of Federal Network Systems at Bolt, Beranek and Newman (BBN) Company. Admiral Jordan's expertise is in naval operational issues—particularly undersea warfare issues. At BBN, Admiral Jordan is responsible for internetworking technology initiatives for the civil and federal government. During his 31-year naval career, Admiral Jordan was deputy director of Space and Electronic Warfare, director of Antisubmarine Warfare, and director of the Command and Control (C2) Plan and Program Division for the Chief of Naval Operations. Admiral Jordan was commander of the Naval Surface Force, U.S. Atlantic Fleet, and also held various commanding positions in destroyer squadrons.

Dr. David V. Kalbaugh is head of the Power Projection Systems Department at the Johns Hopkins University/Applied Physics Laboratory (JHU/APL). Dr. Kalbaugh has an extensive background in missile and air systems. He joined JHU/APL in 1969 and has been involved in the development of the Tomahawk cruise missile system since its inception in the early 1970s. In addition, he was deputy director of the Navy theater ballistic missile defense (TBMD) Cost and Operational Effectiveness Analysis, and deputy director of the Navy TBMD Concept Evaluation and Integration Study sponsored by the Program Executive Office for Theater Air Defense. Dr. Kalbaugh also taught part-time for more than a decade in JHU's Whiting School of Engineering.

Dr. Annette J. Krygiel just completed serving as a distinguished visiting fellow at the Institute for National Strategic Studies of the National Defense University, where she investigated the problem of large-scale system integration. Dr. Krygiel's expertise is in the management of large-scale systems, particularly in regard to software development. Prior to being appointed to the Institute for National Strategic Studies, she was director of the Central Imagery Office. Dr. Krygiel was chief scientist at the Defense Mapping Agency (DMA). She has been a participant in NRC studies, including the work of the Panel on Distributed Geolibraries: Spatial Information Resources.

Ms. Teresa F. Lunt is principal scientist in the Secure Document Systems Area at Xerox Palo Alto Research Center (PARC). Prior to joining Xerox PARC, Ms. Lunt was associate director of the Computer Science Laboratory at Stanford Research Institute (SRI) International, where she was responsible for research in distributed computing, networking, information assurance, and network security. Ms. Lunt's expertise is in computer science, information management systems, and computer security. Fomerly, she was assistant director of distributed systems and program manager for information survivability in the Information Technology Office at the Defense Advanced Research Projects Agency (DARPA).

Dr. Douglas R. Mook is director of Advanced Systems at Sanders, a Lockheed Martin Company. Dr. Mook has experience in acoustic processing and sensor fusion. He is responsible for several key DOD programs including the U.S. Navy's Advanced Acoustics Communications Advanced Technology Demonstration, the U.S. Army's Federated Laboratories Digital Battlefield programs for communications and sensors, and DARPA's Unattended Ground Sensors programs. Dr. Mook is a member of the Navy's Fleet Ballistic Missile Submarine Security Review Panel and has been a member of the Army Research Laboratory Restructuring Panel and the Army Digital Battlefield Definition Panel.

Dr. Donald L. Nielson recently retired as director of the Computing and Engineering Sciences Division and vice president at SRI International. He has extensive experience in information technology, including information transport systems, distributed processing, artificial intelligence-aided language and reasoning systems, terminal systems and human-computer interaction, computer and communications security, information media and standards, telecommunications sciences, and image processing. He has also studied the problems of inserting advanced commercial off-the-shelf information technology into field-deployable military systems. Dr. Nielson has served on a number of government advisory committees and panels, including the Technical Advisory Committee to DARPA, Defense Information Systems Agency (DISA), and the Air Force Scientific Advisory Board.

Dr. Stewart D. Personick is E. Warren Colehower Chair, a professor of telecommunications, and director of the Center for Telecommunications and Information Networking at Drexel University. Dr. Personick, a member of the NAE, has an extensive background in telecommunications, computer operations and security, and optical communications technology and applications. As the first director of Drexel's Center for Telecommunications and Information Networking, he created four initial programs: Networks That Work, Trustworthy Networks, Next Generation Wireless, and Optical Networking. Dr. Personick is retired from Bell Communications Research, Incorporated (Bellcore), where he served as vice president of information networking. Dr. Personick is a current member of the NRC's Board on Army Science and Technology.

Dr. Joseph B. Reagan is a retired vice president and general manager of R&D at Lockheed Martin Missile and Space and was a corporate officer of the Lockheed Martin Corporation. Dr. Reagan, a member of the NAE, has an extensive background in defense technology development, particularly in the area of space and missile technologies. Dr. Reagan joined Lockheed 40 years ago as a scientist, where he led the Space Instrumentation Group for 10 years and was responsible for the development and on-orbit deployment of over 20 scientific payloads for the National Aeronautics and Space Administration (NASA) and the DOD. His research interests included the areas of space sensors, radiation belt and solar particles, nuclear weapon effects, and the effects of radiation particles on spacecraft systems. Today, Dr. Reagan is involved in activities that foster the improvement of science and mathematics education in the United States and is vice chair of the NRC's Naval Studies Board.

Admiral Charles R. Saffell, a retired Rear Admiral, U.S. Navy, is vice president for C4ISR systems at Titan Technologies and Information Systems Corporation, a division of the Titan Corporation, where his areas of concentration are commercial off-the-shelf solutions to C4ISR requirements, end-to-end technical solutions to current and future C4ISR challenges, strategic partnerships and acquisitions, international C4ISR, and information technology (IT) applications development. Prior to his joining Titan, Admiral Saffell was the commander of Amphibious Group Three and deputy director of the C4 Directorate of the Joint Staff. Admiral Saffell was also one of the conceivers of the Joint Chiefs of Staff (JCS) *Joint Vision 2010* that has restructured warfighting operations and doctrine.

Dr. Nils R. Sandell, Jr., is president and CEO at ALPHATECH, Inc. He is also a co-founder of the company. Dr. Sandell has an extensive background in automatic target recognition (ATR) and sensor management. At ALPHATECH, he is currently responsible for projects developing, planning, and scheduling algorithms for airborne reconnaissance platforms and tracking and sensor resource management algorithms for ground-moving target indication and synthetic aperture radars. A former associate professor at the Massachusetts Institute of Technology, Dr. Sandell lectured in areas of estimation and control theory, stochastic processes, and computer systems.

Admiral William D. Smith is a retired Admiral, U.S. Navy. Admiral Smith retired in 1993 after 38 years of active duty service. He has extensive experience in Navy planning, programming, budgeting, and operational issues. His last assignment was as U.S. Military Representative to the NATO Military Committee in Brussels, Belgium. Admiral Smith has served in a number of high-ranking capacities for the Chief of Naval Operations. From 1987 to 1991, Admiral Smith served as Deputy Chief of Naval Operations for Logistics and Navy Program Planning. From 1985 to 1987, he was director, Fiscal Management Division/ Comptroller of the Navy. Admiral Smith is a senior fellow at the Center for Naval Analyses.

Dr. Michael G. Sovereign is professor emeritus of C3 at the Naval Postgraduate School (NPS), having retired on January 1, 1999. His background is in C3, joint warfare analysis, and acquisition cost analysis. For the past year, he has served as visiting research professor for Headquarters, U.S. Commander-in-Chief Pacific Fleet, where his responsibilities included conducting research on the Navy's Virtual Information Center workshops and other experiments aimed at addressing joint C4ISR issues. Before joining NPS, Dr. Sovereign was senior principal scientist at SHAPE Technical Center (now a NATO C3 agency), where he participated in major replanning of NATO C3 systems. Dr. Sovereign has published numerous articles that have spanned a wide range of subject matter, including the economics of instructional media, defense logistics, and economics.

Mr. H. Gregory Tornatore is the program area manager for Defense Communications Programs at the Johns Hopkins University/Applied Physics Laboratory (JHU/APL). His areas of expertise include military C3, wide-area surveillance, over-the-horizon sensors and targeting, communications networks and architectures, high-frequency radar, and ionospheric propagation. Mr. Tornatore also chairs the APL's Internal Research and Development Command and Control Thrust Area, responsible for the application of new technology to DOD C3 problems. Mr. Tornatore has been employed by JHU/APL since 1977 and has been a member of the Principal Professional Staff since 1980. Prior to joining JHU/APL, Mr. Tornatore was employed at the Electro-Physics Laboratory, ITT Avionics Division.

General Paul K. Van Riper, a retired Lieutenant General, U.S. Marine Corps, is a private consultant. General Van Riper recently retired from the Marine Corps after 41 years of active and reserve service. His expertise is in military affairs and operational issues, as well as in the importance of science and technology for the capabilities of the naval forces. Currently, he is a senior fellow with the Center for Naval Analyses participating in a wide array of defense and security-related seminars, conferences, and studies. He had a long and distinguished military career commanding or assigned to ground-combat units and is familiar with all aspects of Marine Corps operations. For 2 years prior to his retirement, General Van Riper served as Commanding General of the Marine Corps Combat Development Command. He is a member of the NRC's Naval Studies Board.

Dr. Bruce Wald is the founder of Arlington Education Consultants, which serves both government and industry. He also holds an adjunct appointment with the Center for Naval Analyses. Dr. Wald has extensive expertise in space and information technology, electronic warfare, and national security implications. Dr. Wald is the former associate director of Research and director of Space and Communications Technology at the Naval Research Laboratory. Dr. Wald has been a member of numerous government and industry advisory boards and panels. He is a member of the NRC's Naval Studies Board.

Admiral Raymond M. Walsh is a retired Rear Admiral, U.S. Navy. Admiral Walsh's background is in DOD/Navy financial management and policy formulation. Currently, Admiral Walsh is a senior systems engineer at Basic Commerce and Industries, Incorporated (BCII). Prior to joining BCII, he was a lead analyst at Sonalysts, Incorporated. Admiral Walsh's broad range of experience includes the command of two surface combatants as a naval surface warfare officer and as an operations analyst ashore involved with Navy planning, programming, and budgeting processes. Admiral Walsh was also the director of the Operations Division for the Office of Budget and Reports under the Navy Comptroller, where he was the responsible official for all Navy operating budget accounts. Admiral Walsh recently served on the NRC's Committee on Shore Installation Readiness and Management.

Ms. Mitzi M. Wertheim is a consultant to Enterprise Solutions at the Center for Naval Analyses. Her expertise is in the application of business process reengineering methods and teaching large corporations to increase service while reducing cost. In recent years, her research interests have focused on naval career, education, and training issues. Before joining CNA, Ms. Wertheim was vice president of Enterprise Solutions at SRA International, Incorporated. Her responsibilities included identifying linkages and interdependencies in organizations and then leveraging IT to achieve business objectives. From 1977 to 1981, Ms. Wertheim was the deputy undersecretary of the Navy. Ms. Wertheim is involved with a number of organizations, including the Council of Foreign Relations and the Advisory Board of the Defense Budget Group. She is a founder and executive committee member of the MIT Seminar XXI. Ms. Wertheim is a member of the NRC's Naval Studies Board.

Mr. Geoffrey A. Whiting is director of Maritime Systems at Sanders, a Lockheed Martin Company. Mr. Whiting has a broad understanding of C4ISR and shipboard defense systems. As director of Maritime Systems, Mr. Whiting is responsible for maritime tactical signal exploitation systems for U.S. and foreign defense customers and for the development of lower life-cycle costs and increased functionality that optimize maritime product lines. Mr. Whiting's professional experience includes more than 25 years with the U.S. Navy in operational, intelligence, and technical positions.

Mr. Dell P. Williams III is senior technical advisor to the president and CEO of Teledesic Corporation. He is responsible for technical oversight of the development of the Teledesic network, a constellation of several hundred low-Earth-orbit satellites providing worldwide access to "fiber-like" telecommunications services. Mr. Williams has an extensive background in commercial communications, systems engineering, space systems, and information assurance. Prior to his work on the Teledesic network, Mr. Williams was vice president of Electronic Defense Programs at ARGO Systems, a wholly owned Boeing subsidiary. He also was director of Advanced Programs at Lockheed Missiles and Space Company and director of Space Systems at NASA Headquarters.

H

Acronyms and Abbreviations

AA	antiair
AADC	area air defense commander
AAMDC	Army Air and Missile Defense Command
AAW	antiair warfare
AAWC	antiair warfare commander/coordinator
ABCCC	Airborne Battlefield Command and Control Center
ABMOC	Air Battle Management Operations Center
AC	Attack Center
AC2ISRC	Air Combat Command's Aerospace Command and Control, Intelligence, Surveillance, and Reconnaissance Center
ACMC	Assistant Commandant of the Marine Corps
ACOS	Assistant Chief of Staff
ACS	acoustic cross section
ACTD	advanced concept technology demonstration
ACTDS	Automatically Cued Target Detecting System
ADC	analog-to-digital converter
ADLA	autonomous drifting line array
ADNS	Automated Digital Network System
ADTOC	Air Defense Tactical Operations Center
AEF	aerospace expeditionary force
AFAPD	Air Force Application Program and Development
AFB	Air Force base
AF EFX	Air Force expeditionary force experiment
AFIWC	Air Force Information Warfare Center

APPENDIX H *475*

AFOSR	Air Force Office of Scientific Research
AFRL	Air Force Research Laboratory
AFRTS	Armed Forces Radio and Television Service
AFSAB	Air Force Scientific Advisory Board
AGM	air-to-ground missile
AGS	advanced gun system
AICE	Agile Information Control Environment
AIM	antenna interface module
AIP	ASARS Improvement Program
AJ	anti-jam(ming)
ALAM	advanced land-attack missile
ALR	automatic landmark recognition
AMDTF	Air and Missile Defense Task Force
AMRAAM	advanced medium-range air-to-air missile
AMRFS	Advanced Multifunction Radio Frequency System
ANDVT	Advanced Narrowband Digital Voice Terminal
AOC	Air Operational Center
APL	Applied Physics Laboratory
ARG	amphibious ready group
ARQ	automatic retransmission queue
ASARS	Advanced Synthetic Aperture Radar System
ASCIET	All Service Combat Identification Evaluation Team
ASD	Assistant Secretary of Defense
ASN	Assistant Secretary of the Navy
ASN CIO	Assistant Secretary of the Navy Chief Information Officer
ASN RDA	Assistant Secretary of the Navy for Research, Development, and Acquisition
ASW	antisubmarine warfare
ATACMS	Army Tactical Cruise Missile System
ATC	air traffic control
ATDL	advanced tactical data link
ATM	asynchronous transfer mode
ATO	air tasking order
ATOS	Aircraft Tape Operating System
ATR	automatic target recognition
ATRC	Aegis Training and Readiness Command
AWACS	Airborne Warning and Control System
AWE	advanced warfighting experiment
BADD	battlefield automated data distribution
BAT	brilliant anti-armor technology
BDA	battle/bomb damage assessment

BF	battle force
BG	battle group
BGSIT	battle group systems interoperability testing
BI	battlespace infosphere
BLOS	beyond line of sight
BMC4I	battle management command, control, communications, computing, and intelligence
BMDO	Ballistic Missile Defense Organization
BOR	Board of Representatives
BPSK	binary phase shift key
C&D	command and decision
C2	command and control
C2ISR	command, control, intelligence, surveillance, and reconnaissance
C2P	command and control processor
C3	command, control, and communications
C3I	command, control, communications, and intelligence
C4I	command, control, communications, computing, and intelligence
C4ISR	command, control, communications, computing, intelligence, surveillance, and reconnaissance
CAOC	Combined Air Operations Center
CAP	Command Air Patrol
CAPI	cryptographic application programming interface
CBT	computer-based training
CCD	camouflage, concealment, and deception; charge-coupled device
CCM	counter-countermeasure
CDC	Combat Development Command
CDL	common data link
CDMA	code division multiple access
CEC	cooperative engagement capability
CEP	circular error probable
CERT	Computer Emergency Response Team
CFO	chief financial officer
CG	guided-missile cruiser
CHBDL	common high-bandwidth data link
CHENG	Chief Engineer (Navy)
CHNAVMAT	Chief of Naval Materiel
CIC	Combat Information Center
CINC	commander-in-chief
CINCLANTFLT	Commander-in-Chief, U.S. Atlantic Fleet

APPENDIX H 477

CINCPAC	Commander-in-Chief, Pacific Command
CINCPACFLT	Commander-in-Chief, U.S. Pacific Fleet
CIO	Chief Information Officer
CISN	Communications, Information Systems, and Networks
CIWS	close-in weapon system (Phalanx)
CJCS	Chairman, Joint Chiefs of Staff
CJTF	command joint task force
CMC	Commandant of the Marine Corps
CMD	cruise missile defense
CNA	Center for Naval Analyses
CNET	Chief of Naval Education and Training
CNO	Chief of Naval Operations
CoABS	Control of Agent-based Systems program
COE	common operating environment
COMNAVAIRLANT	Commander, Naval Air Force, Atlantic
COMNAVAIRPAC	Commander, Naval Air Force, Pacific
COMNAVSUBLANT	Commander, Naval Submarine Force, Atlantic
COMNAVSUBPAC	Commander, Naval Submarine Force, Pacific
COMSEC	communications security
CONOPS	concept(s) of operations
CONUS	continental United States
COP	common operational picture
COTS	commercial off-the-shelf
CPoF	Command Post of the Future program
CRD	capstone requirements document
CSMA/CA	carrier sense multiple access/collision avoidance
CTAPS	Contingency Tactical Air Planning System
CTD	common tactical data set
CTP	common tactical picture; consistent tactical picture; coherent tactical picture
CTS	clear to send
CUDIXS	Common User Digital Information Exchange System/Subsystem
CV	aircraft carrier
CVBG	carrier battle group
CVN	nuclear-powered aircraft carrier
CVX	next-generation aircraft carrier (aircraft carrier, experimental)
CWO	chief warrant officer
DAC	digital-to-analog converter
DAMA	demand assigned multiple access
DARPA	Defense Advanced Research Projects Agency

DCNO	Deputy Chief of Naval Operations
DDB	Dynamic Data Base program
DDS	data distribution system
DEP	distributed engineering plant
DES	Data Encryption Standard
DHCP	dynamic host configuration protocol
DIA	Defense Intelligence Agency
DIAP	Defense-wide Information Assurance Program
DII	defense information infrastructure
DII COE	defense information infrastructure common operating environment
DISA	Defense Information Systems Agency
DISN	Defense Information Systems Network
DLA	Defense Logistics Agency
DMA	Defense Mapping Agency
DMDC	Defense Manpower Data Center
DME	distance measuring equipment
DMIF	Dynamic Multiuser Information Fusion program
DMR	digital modular radio
DNS	domain name service
DOC	Department of Commerce
DOD	Department of Defense
DOT	Department of Transportation
DRM	design reference mission
DSB	Defense Science Board
DSC	Decision Support Center
DSCS	Defense Satellite Communications System
DSL	digital subscriber line
DSMAC	digital scene matching area correlation
EAC	Evaluation Analysis Center
EAF	expeditionary aerospace force
ECM	electronic countermeasures
EFX	expeditionary force experiment
EHF	extremely high frequency (30 GHz to 300 GHz)
ELB	extending the littoral battlespace; extended littoral battlespace
ELF	extremely low frequency
ELINT	electronic intelligence
EO	electro-optical/optics
EPLRS	Enhanced Position Location Reporting System
ERDB	Emerging Requirements Database
ERGM	extended-range guided missile

APPENDIX H 479

ESC	Electronic Systems Command
ESM	electronic support measure
ESSM	evolved sea sparrow missile
ESWS	enlisted surface warfare specialist
ETOC	Education and Training Oversight Council
EW	electronic warfare/electronic warfare technician
FAAD	forward area air defense
FAADC2I	Forward Area Air Defense Command, Control, and Intelligence (Army)
FAC	forward air control(ler)
FARA	Federal Acquisition Reform Act
FBCB2	Force XXI Battle Command Battalion/Brigade and Below program (Army)
FBE	fleet battle experiment
FEC	fast Ethernet channel
FEWSG	Fleet Electronic Warfare Support Group
FIWC	Fleet Information Warfare Center
FLIR	forward-looking infrared
FMF	fleet Marine force
FO	forward observer
FOPEN	foliage penetration/penetrating
FOR	field of regard
FOV	field of view
FPA	focal plane array
FSK	frequency shift keying
FTN	Fault Tolerant Network program
FTP	file transfer protocol
FTX	field training exercise
FYDP	Future Year Defense Program
GAO	General Accounting Office
GBR	ground-based radar
GBS	Global Broadcast System
GCCS	Global Command and Control System
GCCS-M	Global Command and Control System-Maritime
GCSS	Global Combat Support System
GDOP	geometric dilution of precision
GEO	geosynchronous Earth orbit
GIG	global information grid
GLONASS	Global Navigation Satellite System
GMTI	ground moving-target indicator/indication
GPRS	General Packet Radio Service

GPS	Global Positioning System
GSMC	Global System for Mobile Communications
GTN	Global Transportation Network
HARM	high-speed antiradiation missile
HCI	human-computer interface
HDR	high data rate
HEL	high-energy laser
HF	high frequency
HITL	hardware-in-the-loop
HPD	high-power discriminator
HQMC	Headquarters, Marine Corps
HRMTI	high-resolution moving-target indicator
HRR	high-range resolution
HTML	Hyper Text Markup Language
HVT	high-value target
I3	integrated imagery and intelligence
I&W	indications and warnings
IA	information assurance
IBDL	interbattery data link
IBS	Integrated Broadcast System
IC	integrated circuit
ID	identification
IDA	Institute for Defense Analyses
IDM	information dissemination management
IEEE	Institute of Electrical and Electronics Engineers
IFF	identification friend or foe
IFFN	identification, friend, foe, or neutral
IFOV	instantaneous field of view
III	integrated information infrastructure
IJMS	interim JTIDS message specification
IMINT	image intelligence
IMT	information management technology
IMU	inertial measurement unit
INFOSEC	information security
INMARSAT	International Marine/Maritime Satellite
INS	Inertial Navigation System
IO	information operations
IOC	initial operational capability
IOG	independent operations group
IP	Internet Protocol
IPB	intelligence preparation of the battlefield

APPENDIX H 481

IPSec	IP Security
IPT	integrated products team
IR	infrared
IRMC	Information Resource Management College
IRST	infrared search and track
IS	intelligence specialist
ISR	intelligence, surveillance, and reconnaissance
ISS	information superiority and sensors
IT	information technology
IT-21	information technology for the 21st century
ITAA	Information Technology Association of America
ITI	information technology infrastructure
ITIA	Information Technology Infrastructure Architecture
ITM	information technology management
ITMRA	Information Technology Management Reform Act
ITSG	Information Technology Standards Guidance
IW	information warfare
IWAR	integrated warfare architecture
IXS	Information Exchange System/Subsystem
JASSM	joint air-to-surface standoff missile
JBI	joint battlespace infosphere
JCIT	Joint Combat Information Terminal
JCMT	Joint Collection Management Tasking
JCS	Joint Chiefs of Staff
JCTN	Joint Composite Tracking Network
JDAM	joint direct attack munition
JDISS	Joint Deployable Intelligence Support System
JDN	Joint Data Network
JEFX	joint expeditionary force experiment
JFACC	Joint Forces Air Component Command; Joint Forces Air Component Commander program
JFC	joint force commander
JHU/APL	Johns Hopkins University/Applied Physics Laboratory
JICO	joint interface control officer
JITC	Joint Interoperability Testing Center
JMCIS	Joint Maritime Command Information System
JMCOMS	Joint Maritime Communications System
JPN	Joint Planning Network
JROC	Joint Requirements Oversight Council
JSF	joint strike fighter
JSIPS-N	Joint Services Imagery Processing System-Navy
JSOW	joint standoff weapon

JSTARS	Joint Surveillance and Target Attack Radar System
JTA	Joint Technical Architecture
JTACMS	Joint Tactical Missile System
JTAMDO	Joint Theater Air and Missile Defense Organization
JTF	joint task force
JTIDS	Joint Tactical Information Distribution System
JTRS	Joint Tactical Radio System
JWCA	joint warfare capabilities assessment
JWICS	Joint Worldwide Intelligence Communications System
LAAD	low-altitude air defense
LACMD	land-attack cruise missile defense
LAMPS	Light Airborne Multipurpose System
LAN	local area network
LANDSAT	land remote-sensing satellite
LANTIRN	low-altitude navigation and targeting infrared for night
LASM	land-attack standard missile
LDO	limited duty officer
LDR	low data rate
LEO	low Earth orbit
LGB	laser-guided bomb
LGM	laser-guided munitions
LHA/LHD	amphibious assault ship (general purpose/multipurpose)
LHC	long-haul communications
LOS	line of sight
LPD	low probability of detection; amphibious assault transport dock
LPI	low probability of interception
MAGTF	Marine air-ground task force
MANET	mobile ad hoc network
MARCORPS	Marine Corps
MBV	model-based vision
MCBL	Marine Corps Battle Laboratory
MCCDC	Marine Corps Combat Development Command
MCM	mine countermeasures
MCSYSCOM	Marine Corps Systems Command
MCWL	Marine Corps Warfighting Laboratory
MDV	minimum Doppler velocity; minimal detectable velocity
MEF	Marine expeditionary force

MEMS	microelectromechanical systems
METOC	meteorology and oceanography
MEU	Marine expeditionary unit
MFR	multifunction radar
MIDS	Management Information and Data System; Management Information and Distribution System; modern intelligence databases
MILSTAR	Military Strategic, Tactical, and Relay (SATCOM)
MIT	Massachusetts Institute of Technology
MIT/LL	Massachusetts Institute of Technology, Lincoln Laboratory
MLS	multilevel secure
MMIC	monolithic microwave integrated circuit
MOE	measure of effectiveness
MOOTW	military operations other than war
MOP	measure of performance
MRC	major regional conflict
MSS	maximum segment size (TCP)
MSTAR	Moving and Stationary Target Acquisition and Recognition program
MTI	moving-target indicator
MTIm	moving-target imaging
MTW	major theater war
N2	Director of Naval Intelligence
N3/N5	Deputy Chief of Naval Operations (Plans, Policy, and Operations)
N4	Deputy Chief of Naval Operations (Logistics)
N6	Deputy Chief of Naval Operations (Space, Information Warfare, and Command and Control)
N7	Director of Naval Training
N8	Deputy Chief of Naval Operations (Resources, Warfare Requirements, and Assessments)
N86	Deputy Chief of Naval Operations (Surface Warfare Division)
N88	Director, Air Warfare Division
NASA	National Aeronautics and Space Administration
NAVAIR	Naval Air Systems Command
NAVCOMPT	Navy Comptroller
NAVMACS	Naval Modular Automated Communications System
NAVSEA	Naval Sea Systems Command
NAVSPACOM	Naval Space Command
NAVSPASUR	Navy Space Surveillance System

NAVSUP	Naval Supply Systems Command
NCA	National Command Authority
NCII	Naval Command and Information Infrastructure
NCNF	network-centric naval forces
NCO	network-centric operations
NCTR	naval commercial traffic regulator
NCW	network-centric warfare
NDU	National Defense University
NEC	Navy enlisted classification
NES	network encryption system
NFS	naval fire support
NGO	nongovernmental organization
NILE	NATO Improved Link Eleven (Link 22)
NIMA	National Imagery and Mapping Agency
NIPRNET	sensitive but unclassified Internet Protocol Router Network
NIST	National Institute of Standards and Technology
NITF	National Imagery Transmission Format (standard)
N/MCI	Navy/Marine Corps intranet
NORAD	North American Aerospace Command
NPGS	Naval Postgraduate School
NRC	National Research Council
NRL	Naval Research Laboratory
NRO	National Reconnaissance Office
NSA	National Security Agency
NSB	Naval Studies Board
NSF	National Science Foundation
NSTISSI	National Security Telecommunications and Information Systems Security Instruction
NTDR	near-term digital radio
NTM	National technical means
NWC	Naval War College
NWDC	Navy Warfare Development Command
OA	operational architecture
OASD (C3I)	Office of the Assistant Secretary of Defense (Command, Control, Communications, and Intelligence)
OCMD	overland cruise missile defense
OLCD	object level change detection
OMB	Office of Management and Budget
OMFTS	Operational Maneuver From the Sea
OMN	operations and maintenance, Navy

ONR	Office of Naval Research
OODA	observe, orient, decide, and act
OOTW	operations other than war
OPA	optical phased array
OPM	Office of Personnel Management
OPN	other procurement, Navy
OPNAV	Office of the Chief of Naval Operations
OQPSK	offset quadrature phase shift key
ORD	operational requirements document
OS	operations specialist
OSD	Office of the Secretary of Defense
OSI	Open Systems Interconnection (model)
OSPF	open shortest path first
OTAR	over-the-air rekeying
OTC	officer in tactical command
OTH	over-the-horizon
OUSD (A&T)	Office of the Under Secretary of Defense (Acquisition and Technology)
OUSD (P&R)	Office of the Under Secretary of Defense (Personnel and Readiness)
P3I	preplanned product improvement
PADIL	Patriot data link
PCS	portable control station
PDA	personal digital assistant
PE	program element
PEO	program executive office(r)
PEO C3S	Program Executive Office for Command, Control, and Computer Systems
PEO TSC	Program Executive Office for Theater Surface Combatants
PK	public key
PKI	public-key infrastructure
PLI	position location information
PLRS	Position Location and Reporting System
PM	program manager
PMR	programmable modular radio
POC	point of contact
POM	program objective memorandum
PPBS	Planning, Programming, and Budgeting System
PPDB	point positional database
PPDL	point-to-point data link
PPLI	precise position location and identification

PSA	Principal Staff Assistant
PTP	point to point
QOL	quality of life
QOS	quality of service
R&D	research and development
R2	reporting responsibility
RAM	rolling airframe missile
RCS	radar cross section
RDA	research, development, and acquisition
RF	radio frequency
RFP	request for proposal
RFS	request for service
RM	radioman
RMS	Requirements Management System
ROEs	rules of engagement
ROV	remotely operated vehicle
RSVP	Resource Reservation Protocol
RTR	Real Time Targeting and Retargeting program
RTS	request to send
S&T	science and technology
SA	selective availability
SAAWC	sector antiair warfare coordinator/center
SALSA	Software Architecture and Logic for Secure Cooperative Applications
SALTS	Streamlined Automated Logistics Transmission System
SAM	surface-to-air missile
SAR	synthetic aperture radar
SAS	synthetic aperture sonar
SATCOM	satellite communications
S-Band	frequency band, 1550 MHz to 5200 MHz
SD	spiral development
SEAD	suppression of enemy air defenses
SECNAV	Secretary of the Navy
SEW	space, electronic warfare
SFW	sensor-fused weapon
SHF	superhigh frequency
SIAP	single integrated air picture
SIGINT	signal intelligence
SINCGARS	Single Channel Ground and Airborne Radio System

APPENDIX H   487

SINTRA	Secure Information Through Replicated Architecture
SIPRNET	Secret Internet Protocol Router Network
SLA	Service-level agreement
SLAM-ER	standoff land attack missile-expanded response
SM-3	standard missile 3
SNMC	Simple Network Management Control
SNMP	Simple Network Management Protocol
SONET	synchronous optical network
SPAWAR	Space and Naval Warfare Systems Command
SPMAGTFX	Special Purpose Marine Air-Ground Task Force-Experimental
SSBN	nuclear-powered ballistic missile submarine
SSC	SPAWAR Systems Center
SSG	strike and surface fire
SSL	Secure Socket Layer
SSN	nuclear-powered attack submarine
ST	sonar technician
STAP	space-time adaptive processing
STOL	short takeoff and landing
STOM	ship to objective maneuver
STOVL	short takeoff and vertical landing
STOW	synthetic theater-of-war
STU III	secure telephone unit-third generation
SUBPAC	Submarine Forces, Pacific
SURFPAC	Surface Forces, Pacific
SUW	surface warfare
SYSCOM	Systems Command
TACAN	tactical air and navigation
TACC	Tactical Air Control (Command) Center
TACFIRE	Tactical Fire Direction System (Army)
TACINTEL	tactical intelligence
TAD	theater air defense
TADIL	tactical digital information link
TADIL A	Tactical Digital Information Link (Link 11)
TADIL B	Tactical Digital Information Link (Link 11B)
TADIL C	Tactical Digital Information Link (Link 4A)
TADIL J	Tactical Digital Information Link (Link 16)
TAMD	tactical (theater) air and missile defense
TAN	terrain-aided navigation
TAOC	Tactical Air Operations Center (USMC)
TAP	tactical awareness packet; terminal awareness packet
TAV	total asset visibility

TBMCS	Theater Battle Management Core System
TBM(D)	tactical (theater) ballistic missile (defense)
TCDL	tactical common data link (U.S. DOD joint)
TCP	Transmission Control Protocol
TCT	time-critical target
TDDS	Tactical Data Dissemination System
TDMA	time division multiple access
TDN	Tactical Data Network
TDS	Tactical Direction System
TEL	transporter-erector-launcher
TERCOM	terrain-contour matching
TESS	The Enhanced Surveillance System
TFNF	technology for future naval forces
THAAD	theater high-altitude area defense
TIBS	Tactical (Theater) Information Broadcast System
TISS	Thermal Imaging Sensor System
TLE	target location error
TMAD	tactical (theater) missile and air defense
TnT	tactical/nontactical
TOA	total obligational authority
TOC	Tactical Operations Center
TOF	time of flight
TPED	tasking, processing, exploitation, and dissemination
TPFDD	time-phased force deployment data
TPFDL	Time Phase Force Deployment List
T/R	transmitter/receiver
TRADOC	Training and Doctrine Command (Army)
TRANSIT	Navy Satellite Navigation System
TRAP	Tactical Receive Applications Program
TRE	tactical receive equipment
TSB	time slot block
TSC	tactical support center; Theater Surface Combatant
TSR	time slot reallocation
TTPs	tactics, techniques, and procedures
TYPE CDR	functional type commander
UAV	unmanned aerial vehicle
UCAV	uninhabited combat air vehicle
UHF	ultrahigh frequency
UNSECNAV	Under Secretary of the Navy
URL	unrestricted line (USN officer designation); uniform resource locator (World Wide Web address)
USA	United States Army

APPENDIX H   *489*

USACOM	United States Atlantic Command
USAF	United States Air Force
USD AT	Under Secretary of Defense for Acquisition and Technology
USJFCOM	United States Joint Forces Command
USMC	United States Marine Corps
USMTF	uniform standard message transfer format
USN	United States Navy
USPACOM	United States Pacific Command
USW	undersea warfare
UUV	unmanned underwater vehicle
VCJCS	Vice Chairman, Joint Chiefs of Staff
VCNO	Vice Chief of Naval Operations
VHF	very high frequency
VLS	vertical launch system
VMF	variable message format
VOR	voice-operated relay; VHF omni-range navigation system
VPN	virtual private network
VRC	vehicular radio communications
VSR	volume search radar
VTC	video teleconference
VTOL	vertical takeoff and land(ing)
VTT	visual true type
WDM	wave-division-multiplexed
WIPT	working integrated process team
WMD	weapon of mass destruction
WOC	War Operations Center
WWMCCS	Worldwide Military Command and Control System
XML	Extensible Markup Language